MOLECULAR BIOLOGY OF CIRCADIAN RHYTHMS

MOLECULAR BIOLOGY OF CIRCADIAN RHYTHMS

Edited by
AMITA SEHGAL
Howard Hughes Medical Institute
Department of Neuroscience
University of Pennsylvania School of Medicine
Philadelphia, Pennsylvania

WILEY-LISS
A John Wiley & Sons, Inc., Publication

Published by John Wiley & Sons, Inc., Hoboken, New Jersey.
Published simultaneously in Canada.

For general information on our other products and services please contact our Customer Care Department within the U.S. at 877-762-2974, outside the U.S. at 317-572-3993 or fax 317-572-4002.

Wiley also publishes its books in a variety of electronic formats. Some content that appears in print, however, may not be available in electronic format.

Library of Congress Cataloging-in-Publication Data:

Molecular biology of circadian rhythms / edited by Amita Sehgal.
p. ; cm.
Includes bibliographical references and index.
ISBN 0-471-41824-2 (cloth : alk. paper)
1. Circadian rhythms. 2. Molecular biology.
[DNLM: 1. Circadian Rhythm—genetics. 2. Circadian Rhythm—physiology. 3. Molecular Biology. QH 527 M7175 2003] I. Sehgal, Amita.
QP84.6.M65 2003
571.7′7—dc21 2003000610

Printed in the United States of America

10 9 8 7 6 5 4 3 2 1

CONTRIBUTORS

John D. Alvarez, M.D., Ph.D., Department of Pathology, University of Pennsylvania School of Medicine, Philadelphia, Pennsylvania

Juan Capel, Ph.D., Departmento de Biologia Aplicada, Universidad de Almeria, Almeria, Spain

Anthony R. Cashmore, Ph.D., Plant Science Institute, Department of Biology, University of Pennsylvania, Philadelphia, Pennsylvania

David F. Dinges, Ph.D., Division of Sleep and Chronobiology, Department of Psychiatry, University of Pennsylvania School of Medicine, Philadelphia, Pennsylvania

Jadwiga M. Giebultowicz, Ph.D., Department of Biology, Oregon State University, Corvallis, Oregon

Jose A. Jarillo, Ph.D., Departmento de Biotechnologia, Instituto Nacional de Investigion y Technologia Agraria y Alimentaria (INIA), Madrid, Spain

Gerard A. Kerkhof, Ph.D., Department of Psychology, University of Amsterdam, Roetersstraat 15, Amsterdam, The Netherlands

Peter McNamara, Ph.D., Phenomix Corporation, La Jolla, California

Nirinjini Naidoo, Ph.D., Center for Sleep and Respiratory Neurobiology, Department of Medicine, University of Pennsylvania School of Medicine, Philadelphia, Pennsylvania

Jeffrey L. Price, Ph.D., School of Biological Sciences, University of Missouri-Kansas City, Kansas City, Missouri

Amita Sehgal, Ph.D., Department of Neuroscience, Howard Hughes Medical Institute, University of Pennsylvania School of Medicine, Philadelphia, Pennsylvania

Hans P.A. Van Dongen, Ph.D., Department of Psychiatry, University of Pennsylvania School of Medicine, Philadelphia, Pennsylvania

Julie A. Williams, Ph.D., Department of Neuroscience, Howard Hughes Medical Institute, University of Pennsylvania School of Medicine, Philadelphia, Pennsylvania

CONTENTS

PREFACE

It appears to finally be "time"—time to compile what we know about the molecular basis of circadian rhythms. In the mid-1990s, this would have seemed a preposterous idea; the fact that it is now being done is a testament to the progress made in the field in recent times. Up until 1994 there were only two known so-called "clock genes," one in Drosophila and the other in Neurospora. Since then, not only have many new components been identified in these two systems, but molecular circadian models have been rapidly developed for other systems such as cyanobacteria and mammals. In addition, there is increasing awareness of the overwhelming presence of circadian control in normal physiology. Clearly, temporal aspects of physiology must be considered for all biological processes. As a result, the circadian field has also grown in size, populated now not only by diehard chronobiologists, but also by many researchers who discovered that their favorite molecule or process is regulated in a circadian fashion.

My own motivation to put this book together came, at least in part, from the number of circadian biology queries I started receiving from students (graduate, undergraduate, and, in a couple of cases, even high school students). These students were basing their preliminary proposals or dissertation topics or research projects on the analysis of some molecular aspect of circadian rhythms. I must add that I was suitably impressed with the depth of the questions that these students were asking and with the effort they had made to read the primary literature. Another reason I thought a book like this might be useful is because I recently collaborated with several groups whose work lead them into the area of circadian rhythms and who would have appreciated the introduction to molecular circadian biology that I hope this book will provide. Finally, it appears that many graduate and even undergraduate institutions are now offering classes in chronobiology, either as part of a broader course or even as a course by itself. Now that the molecular biology of the system is such an integral part of circadian studies, a book on this topic may be beneficial. While the writing of this book was under way, I was also approached by another publisher, seeking to put a similar book together. Clearly, many people now perceive the need for a textbook explaining the intricacies of circadian function. I anticipate that this book will be useful to all the scientists mentioned above, in particular to undergraduate and graduate students.

Having decided to compile this book, it took me only a couple of minutes to decide that it should reflect the truly interdisciplinary nature of this field. Molecular approaches to circadian rhythms have been used in several, very diverse species, and the field as a whole has benefited tremendously from research done in each of these. In addition, the overall approaches used in all these species are very similar. Thus, the book starts with an introduction outlining the general properties of circadian rhythms and the description of commonly used terms (jargon) and then introduces the methodology used to study the molecular biology of clocks. It then provides an up-to-date (as much as is possible in this fast moving field) account of the research done in six organisms and systems commonly used for molecular circadian biology. The last section is devoted to the organization of circadian systems in complex organisms and the overall impact on physiology. The emphasis in this section is on mammals, although much of the discussion is also applicable to other organisms.

All the chapters have been written by people with expertise in the molecular analysis of circadian rhythms. Where possible, they have been written by researchers working on the specific organism they are writing about. In other cases, individuals working with other species have gone to tremendous effort to thoroughly research the system they were assigned and provide a comprehensive account of it. I am extremely grateful to all the contributors and would like to take this opportunity to thank each one of them. They all cheerfully honored their commitments and endured endless messages and questions from me. Special thanks goes to Jay Dunlap, Jennifer Loros, Susan Golden, and Carla Green for providing expert opinions on the Neurospora and cyanobacteria chapters.

My secretary, Irene Stevens, scanned many of the figures and kept track of copyright permits, for all of which I thank her. I would also like to thank members of my laboratory for their patience when my time was so heavily occupied by this book. Last, but definitely not least, I am indebted to my family, my husband Jeff and my daughters Natania and Anjalie, for their endless support. They have kept me sane throughout my research career, including this most recent period when I was writing and compiling this book.

Amita Sehgal
Philadelphia, Pennsylvania

ACRONYMS*

ACC	aminocyclopropane—carboxylic acid
ACTH	adrenocorticotropic hormone
ADH	antidiuretic hormone
AMP	adenosine monophosphate
APRR	arabidopsis pseudoresponse regulator
ASPS	advanced sleep phase syndrome
AVP	arginine vasopressin
BAC	bacterial artificial chromosome
bHLH	basic helix—loop helix
BMAL	brain—muscle ARNT-like (ARNT = arylhydrocarbon receptor nuclear transcriptor)
BP	blood pressure
BRET	bioluminescene resonance energy transfer
CAB	chlorophyll A/B
CAM	crassulacean acid metabolism
CaMK	calcium/calmodulin-dependent kinase
CAT	catalase; chloramphenicol acetyltransferase
CBT	core body temperature
CCG	clock-controlled gene
CCR	carbon catabolite repression
CED	*cis*-epoxycarotenoid dioxygenase
CiK	circadian input kinase
CK	casein kinase
CLD	cytoplasmic localization domain
CLIF	cycle-like factor
ClkRE	clock-responsive element
CLOCK	circadian locomotor output cycle(s) kaput
COP	constitutive photomorphogenic
CRE	cyclic AMP (cAMP)-response element (CREB = CRE binding)
CRF	corticotropin releasing factor
CRH	corticotropin releasing hormone
CT	circadian time
DA	dopamine
DBP	D-element binding protein
DBT	double-time
DD	constant darkness (dark-dark); differential display
DEC	differentiated (human) embryo chondrocyte
DET	deetoliate(d)
DMH	dorsal medial (nucleus of the) hypothalamus
DREG	Drosophila rhythmically expressed gene
DSPS	delayed sleep phase syndrome
EGFR	epidermal growth factor receptor
EMS	ethylmethanesulfonate
EMSA	electrophoretic mobility shift assay
ENU	*N*-ethyl-*N*-nitrosurea
EPAS	endothelial PAS (see PAS)
ERG	electoretinogram
EYFP	enhanced yellow fluorescent protein
FAD	flavine-adenine dinucleotide

* This list does not include very common terms (e.g., DNA, IR, UV).

FASPA	familial advanced sleep phase syndrome
FSH	follicle stimulating hormone
FLC	flowering locus C
FMN	flavine mononucleotide
FT	flowering (locus) T
GA	gibberellic acid
GABA	(gamma)-aminobutyric acid
GBF	G-box factor
GDRDA	genetically directed representational difference analysis
GFP	green fluorescent protein
GHRH	growth hormone releasing hormone
GHT	geniculohypothalamic tract
GnRH	gonadotropin releasing hormone
GRP	gastrin releasing peptide
GSK	glycogen synthase kinase
GST	glutathione-*S*-transferase
HEK	human embryonic kidney
HIOMT	hydroxyindole-*O*-methyltransferase
HKR	histidine-kinase-related
HLF	hepatic leukemia factor
HR	heart rate
IAA	indole-3-acetic acid
IGF	insulinlike growth factor
IGL	interogeniculate leaflet
IMLCSC	intermediolateral column of the spinal cord
IRBP	interophotoreceptor retinoid binding protein
LD	light—dark; long day
LDH	lactate dehydrogenase
LGN	lateral geniculate nucleus
LHC	light harvesting complex
LL	constant light (light-light)
LNv	ventrolateral neuron
LOV	light, oxygen, voltage
LRE	light-responsive element
MAP	mean arterial pressure
MAPK	mitogen-activated protein kinase
MOP	member of PAS (see PAS)
MPN	medial preoptic nucleus
MTHF	methenyltetrahydrofolate
NAD	nicotinamide adenine dinucleotide
NAT	*N*-acetyltransferase
NCAM	neural cell adhesion molecule
NLS	nuclear localization signal
NMDA	*N*-methyl-D-aspartate
NPAS	neuronal PAS (see PAS)
ORF	open reading frame
PACAP	pituitary adenylate cyclase activating polypeptide
PAH	planar aromatic hydrocarbon
PAI	plasminogen activator /inhibitor
PAR	proline and acid residue(s)
PAS	*P*ER, *a*rylhydrocarbon receptor of mammals and *s*ingle-minded gene in Drosophila
PCR	polymerase chain reaction
PDF	pigment dispersing factor
PEPC	phosphoenolpyruvate carboxylase
PEX	period extender
pl	isoelectric point
PI	phosphoinositide
PIF	phytochrome interacting factor
PK	prokineticin; protein kinase (PKA = protein kinase A)
POMC	proopiomelanocortin
PPAR	peroxisome proliferator-activated receptor
PRC	phase-response curve
PRR	pseudoresponse regulator
PS	phytochrome substrate
PSA	poly(sialic acid)
PT	pars tuberalis
PTH	parathyroid hormone
PTX	pertussis toxin
PVN	paraventricular nucleus
PVT	paraventricular (nucleus of) thalamus
QTL	quantitative trait locus
RA	retinoic acid
RAR	retinoid acid receptor
RCA	rubisco activase
RD	retinal degeneration (RDS = RD slow)

REM	rapid eye movement (sleep stage)
REMI	restriction-enzyme-mediated integration
RHT	retinohypothalamic tract
RGC	retinal ganglion cell
ROR	retinoid-related orphan receptor
RPA	RNase protection assay
RR	response regulator
RXR	retinoid X receptor
SAMDC	*S*-adenosylmethionine decarboxylase
SCG	supracervical ganglion
SCN	suprachiasmatic nucleus (mSCN, vSCN = medial, ventrolateral SCN)
SD	short day
SHM	serine hydroxymethyltransferase
SON	supraoptic nucleus
SPZ	subparaventricular zone
SSLP	simple sequence length polymorphism
SW	slow wave (sleep); sleep—wake (cycle)
TEF	thyrotroph embryonic factor
TGF	transforming growth factor
THR	thyroid hormone receptor
TRH	thyrotropin releasing hormone
TSH	thyrotropin secreting hormone
VIP	vasoactive intestinal peptide
WT	wild type
YAC	yeast artificial chromosome
ZEP	zeaxanthine epoxidase
ZT	Zeitgeber time

Part I

INTRODUCTION TO CIRCADIAN RHYTHMS

1

GENERAL CONCEPTS

Amita Sehgal

■ BASIC ATTRIBUTES OF CIRCADIAN RHYTHMS

The word *circadian* derives from the Greek words *circa* (about) and *dian* (day). Circadian rhythms then are cycles that occur with a periodicity of ~24 hours. Given that we live in a cyclic environment created by the rotation of the Earth, it is not surprising that adaptation has involved the circadian cycling of physiological and behavioral activities. What is amazing is the extent to which circadian rhythms pervade life on Earth. Previously thought to be a characteristic of eukaryotic organisms, circadian rhythms are now known to occur in prokaryotes such as cyanobacteria. Indeed, as we shall see in Chapter 6, much of the cyanobacteria genome, and by inference much of its physiology, is expressed in a cyclic fashion. Clearly, almost all organisms have evolved ways of coordinating their physiology such that different functions occur at different times of day. As we are becoming more aware of temporal aspects of regulation, we are realizing that there is scarcely a physiological process that does not involve some cycling component.

Example of Circadian Rhythm: Rest–Activity Cycle

Human sleep–wake cycles constitute perhaps the best known example of a behavior that occurs with a ~24-hour periodicity. Many other organisms display similar behavior, which is commonly referred to as *rest–activity cycles*, taking into account the fact that "rest" is not always "sleep." A record depicting rest–activity cycles in the fruitfly, *Drosophila melanogaster*, is shown in Figure 1.1. Each horizontal line in this record represents a 24-hour day, and the vertical lines correspond to the activity of a single fly. The activity of this fly was monitored

Molecular Biology of Circadian Rhythms. Edited by Amita Sehgal
ISBN 0-471-41824-2

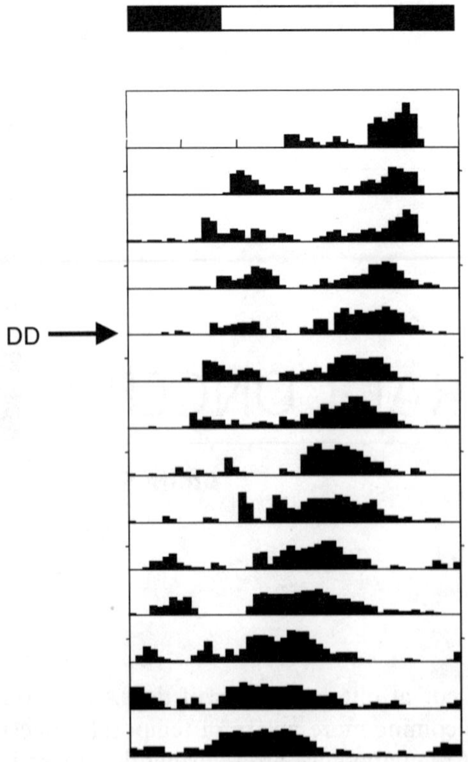

Figure 1.1. The circadian rhythm of rest–activity in a fruitfly. The "activity record" (also called an *actogram*) shown here depicts the rest:activity cycle of a single fly over a 14-day period. Each horizontal line represents 24 hours or a single day. The activity of the fly was measured by isolating it in a glass tube and monitoring its movement along the tube, based on the deflection of an invisible light beam projected into the tube (see also Chapter 3). The vertical ticks (shading) in the record correspond to activity. This particular fly was monitored in light–dark cycles for 4 days and thereafter in constant darkness. The position of the arrow indicates the transfer to constant darkness (DD). The light–dark bar at the top indicates the timing of the light–dark cycle. As is evident from the record, the circadian rhythm of rest:activity persists in constant darkness, but it now "freeruns" with its endogenous periodicity instead of being driven by the environment.

for a total of 14 days; the first 4 days were measured in the presence of light–dark cycles (12 hours of light followed by 12 hours of dark each day) and the last 10 in constant darkness. Activity of this fly shows daily consolidation, such that it occurs at the same time each day. This is true even when the light–dark cycles are removed. In addition, it is clear that this recurring pattern, or rhythm, occurs with a periodicity that is close to 24 hours. This record is referred to in the following sections to illustrate various properties of circadian rhythms.

Control of Circadian Rhythms by Endogenous Clocks

The issue then is how these rhythms are generated. The first indication that rhythms may be generated within an organism, as opposed to being driven by the cyclic environment, came from the observations of the French astronomer De Mairan in 1729. In observing the leaf movements of a *Mimosa* plant, he noticed that the 24-hour periodicity of the movement was maintained in constant darkness. Despite this observation, however, it was a long time before the endogenous nature of these rhythms was accepted.

Critics of the endogenous control idea argued that even under so-called constant conditions (e.g., constant darkness) it was not clear that all environmental influences relating to the Earth's rotation were absent. In the 1930s Kalmus and Bunning argued for an endogenous mechanism based on the finding that the periodicity of the rhythm varied with temperature fluctuations that were clearly independent of the Earth's rotation. Although their findings of temperature dependence were subsequently disputed by research showing that circadian rhythms are temperature-compensated (see discussion below), they were obviously right about endogenous control. In the 1940s the idea that endogenous clocks control circadian rhythms gained wider acceptance. What helped to reinforce this was the finding that the periodicity of circadian rhythms under noncycling environmental conditions varies from species to species. More importantly, the period in most cases did not precisely match that of any environmental cycle, indicating that it was most likely endogenously generated. For instance, in the record shown in Figure 1.2, note how, in constant darkness, activity each day begins and ends a little earlier than it did the day before. This drift occurs because the periodicity of this rhythm, and thereby that of the underlying clock, is a little less than 24 hours. In the presence of light–dark cycles, the drift does not occur because, as discussed below, light *resets* the clock.

Despite the general acceptance of the idea, some skeptics remained, and this skepticism inspired Hamner and others to conduct an interesting experiment in the early 1960s. This experiment was performed at the South Pole with a number of organisms, all of which were maintained on a rotating turntable. Although the turntable rotated with the same periodicity as the Earth, it did so in the opposite direction. Under these conditions, the circadian rhythm of eclosion in *Drosophila pseudoobscura* persisted, as did a number of other physiological rhythms.

It should be noted that not all rhythmic functions are controlled by an endogenous clock. Some are driven by the day–night cycle and therefore do not persist in constant conditions. These are referred to as *diurnal rhythms*, to distinguish them from circadian rhythms, which, by definition, persist in noncycling environmental conditions. In humans, rhythmic processes include sleep–wake cycles, release of several hormones, regulation of body temperature, blood pressure, urine production, cholesterol metabolism, respiratory function, and incidence of asthma and heart attacks. While some of these are clearly under circadian control, others may turn out to be dependent on other factors (e.g., behavioral state). For instance, the release of some hormones appears to be rhythmic because it is influenced by sleep, which normally occurs in a cyclic fashion, rather than by clock function (see Chapter 12).

Circadian Rhythm Properties

This periodicity of ~24 hours is one attribute of circadian rhythms. A second is that these rhythms can be synchronized or reset by environmental cues. The dominant environmental signal that affects rhythms is light, with most rhythms synchronized to the day–night cycle. Jurgen Aschoff used the term *zeitgeber* (“timegiver” in

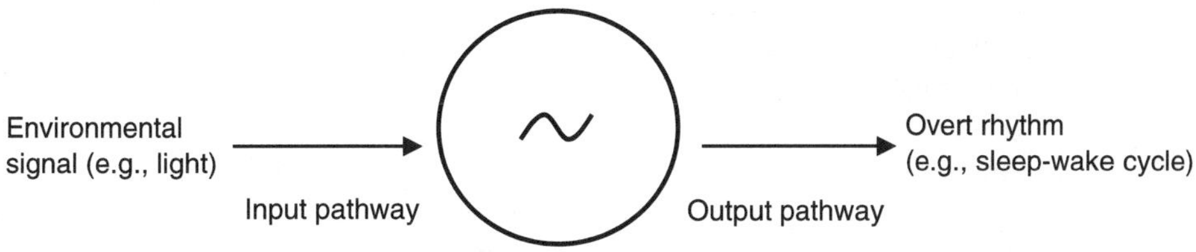

Figure 1.2. A simple circadian system. The very basic circadian system is thought to contain a clock, an input pathway that transmits environmental signals to the clock, and an output pathway that carries signals away from the clock and results in the manifestation of overt rhythms. In practice, there can be multiple input and output pathways to serve a single clock. In addition, the system most likely does not work in the linear fashion depicted here. Outputs can feedback on the clock and perhaps even on input and the clock can feedback on the input pathway.

German) to denote a periodic environmental signal that synchronizes or sets the time of a circadian rhythm. The process of synchronization itself is called *entrainment*. Since most rhythms are synchronized to environmental cycles, chronobiologists refer to time as defined by these environmental cycles, thereby calling it *zeitgeber time* (ZT). When an organism is maintained in a 24-hour cycle that consists of 12 hours of light, followed by 12 hours of darkness, zeitgeber time or ZT0 corresponds to "lights on" and ZT12 is "lights off." Thus, all timepoints between 0 and 12 refer to daytime hours, while those between 12 and 24 (which is the same as "0") refer to nighttime hours. This terminology will be used throughout this book.

Light then is a major zeitgeber that entrains circadian rhythms. The time at which a rhythmic activity/process occurs defines the *phase* of that particular rhythm. In the record shown in Figure 1.1, the time at which the fly is active indicates the phase of its rest–activity rhythm. Likewise, the time at which it sleeps, or rests, can also be used as a measure of phase. Typically, the parameter that is most constant from day to day (whether wakeup time or sleep time) is used to define the phase of a rhythm. It is clear that nocturnal and diurnal animals display opposite phases in their sleep–wake rhythms. A change in the timing of the zeitgeber, as might be caused by travel to a different time zone, results in a *shift* in the phase of the rhythm. However, as we all well know, this shift does not occur immediately, thereby resulting in the phenomenon commonly known as "jet lag." During the process of adjustment to a new cycle a rhythm displays "transient" behavior, consisting of erratic phase and periodicity.

To reiterate an important property of circadian rhythms, they can persist in the absence of cyclic environmental cues. Under these conditions they are said to be *freerunning*. The endogenous nature of a rhythm and its periodicity have to be determined under freerunning conditions. Again, note how in Figure 1.1, the pattern of the rhythm in constant darkness (freerun) is somewhat different from that in the presence of light–dark cycles. This is because in constant darkness it runs with its endogenous periodicity. Under these conditions, we speak in terms of *circadian time* (CT), which is the time of the internal clock. Since the periodicity of the internal clock is frequently a little different from 24 hours, CT is not the same as ZT. However, on the first and second days of constant darkness, when the endogenous rhythm has not drifted significantly from the environmental cycle, CT may approximate ZT for an organism that has a periodicity close to 24 hours, and researchers may use it as such.

Finally, the periods of circadian rhythms are temperature compensated. This was demonstrated in some classic experiments done by Colin Pittendrigh in the 1950s. He determined the effects of temperature on the eclosion (hatching of adult flies from pupae) rhythm in *Drosophila pseudoobscura* flies. In his first series of experiments he improvised "incubators"—a pressure cooker maintained in a cold-water creek and an outhouse darkroom, to maintain the flies at cold and warm temperatures respectively. He found that the periodicity of the rhythm was relatively the same (within an hour) at the two temperatures. In subsequent experiments done at Princeton University, he confirmed these findings, showing that the periodicity was essentially the same at 26 and 16°C.

Thus, the circadian period remains constant over a wide temperature range. Assuming that the underlying basis of overt rhythms is some biochemical mechanism, as indeed must be the case, then normally the changes in protein function at different temperatures would alter the rate of the reaction and thereby the periodicity of the rhythm. The rate of most physiological processes doubles with a 10°C rise in temperature. The fact that this does not happen to circadian periodicity indicates that there is some mechanism that compensates for these temperature-induced changes. Clearly, it is important for organisms to maintain a periodicity that approximates that of the day–night cycle despite changes in environmental temperature. While homeotherms can obviously do this with no problem, organisms that do not maintain a constant body temperature had to evolve mechanisms for this purpose.

Circadian Clock Input/Output

Circadian clocks are part of a system that includes input and output components (Fig. 2.2). If rhythms are generated by an endogenous clock—and, indeed, at this point we can say that they are, based not only upon the investigations mentioned above, but also upon a vast amount of genetic and molecular data that are the subject of this book—then, if follows that the properties of rhythms arise from inherent characteristics of the clocks that produce them. In fact, clocks are entrained by external/environmental signals through what is termed an *input* or *entrainment* pathway. The implication is that the clock itself may not be in contact with the environment, but can communicate with it through a distinct pathway(s). Since the predominant entraining cue is light, the input pathway is usually perceived as containing a photoreceptor. As we shall see in subsequent chapters, the identification of the photic input pathway is an intense area of research in every organism used for the study of circadian biology.

Time-of-day signals generated by the clock must be transmitted to other parts of the organism to drive physiology/behavior in a rhythmic fashion. The pathway that carries signals away from the clock and produces overt rhythms is called the *output pathway*. Since outputs can vary greatly, even within the same organism, it follows that there must be more than one output pathway. In fact, we now know (as described below) that many of these outputs are actually generated not just by independent output pathways, but by separate clocks within an organism.

Together, the clock, input pathway and output pathway are termed a *circadian system*. Although the components of this system are typically depicted in a linear diagram, this is an oversimplification of the mechanisms involved. In many cases, the outputs can feed back and affect the clock (e.g., forced activity can reset the clock), and the clock may feedback to regulate input, resulting in the response to light being stronger at certain times of day. The response to light is said to be "gated" by the clock.

Clock Entrainment to Input Components (Light)

As mentioned above, most biological clocks have periodicities that are not precisely 24 hours. Aschoff suggested that the periodicity of the same organism may be different under different conditions, depending on the functional state of the organism or even the nature of the "constant" environment (e.g., constant light vs. constant dark). He was intrigued by the fact that light-active (diurnal) animals show an increase in spontaneous frequency, as measured by increased bouts of activity, with increasing intensity of light. This increased frequency of the oscillation results in an overall shorter period. Night-active (nocturnal) animals, on the other hand, decrease their spontaneous frequency with increasing intensity of light, thereby displaying longer periods. On the basis of these observations, Aschoff proposed a circadian rule according to which diurnal animals have shorter than 24-hour rhythms in LL (constant light) and longer than 24-hour rhythms in DD (constant darkness). The reverse would be true in nocturnal animals.

However, as Aschoff himself recognized, there are exceptions to this rule. Drosophila are diurnal and yet display periodicities that are slightly less than 24 hours (~23.8 hours) in DD.

Regardless of the periodicity of the endogenous rhythm all organisms are synchronized to the 24-hour environmental cycle. This means that they must be reset every day because otherwise they would eventually drift out of synchrony with the day–night cycle. Take, for instance, a rhythm that has a period of 23.8 hours. If it were to freerun, then on each successive day it would be an additional 0.2 hour out of phase with the environment. In 2 months it would have a reverse phase, relative to the environment, from the one with which it started. A rhythm such as this, which has a shorter period than the environmental cycle, is delayed by 0.2 hour each day. This is illustrated by the example shown in Figure 1.1, where the endogenous periodicity of the fly is ~23.5 hours, but it synchronizes to a 12–12 light–dark cycle, because it is reset each day. A longer period rhythm must advance each day by the amount of time its periodicity differs from that of the day–night cycle. The manner of the resetting can differ between rhythms. A rest–activity rhythm could advance each day by initiating either the rest or the active phase earlier than dictated by the internal clock. These earlier onsets are typically driven by something in the environment (e.g., sunrise or sunset).

As will be seen in subsequent chapters, even organisms that lack clock function, through genetic manipulation, are driven by the environmental cycle to display 24-hour rhythms. The difference between these rhythms and those in clock-containing organisms is that they cannot persist when the rhythmic signals from the environment are removed. In addition, they cannot anticipate the daily transitions in the environmental cycles. For instance, in a diurnal animal that lacks a clock, activity is usually triggered by the light signal. The animal does not become active in anticipation of the dark–light transition. In a nocturnal animal light would have the opposite effect. This effect of light that is independent of the endogenous clock is termed "masking," so called because it can mask the manifestation of the endogenous rhythm. Thus, even in an animal that has a clock, masking effects of light can confound analysis of the endogenous rhythms, which are therefore best assayed under freerunning conditions (i.e., constant darkness and constant temperature). It should be noted that although masking, as described here, denotes clock-independent effects of light on the activity rhythm, it can, in principle, apply to any influence that obscures a rhythmic function.

Entrainment to Light Pulses

In addition to being entrained to day–night cycles, circadian rhythms can be entrained by pulses of light. This type of entrainment is called *nonparametric*, distinguishing it from *parametric* entrainment which involves synchronizing to cycles of a stimulus. In general, a pulse of light delivered when an organism expects to see light ("subjective day", as it is called in constant darkness) does not reset the rhythm. However, pulses at night reset the rhythm with those in the first part of the night having effects opposite those in the second half of the night. A pulse in the early night delays the phase of the rhythm. It can be thought of as analogous to extended daylight, which has the effect of delaying the onset of the night function (e.g., in a diurnal animal it might delay the onset of sleep or rest). Conversely, a pulse in the second half of the night advances the phase of the rhythm. Again, the best analogy perhaps is with a shift in the natural day–night cycle. In this case, it would be like dawn occurring earlier, which would lead to earlier onset of the next day's activities. The idea is that a pulse in the early night resets the cycle to dusk, while one in the late night resets it to dawn. In practice, of course, the resetting is not quite so perfect. Delays are usually of larger magnitude and easier to achieve than advances.

A graph that plots the magnitude of the shift relative to the time of day that the shifting stimulus is delivered is called a *phase-response curve* (PRC). A typical such curve is shown in Figure 1.3a. As mentioned above, there is no shift during the subjective day, which is thereby called the "dead zone." In the middle of the night, the curve crosses over from delays to advances. In this type of PRC, also called a "Type 1 PRC," pulses of light at the crossover point do not produce any shift in the rhythm. Some organisms are more sensitive to light and display a Type 0 PRC (Fig. 1.3b). A Type 0 PRC is characterized by larger magnitude shifts and the absence of any crossover point in the middle of the night. This lack of a crossover typically gives a false impression of discontinuity in the PRC; in a rhythm with a 24-hour period, a 12-hour delay is indistinguishable from a 12-hour advance, so a shift from plotting strong delays to weaker advances produces the apparent discontinuity in the plot. Although previously regarded as very distinct entities, there is now increasing evidence that with strong light pulses (either of higher intensity or increased duration) a Type I PRC can be turned into a Type 0.

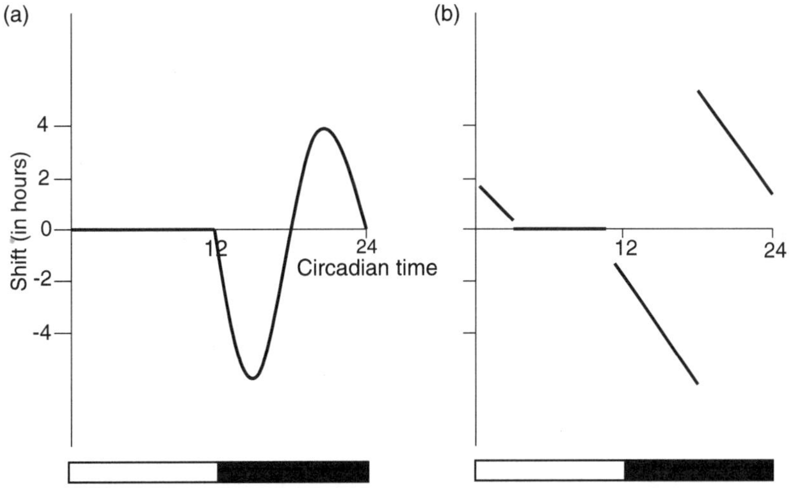

Figure 1.3. Entrainment to pulses of light. Based on the response of an organism to resetting stimuli delivered at different times of day, phase response curves (PRCs) can be constructed. This type of curve basically demonstrates the effect of time of day on the magnitude of the phase shift produced by a specific stimulus. The best documented PRCs are those in response to light stimuli and these are of two major kinds, shown in (a) and (b). The light–dark cycle used to entrain the organism is indicated at the bottom of each plot. The number of hours by which the phase of the clock shifts (*y* axis) in response to a light pulse delivered at different times of day (*x* axis) is plotted. In a Type 1 PRC (a), pulses of light delivered during a time when an organism would expect to see light (daytime hours) have no effect on the phase of the rhythm. A pulse of light in the early part of the night delays the phase (indicated by negative numbers), and a pulse in the second half of the night advances the rhythm (indicated by positive numbers), with a gradual transition between phase delays and phase advances. In a Type 0 PRC, the clock is very sensitive to light and shows large shifts, in response to pulses, at all times of the cycle, with no gradual change from phase delays to phase advances.

Multiple Clocks

An organism can have more than one circadian clock. It has now become clear that most multicellular organisms have more than one clock. It appears that with increasing complexity of physiological function, there is increasing need to have clocks specific for different organs/tissue systems. The question then becomes: Is there one clock that controls all others in an organism —a master clock, if you will—or are all clocks equal? Among the animal systems, the master or central clock (if there is one) is usually located in the brain (Fig. 1.4). Clocks in other parts of the body are referred to as *peripheral oscillators* to distinguish them from the central clock. The degree of autonomy that these oscillators possess varies from species to species and also from oscillator to oscillator. Insects contain many clocks that are autonomous with respect to timekeeping as well as with respect to photoreception. However, there is at least one oscillator in Drosophila, the insect of choice for molecular analysis

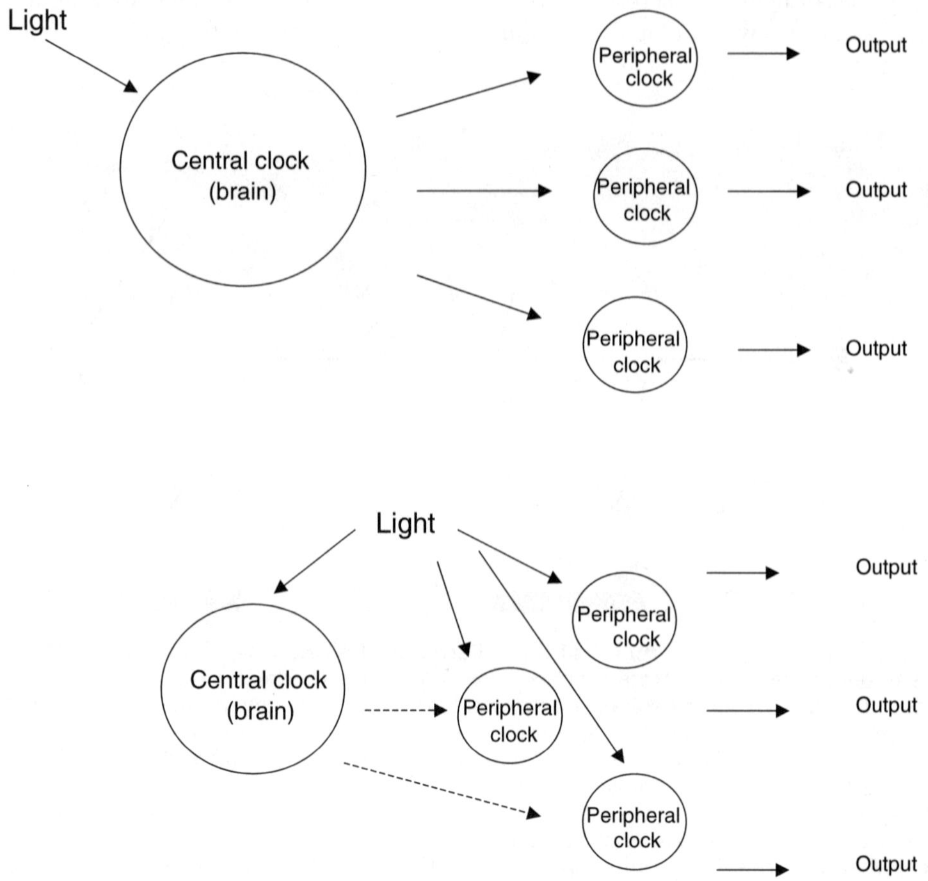

Figure 1.4. Organization of clocks in complex organisms. In mammals, a light-responsive central clock in the brain regulates clock activity in clocks found in peripheral tissues (top). The peripheral clocks are not directly photosensitive although they can entrained by other signals (e.g., hormones). In other organisms, peripheral clocks can be photosensitive and vary in their dependence upon the central clock.

of clock function, that is controlled by the central clock although it may also be capable of direct photoreception. Likewise, most oscillators in the nonmammalian vertebrate, zebrafish, are autonomous and have their own photoreceptors, although interactions with the central clock likely occur. Other nonmamalian vertebrates, that do not have the transparent skin of the zebrafish, may not show this extent of autonomy in peripheral oscillators.

In mammals, there is a master clock that is housed in the hypothalamus and is called the *suprachiasmatic nucleus* (SCN). Thus, clock functions appear to have become centralized over the course of evolution as have other body functions (Fig. 1.4). The SCN can sustain clock activity for extended periods of time in culture. It is also the only clock in the body, with the exception of the eye clock, which receives photic information. Details of this can be found in Chapter 4.

Although peripheral oscillators in mammals cannot perceive light, they can be entrained by other physiological stimuli. Most notable are hormones, which might be controlled by the SCN, but may also act independently on peripheral oscillators. In addition, retinoids will reset a clock in the vasculature, and restricted feeding can entrain the clock in the liver without affecting the SCN. Clearly, the organization of circadian systems in terms of central and peripheral oscillators, the interactions between the different components and the mechanisms that entrain them are all important aspects of physiology. These issues are discussed in the chapters on peripheral oscillators and hormonal rhythms. Specific details that pertain to individual organisms can be found in the chapters on these respective organisms.

■ CLOCK MECHANISMS

Since the general acceptance of the idea that clocks are endogenous, a major question has been: How, within an organism, do you make a clock? Over the years, a number of mathematical models have been developed to explain how this might occur. One aspect that these models have in common is the presence of a feedback loop. Overt cycles must have their basis in an endogenous cycle or loop that can sustain oscillations indefinitely in the absence of environmental cues. The question then is how these loops are generated. Do they operate at the level of the organism, the tissue, or the cell? As outlined below and described in detail in other chapters in this book, in all organisms examined, the basic clock mechanism (i.e., the oscillatory loop) is, in fact, intracellular. While an intracellular clock might be expected of a cyanobacterium or of the bread mold Neurospora, it comes as a surprise when it turns out to be the case in a mammal. This is not to say that timekeeping is performed by a single cell. In multicellular organisms, a clock tissue (central or peripheral) is likely to be composed of a cluster of cells, each of which contains a loop that oscillates in synchrony with the loops in neighboring cells. An SCN is an example of such a cluster. Rhythmic signals generated by these clusters drive output rhythms.

As one might have gathered from the discussion above, the general nature of the clock mechanism is conserved from cyanobacteria to mammals. However, since circadian rhythms are present in widely separated phyla and the actual molecules involved are different in the diverse phyla, the prevailing notion is that clocks evolved several times. The fact that the same mechanism evolved in each case attests to the efficacy of this particular mechanism. In addition, the availability of multiple, biological systems that use the same mechanism has allowed a powerful, multidisciplinary approach to the study of clocks. Particularly important have been studies in model organisms, which offer powerful genetic approaches and simpler systems than more complex vertebrates. A general model for the clock mechanism was first worked out in Drosophila and then in Neurospora. Mammalian studies of clock function built largely on the information that was already available from these other systems, in particular on the framework provided by the Drosophila model. This book highlights this multiple-species approach to the molecular analysis of rhythms by describing the salient features of the clock, input and output pathways in different species and comparing, wherever applicable, the similarities in the systems.

Clock Mechanism at a Molecular Level

The basic clock mechanism is a feedback loop in which oscillating products of specific genes regulate their own expression. These genes are called *clock genes*, and the autoregulation consists of the proteins negatively regulating synthesis of their own mRNAs (Fig. 1.5). Each complete turn of the loop takes ~24 hours to complete, resulting in circadian oscillations of RNA and protein levels. Rhythmic expression of the proteins ensures that the negative feedback on transcription is rhythmic, which, in turn, ensures rhythmic expression of the mRNA. The precise molecules that function in this fashion may differ from one species to another although certain structural features are conserved from cyanobacteria to mammals. Between Drosophila and mammals, the molecules are also conserved.

Given that this is a cyclic process, it is, of course, difficult to ascertain what comes first—the RNA cycling or the protein cycling. It is clear that the mRNAs do not cycle in the absence of rhythmic feedback from the proteins. However, the most

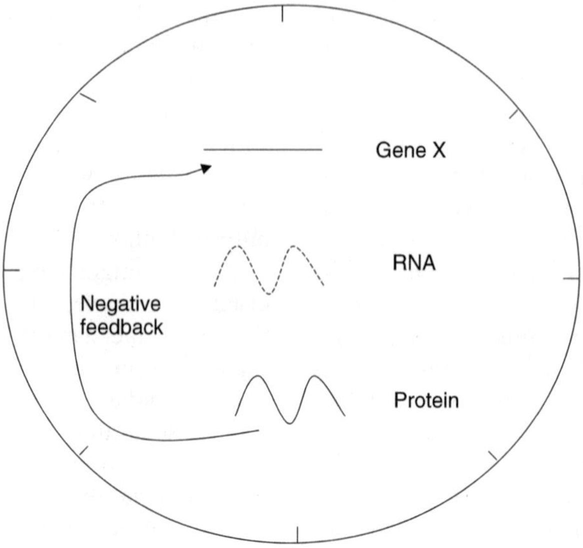

Figure 1.5. The basic model for a 24-hour oscillator. The model is based on work done in many different species ranging from cyanobacteria to mammals and consists of an autoregulatory feedback loop in which a clock gene encodes cycling RNA and protein, and the protein negatively regulates synthesis of its own mRNA. The loop takes ~24 hours to complete (shown within a 24-hour clock).

recent data indicate that cycling of specific clock proteins can be maintained under conditions where their RNAs do not cycle. Posttranslational mechanisms, such as cyclic phosphorylation, appear to contribute to rhythmic expression of clock proteins and may account for oscillations of these proteins under conditions where cycling of the corresponding mRNAs is blocked. It is important to note also that these experiments have not eliminated the cycling of all clock gene mRNAs in an organism, so the cycling of one may be able to compensate for loss of cycling of the others.

Although we have refrained from the use of chronobiology jargon in this book, it is worth introducing some of the terminology used in the field here because it may appear in other readings of the circadian literature. A molecule that is part of the clock is frequently referred to as a *clock component*. A clock component can be a *state variable*, which is a molecule that imparts time-of-day cues through rhythmic changes in its levels or activity, or a *state parameter*, which does not itself cycle, but is required for the appropriate regulation of the state variable. Researchers have developed criteria for state variables, which include a requirement that its levels or activity cycle, that it regulate itself through a feedback mechanism and that mutations in it produce effects on overt rhythms.

Gene Expression

We are starting to also gain an understanding of how it is that clocks communicate their temporal signals to the rest of the organism, although these "output pathway" mechanisms represent the least understood aspect of circadian rhythms. The diversity of overt rhythms, and thereby that of the output pathways that lead to these rhythms, may account, in part, for the dearth of knowledge in this area. Clearly, the molecules and mechanisms that produce rhythmic rest–activity will be very different from those that drive a rhythm in cholesterol metabolism in the liver.

What is clear, however, is that one of the ways clock genes transmit output signals is by driving rhythms of gene expression. Although the number varies from species to species, we now know that a significant proportion of the genome is expressed rhythmically in all organisms that possess clocks. The number as well as the nature of the genes expressed in a rhythmic fashion vary not only from species to species but also from one organ or tissue to another in the same organism. A gene expressed rhythmically in one tissue may not be expressed rhythmically in another tissue, because it serves a different function in the other tissue or because the process with which it is involved does not cycle in the other tissue. The rhythms in gene expression are translated into rhythms in protein expression and function that ultimately give rise to rhythms in physiology and behavior.

The most direct mechanism by which the clock can drive a rhythm in gene expression is by rhythmically regulating promoter activity. Promoters of many genes contain elements recognized by transcription factors that are part of the clock. Rhythmic activity of these factors results in rhythmic expression not only of clock genes but also of genes that are downstream of the clock (Fig. 1.6). These downstream genes may encode other transcription factors or other proteins that affect a cell property or function.

However, it is important to note that clocks can also regulate downstream components through transcription-independent mechanisms. Although the mechanisms are not known yet, there are examples of circadian components that cycle at the protein, but not the RNA level. In addition, in Drosophila, a neuropeptide that is released specifically by brain clock cells cycles at axon terminals. Since expression of this protein does not cycle in the cell body, the interpretation is that the levels do

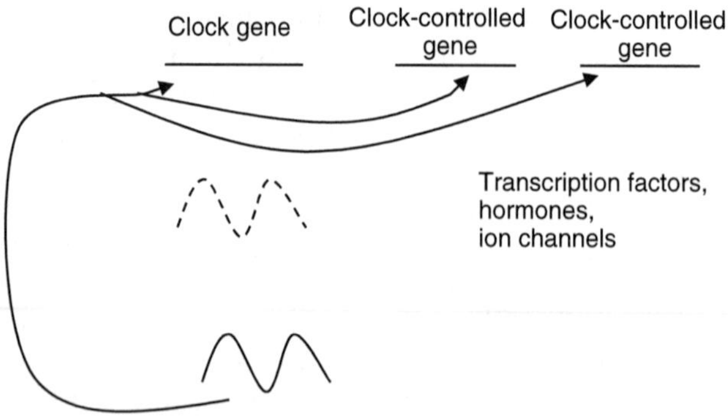

Figure 1.6. Control of downstream genes by the clock. A molecular clock (such as the one shown in Fig. 1.7) can regulate the output pathway by driving cyclic expression of downstream genes. One mechanism by which this occurs is shown here. While regulating its own transcription, the clock protein also regulates the transcription of downstream genes at a specific time of day. This leads to rhythmic expression of those genes.

not cycle, but the release from axon terminals does. A simplistic model for circadian control of a complex function, such as rest–activity behavior, is that oscillations in clock genes result in oscillations in some cellular (neuronal) property. This oscillation in turn drives rhythmic expression of an output (perhaps a hormone) from the cell, and that output mechanism then acts rhythmically on cells controlling rest–activity (Fig. 1.7). This results in the manifestation of rest–activity rhythms. In a simple circadian system the final output rhythm may be produced directly by the clock cell.

Thus the process is more complicated than would appear at first glance. These complexities, most of which are not completely understood yet, are acknowledged here and elsewhere in this book. However, at the same time, it is clear that we have learned a lot about how clocks operate in many different species. Amazingly, most of this progress has occurred very recently (as of 2003). In the late 1990s a book focused on molecular mechanisms underlying clock function might have seemed premature. Now it is perceived as something that is needed. It is important to note, though, that this is still a rapidly moving and growing field and so the material presented here provides the basis, but by no means the complete picture, of molecular clock function.

■ ADAPTIVE SIGNIFICANCE OF CIRCADIAN RHYTHMS

An introduction to circadian rhythms would, of course, be incomplete without some discussion of their adaptive significance. Several possible explanations have been proposed. In 1987 Rapp argued that biological systems are periodic because temporal organization allows for better synchrony or coordination between different processes in the same organism, it allows incompatible processes to be separated from each other in time (this is particularly advantageous in unicellular organisms; see Chapter 6) and it allows for a more precise, efficient system.

Perhaps it is easy to see how adaptation to a cyclic environment would be facilitated by cyclic physiology or behavior. However, most rhythms can be driven by the environmental cycle. As mentioned above, even organisms that lack clocks, or contain

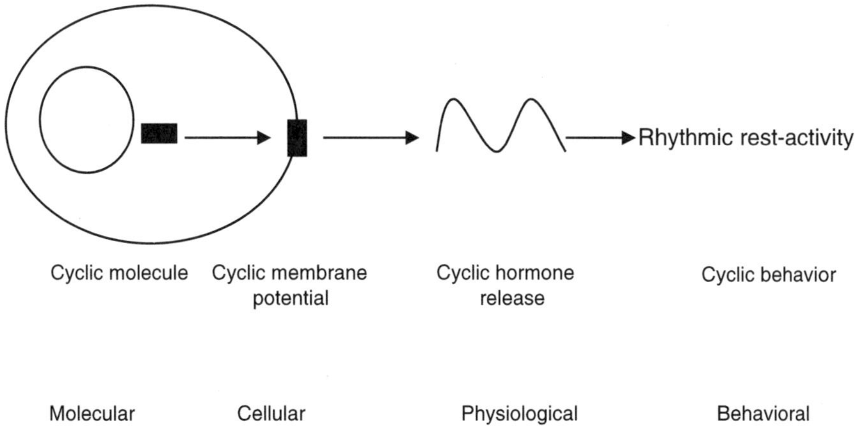

Figure 1.7. Control of an overt, behavioral rhythm by the clock. A simple model is shown in which the clock drives rhythmic expression of a gene that regulates membrane potential. Cycling membrane potential results in rhythmic release of a secreted factor that has an effect on levels of rest and activity.

mutant clocks with aberrant periodicity, can be driven to display rhythmic activity in a 12-hour light–12-hour dark cycle. Given this, why do we then need endogenous clocks? The most commonly cited explanation for endogenous rhythmicity is that it enables organisms to anticipate daily transitions in the environment. It is stated above that changes in the activity patterns of clockless animals are triggered by the environmental stimulus (say, light). An animal with a clock becomes active in anticipation of the dark–light transition, which may confer a selective advantage on it. It is a case of the early bird getting the worm. In addition, a clock allows finer temporal control over physiological processes.

The advent of cellular and genetic tools allowed researchers to experimentally test the importance of clocks. The basic experiment is to remove the clock and assay survival or fitness. Cellularly, this has been done by lesioning the SCN. Although we now know that clocks are found in many tissues outside the SCN, the master clock in mammals is in the SCN. In addition, genetic tools have allowed the generation of animals that lack one or more clock genes in all tissues and so are arrhythmic. Interestingly, early attempts to identify viability or fertility deficits in these animals failed. The advantage conferred by clocks may not be evident under laboratory conditions where the animals are supplied with ample food and water and do not have to fend for themselves. However, more detailed analysis is now revealing deficits associated with loss of clock function. Furthermore, researchers found a small reduction in lifespan in some of the Drosophila mutants that affect circadian period. The differences in lifespan between wild type and mutants were altered somewhat in light–dark cycles of different lengths. However, the reduction in lifespan was seen even when the mutants were maintained in a light–dark cycle that approximated the periodicity of their internal clock. Thus, the advantage of having a ~24-hour rhythm in Drosophila may extend beyond the need to match the periodicity of the environmental cycle. In cyanobacteria, synchrony with the environmental cycle confers a selective advantage. When cyanobacterial strains of different periodicities are grown in the same culture, the ones that grow best, and thereby outcompete their neighbors, have periodicities that match that of the environment. Finally, studies of SCN-lesioned mammals released into the wild have also revealed advantages of having clocks. The lack of a clock appears to make these animals more susceptible to predators.

FURTHER READING

Aschoff, J (1960): Exogenous and endogenous components in circadian rhythms. *Cold Spr Harb SympQuant Biol* **25**: 11–28.

Bunning, E (1973): *The Physiological Clock. Circadian Rhythms and Biological Chronometry*, The English Universities Press.

DeCoursey, PJ, Walker, JK, Smith, SA (2000): A circadian pacemaker in free-living chipmunks: Essential for survival? *J Comp Physiol [A]* **186**: 169–180.

Dunlap, JC, Loros, JJ, DeCoursey, P (eds) (2003): *Chronobiology: Biological Timekeeping*, Sinauer Assoc.

Klarsfeld, A, Rouyer, F (1998): Effect of circadian rhythms and LD periodicity on the lifespan of *Drosophila melanogaster*. *J Biol Rhythms* **13**: 471–478.

Mori, T, Johnson, CH (2001): Circadian programming in cyanobacteria. *Semin Cell Dev Biol* **12**: 271–280.

Pittendrigh, CS (1954): On temperature independence in the clock system controlling emergence time in Drosophila. *Proc Natl Acad Sci* (*USA*) **40**: 1018–1029.

Pittendrigh, CS (1993): Temporal organization: Reflections of a Darwinian clock-watcher. *Ann Rev Physiol* **55**: 17–54.

Rapp, PE (1987): Why are so many biological systems periodic? *Prog Neurobiol* **29**: 261–273.

Saunders, DS (1982): *Insect Clocks*, Pergamon Press, Oxford (Oxfordshire); New York.

2

GENETIC AND MOLECULAR APPROACHES USED TO ANALYZE RHYTHMS

Amita Sehgal and Jeffrey L. Price

Since the approaches used to study the molecular basis of circadian rhythms are, to a great extent, the same in different species, it is worth discussing the general nature of these approaches. Genetic approaches have been by far the most successful at unraveling the intricate mechanisms that underlie clock function. Whether in cyanobacteria, Arabidopsis, Neurospora, Drosophila, or mouse, genetic screens, in which organisms subjected to random mutagenesis are screened for circadian phenotypes, have provided a wealth of information. They have identified genes involved in every aspect of the circadian system. What we have since learned about some of these genes, in particular the genes that are actually part of the clock, has led to a standard set of questions that are now asked about every new candidate circadian gene. These questions include

- Do levels of the RNA and protein cycle?
- Is it expressed in clock cells?
- Is it part of the clock?
- Does it show an acute response to light?
- Does it interact with other clock proteins?

As a result, regardless of the organism under study, these are the questions that must be addressed. Although the precise methodology may differ, the overall approaches are the same and are discussed below.

Molecular Biology of Circadian Rhythms. Edited by Amita Sehgal
ISBN 0-471-41824-2

■ USE OF GENETIC SCREENS TO IDENTIFY CIRCADIAN GENES

Random Mutagenesis Screens

In the absence of any initial knowledge about the molecules that participate in a process, a genetic analysis is frequently the best way to initiate an analysis of the process (see Fig. 2.1). A genetic analysis requires no *a priori* assumptions about the mechanism of the process. Instead, it identifies mutants in which the process (e.g., a circadian rhythm such as locomotor activity) is altered. The identity of the affected genes and the elucidation of the pathways with which they are involved may reveal the mechanism. Conceptually, genetic analysis is analogous to figuring out how an automobile works by removing or altering parts, and then determining the effects of this manipulation on the functioning of other parts and of the car as a whole. The difference is that one does not choose the part to remove as the mutants are generated through random mutagenesis. To maximize the chances of detecting a handful of circadian genes from the vast, total number of genes in the genome, large numbers of mutagenized individuals must be screened.

The methods used for mutagenesis vary from species to species. Classical methods use chemical mutagens, the concentrations of which are titrated to yield maximal mutagenesis without overly compromising the viability or fertility of the organism. Another popular method for mutagenesis, particularly in Drosophila, is transposon mobilization. Transposons are small segments of extraneous DNA that can insert into chromosomes largely, although not entirely, in a random fashion. They can be mobilized by introducing an enzymatic activity (transposase), and they frequently affect gene expression at sites of insertion.

In order to select for germline mutations, it is the progeny of mutagenized individuals that are screened in the assay of interest. The assay historically consists of measurement of an overt rhythm, although researchers have also developed methods to screen for disrupted molecular rhythms. Also, to date, the emphasis has been on freerunning phenotypes, and so the rhythm is assayed under constant conditions. Candidates that show aberrant phenotypes are bred or propagated and subsequent generations are further analyzed.

Mutant Phenotypes

The mutant phenotype may provide some clues regarding the role of the gene in the

Figure 2.1. Isolation and analysis of circadian genes through forward mutagenesis. A schematic representation of the standard protocol is shown. In a forward mutagenesis screen the organism is mutagenized and the progeny are screened for circadian phenotypes. Mutants with interesting phenotypes are rescreened, and the relevant mutations are then mapped and subjected to molecular cloning. Once the gene is in hand, cDNAs (complementary to the mRNA) for it can be obtained, and these provide probes for RNA detection and reagents for the production and analysis of protein. The gene itself, with its promoter, exons, and introns, provides DNA elements that regulate expression of the gene (e.g., the promoter and elements that mediate RNA stability or splicing). The cDNA and genomic clones (either wild-type or mutant forms constructed by the experimenter) can also be expressed in transgenic cells or organisms to test specific hypotheses for the circadian mechanism. For instance, as noted in the text, the promoter could be fused to a reporter like luciferase and expressed in the organism to determine if it drives cyclic expression. It could also be expressed in cultured cells and tested as a target for candidate transcription factors.

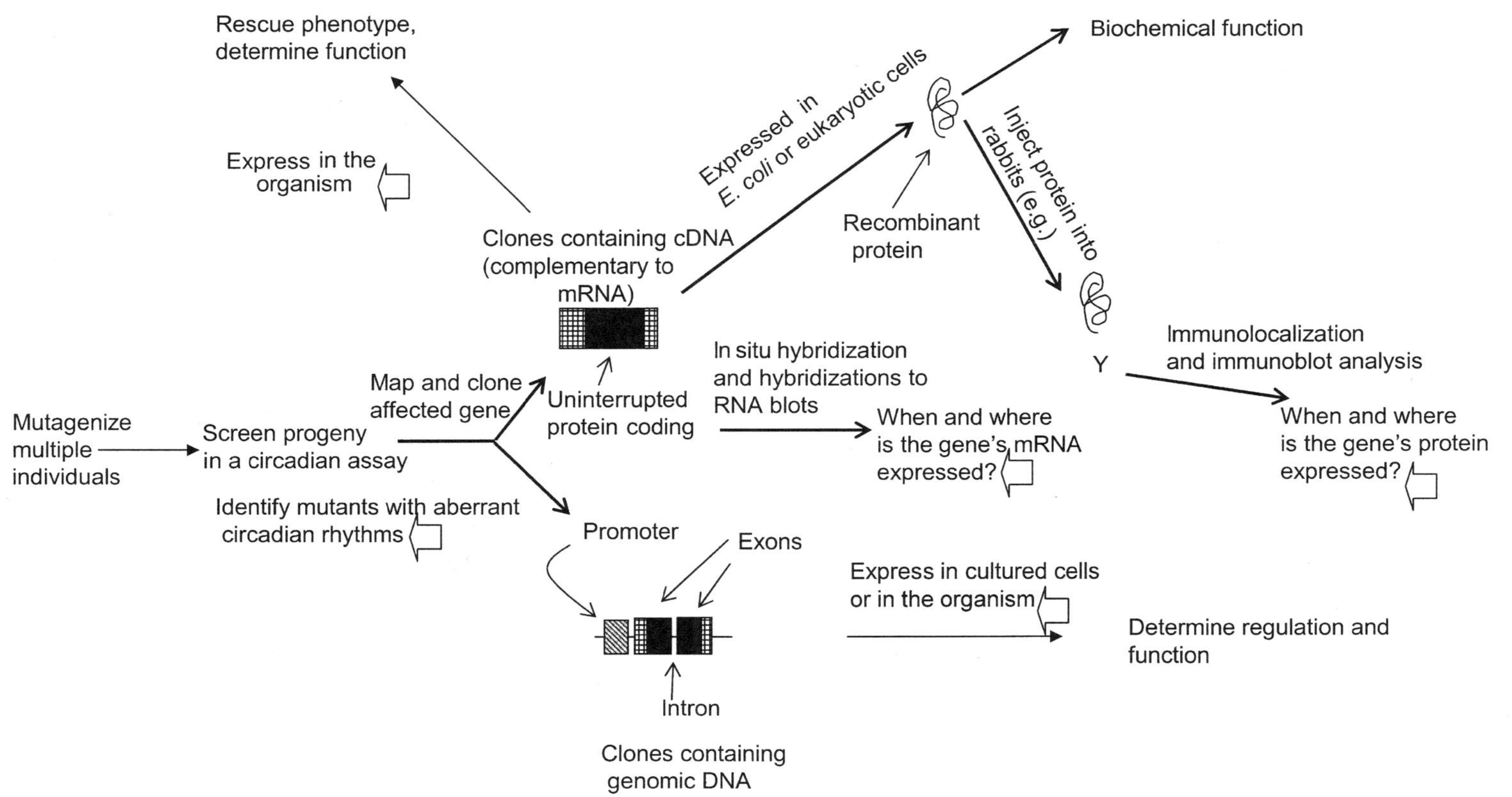

The indicated phenomena (⇦) are examined in wild-type flies and in flies with mutations in this gene or mutations in <u>other</u> genes circadian system. Genetic analysis may elucidate input and output pathways, as well as the mechanism of the circadian oscillator. As discussed in Chapter 1, the circadian paradigm consists of *input pathways*, which keep the clock in phase with environmental cues such as light and temperature cycles (a process termed *entrainment*); a *circadian oscillator*, which is entrained by the input pathway but capable of sustaining an oscillation in the absence of environmental cues; and *output pathways*, which convey the temporal information of the oscillator to the diverse processes it regulates (Fig. 1.2). While this paradigm may be somewhat simpler than the system is in reality, it has been of enormous heuristic value. If a gene is necessary for any of these pathways, a mutation producing a nonfunctional gene product can disrupt overt circadian rhythms such as those of rest–activity (Fig. 1.1). For instance, mutations of a particular gene that is necessary for a clock-controlled process, or for passing temporal information from the clock to the overtly rhythmic process (the "output'), would produce arrhythmia in the output process, while leaving the circadian oscillator intact. In an analogous manner, removal of the clutch will produce a car that cannot move, but the motor will still work. Just as one cannot conclude that a clutch is a component of the motor because the car will not move when it is taken out, one cannot conclude that a gene is essential for the central circadian oscillator because it can produce an arrhythmic output when mutated. In fact, some of the circadian mutants affect just one rhythmic output of the clock and can therefore be assigned to a specific output pathway.

Mutations that produce arrhythmia in all known circadian outputs are more likely to affect the central oscillator mechanism, but they could also affect an output pathway that is common to all known circadian outputs. For instance, if a circadian oscillator communicated temporal information to all its outputs with the rhythmic release of a neurotransmitter, a mutation that eliminated the synthesis of the neurotransmitter could produce arrhythmia in all of the known outputs without an effect on the oscillator mechanism per se. Eventually, molecular assays (discussed below) can be used to determine whether the oscillator mechanism is affected.

If a gene can be mutated to alter the period length of circadian outputs, it cannot be involved with just an output pathway. Its effect on the period length would indicate that it is affecting the basic timekeeping mechanism. This could be through a direct effect on the clock (i.e., the mutation is in a clock component), through an effect on an input component that, by virtue of being upstream of the clock, can affect clock function, or through modulation of an output that feeds back to affect clock function. It can be difficult to distinguish between these possibilities. Mutations in the input pathway may be delineated through assays for entrainment—for example, either entrainment to light–dark cycles or to acute pulses of light. The prediction for a clock component is that a null mutation, which by definition eliminates all function of the encoded protein, should eliminate circadian rhythms. The standard way of determining whether a mutation is a null is by determining whether it produces the same result in complementation tests as a deletion of the gene. For example, the *per*^{0} mutation in Drosophila has been determined to be a null because it uncovers a short-period, long-period or arrhythmic phenotypes when heterozygous with a short-period, long-period, or arrhythmic *per* allele, just as a deletion of *per* does when heterozygous with the same alleles. Although a null mutation in a clock gene is expected to produce arrhythmia, it should be noted that in mammals, redundant or compensatory mechanisms may result in less severe phenotypes (e.g., period alterations).

The final determination of how a mutation affects circadian rhythms must await isolation of the gene affected. At that point, the spatiotemporal expression profile, possible interactions with known clock proteins, and the biochemical function of the protein all contribute toward determining its precise role in the circadian system.

MOLECULAR ANALYSIS OF CIRCADIAN GENES

The methods used to isolate the gene corresponding to a mutant phenotype vary somewhat from species to species. The chromosomal location of the mutation, and the proximity of genetic/molecular markers, can also influence the methodology and ease of mapping it and isolating the relevant gene. Thus, these methods are not discussed at length here, although details for individual genes can be found in subsequent chapters. Once the gene is isolated, it is analyzed through a standard series of procedures outlined in Figure 2.1 and described below. It is important to note, though, that the approaches discussed here are based on what we know of existing clock genes. When the first circadian gene, which was the Drosophila *period* (*per*) gene, was being analyzed, there was no clear course of action. It was the analysis of *per* that paved the way for future molecular studies of circadian rhythms. Expression studies of *per* pinpointed the clock cells in Drosophila, and the discovery of *per* cycling and feedback regulation led to the current model for clock function.

Gene Sequencing

The sequence may predict the biochemical function of the protein. Once a gene is isolated, it is sequenced—these days quite rapidly by automated sequencing facilities. Now that entire genomes of many organisms have been sequenced, the results of a short sequence run can be compared with existing databases to extract the sequence of the full-length gene. While it is still important to sequence the clones isolated, because there are frequently errors in the database sequence, the availability of this information allows researchers to rapidly start designing experiments based on the sequence. The sequence may predict a biochemical function—For instance, it may contain domains found in kinases or transcription factors, and/or it may reveal the presence of a PAS (*P*ER, *a*ryl hydrocarbon receptor of mammals, and *s*ingle-minded gene in Drosophila) domain, which is a protein–protein interaction domain found in many clock proteins. Data obtained from other assays (described below) can be coupled with the sequence information to address the role of the candidate gene. For instance, if the original mutant affects expression of *per* and the gene encodes a transcription factor, then it is worth determining if *per* is a direct transcriptional target. The DNA clones can be used to express the proteins in bacteria or tissue culture cells, so that large quantities of the proteins can be isolated and used for biochemical studies (see text below).

The recombinant proteins produced in bacteria or cultured cells are also injected into mice, rats, or rabbits to produce antibodies that specifically bind to the proteins. The DNA clones provide probes to analyze RNA, while the antibodies can be used to detect protein. Using the cloned genes and antibodies as probes, one can ask when and where the genes are expressed, and how the functions of the multiple clock genes interact to produce circadian rhythms.

Cell Culture

Cell culture systems can be used to study protein function and are becoming increasingly popular for the study of circadian rhythms. As will be discussed in Chapter 4, mammalian cell lines can be induced to display a cycling circadian gene expression, which can also be reset by chemical/humoral signals. Although this is obviously a very useful and easily manipulated system, some caution is required because the physiological effects as well as some of the interactions in these cells may not be relevant for clock function in the intact organism. However, regardless of their limitations for understanding clock physiology in whole organisms, cell culture systems are valuable for assaying biochemical properties of candidate proteins. For instance, phosphorylation assays and transcription assays are effectively performed in cultured cells.

If the RNA levels of the candidate gene are found to cycle (see text below), it is worth investigating the promoter region of the gene in question. Transcription factors involved in clock function have been identified, and the promoter sites recognized by many of these factors are known. The presence of such a recognition site in the promoter region of the gene of interest would suggest that is likely to be a direct target of a clock transcription factor. Alternatively, as mentioned above, the candidate gene may encode a transcription factor that affects expression of a known clock gene. Transcriptional effects can be tested experimentally by fusing the relevant promoter to a reporter gene such as bacterial β-galactosidase or firefly luciferase. Neither of these genes is present endogenously in the cell culture system, and neither would show any regulation by circadian transcription factors unless a circadian promoter is fused to the reporter gene. Activation/repression of the reporter by the candidate transcription factor then establishes the role of the transcription factor and the target promoter sequence. Typically, these experiments are done in cultured cells transfected with the relevant constructs (Fig. 2.2). It should be noted, though, that sometimes activation is not observed, even if the gene is a direct target of the transcription factor, because another factor is also required.

Temporal Analysis

Knowing, as we do now, that levels of many clock and clock-controlled genes cycle, it follows that every new potential circadian gene is assayed for temporal regulation. Tissues are collected at different times of day, and RNA levels are assayed through northern blot analysis, RNase protection assays (Fig. 2.3), or quantitative polymerase chain reaction (PCR) experiments. Likewise protein expression is assayed in tissue obtained at different times of day, usually through western blot analysis (Fig. 2.4). The western blots will frequently also reveal changes in the mobility (i.e., the size of the protein) over the course of the day. For many clock proteins these mobility changes are known to be due to cyclic phosphorylation events, which are part of a mechanism regulating clock protein levels and function.

A key point to remember with all of these studies is that while oscillations of RNA, protein, or phosphorylation levels may be first assayed in the presence of light–dark cycles, they are circadian-relevant only if they also occur in freerunning conditions (e.g., constant darkness and constant temperature). The reason for also assaying them in the presence of light–dark cycles is because molecular oscillations frequently dampen in freerun. Thus, if the assays were done only in freerun, the low amplitude of some oscillations could result in a failure to detect any cycling.

Spatial Analysis

In situ experiments using either labeled RNA probes or antibodies can be used to identify the sites of expression. In these experiments, the RNA or protein is detected in tissue sections or whole mounts

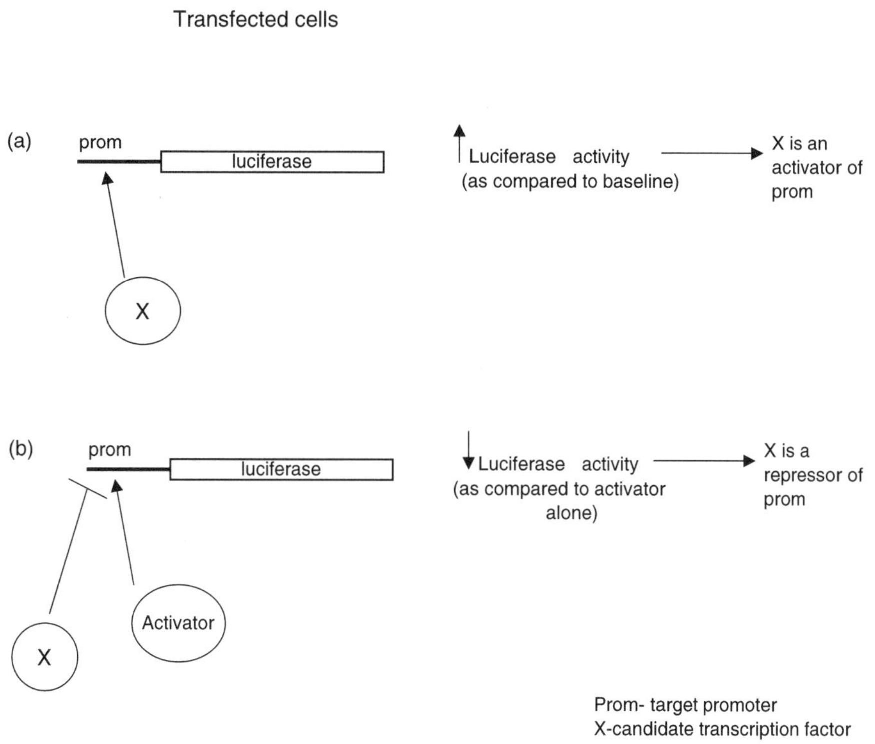

Figure 2.2. Use of cell culture assays to determine the effects of candidate genes on transcription. As described in the text, the promoter of the putative target gene is fused to a reporter and transfected into cultured cells. The reporter shown here is the firefly luciferase protein whose activity can be reliably and easily quantified in a luminometer. Luciferase activity is assayed in cells cotransfected with the candidate transcription factor as well as in control cells (these may be cotransfected with a transcription factor that is not relevant for this process). The level of luciferase activity in control cells serves as a baseline. If the candidate factor is suspected of being a repressor, its effects are best studied under conditions where the target promoter is actively transcribed. Multiple transcription factors may be transfected to determine how the presence of one affects the activity of the other.

of biological specimens, such that the gene product is visualized along with the cellular and subcellular structure of the specimen. Thus, these types of experiments tell us not only whether and when a gene is expressed, but precisely where it is expressed. If a gene is expressed in clock cells it becomes a candidate for a clock gene, although it could turn out to be a component of the input or the output pathway. Conversely, if it is not expressed in clock cells, then it almost certainly is not a clock gene. Its localization may then serve to identify the cells, or the circuits, that are part of the input or output pathways of the circadian system.

Subcellular Localization

The subcellular localization may cycle over the course of a day. It has become clear through analysis of several clock genes that the subcellular localization—in particular the expression in nuclei—is an important, regulated process. Although temporal control of nuclear expression has not been demonstrated in all cases, the fact that

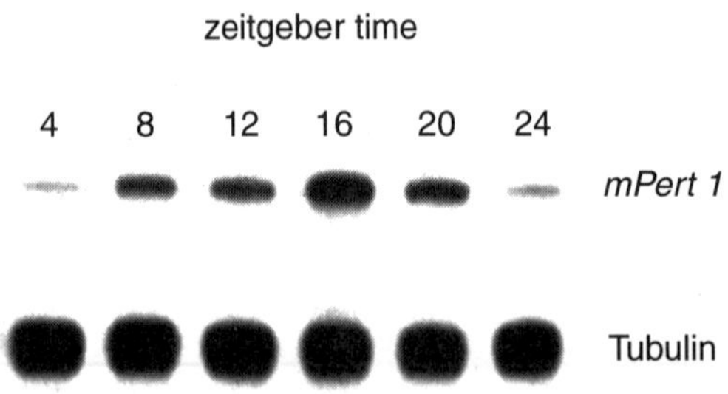

Figure 2.3. Temporal analysis of RNA expression by RNase protection assay (RPA). The data shown here are for the *mPer1* gene in the mouse kidney. Mice were entrained to light–dark cycles for several days and then collected and sacrificed at the times indicated (recall that ZT0 = lights on and ZT12 = lights off). Kidneys were isolated and homogenized. RNA was extracted for each timepoint and subjected to the analysis shown here. The analysis involves the synthesis of a radiolabeled antisense RNA probe from the gene being studied and then the hybridization of this probe to total RNA from the relevant tissue. The mix is then treated with RNases that digest all the single-stranded RNAs present. RNA that corresponds to the gene under study is hybridized to the probe, and thus the radiolabeled probe is "protected" from digestion. Protected fragments are visualized on a denaturing gel. Typically, the same RNA samples are also assayed for an RNA that is expected to be constant in all samples. This serves to control for the amount of RNA in each sample. The control RNA assayed in the experiment shown here is that of the tubulin gene. A radiolabeled antisense probe complementary to the tubulin mRNA was added to the sample at the same time as the *mPer1* probe. As is evident, levels of tubulin are constant at all times of day while levels of *mPer1* cycle.

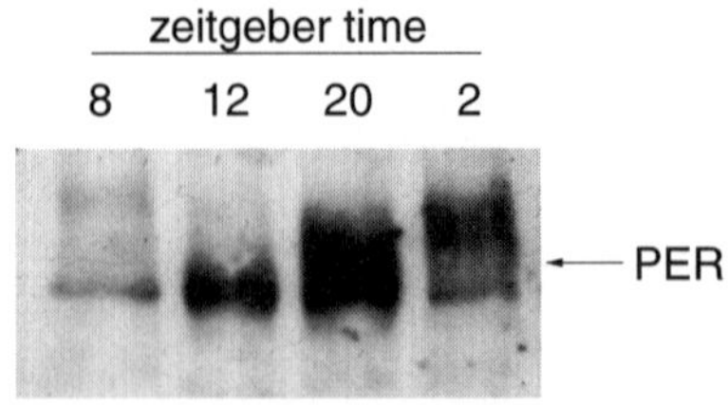

Figure 2.4. Temporal analysis of protein expression by western blot analysis. The data shown are for the Drosophila period protein (PER). Adult Drosophila were entrained to a light–dark cycle for several days, after which they were collected at the timepoints indicated [in zeitgeber time (ZT)]. Heads were isolated and homogenized to prepare protein extracts. The extracts were run on SDS polyacrylamide gels, which separate denatured proteins based on their size. The proteins were then transferred to membranes ("western-blotted") and probed with an anti-PER antibody. The antibody binds to the membrane only where PER antigen is present, and the presence of the bound antibody is detected indirectly by coupling an enzyme, which emits a luminescent signal assayed through autoradiography. As evident from the figure, levels of PER cycle over the course of the day. The phosphorylation state of PER also cycles. The high-molecular-weight bands seen at ZT2 represent phosphorylated forms of PER.

multiple mechanisms are employed to regulate the expression of some clock proteins in nuclei makes it likely that nuclear import/export occurs in a temporal fashion. Thus, when determining the spatial distribution of a new protein, it is also desirable to determine its subcellular localization and, if possible, determine if the protein moves from one compartment to another within the cell over the course of a day.

Assaying Expression in Mutant Backgrounds

Now that multiple clock genes and other components of the circadian system are known, there is a framework in which new components can be placed. In order to identify their relative placement or role, it is important to determine their relationship to known components. One way of doing this is to look for physical interactions, either through yeast two-hybrid systems or coimmunoprecipitation assays (explained below). However, it is sometimes difficult to predict the putative partners, and in many cases the interactions are weak and undetectable in these assays. The other way to go about addressing this question is to use the genetic tools available, namely, the existing clock mutants. By assaying the expression of the new gene in different genetic backgrounds, one can determine the order of gene action in the mechanism. For instance, if a mutation in A alters the expression of B, but a mutation in B does not alter the expression of A, then B is downstream of A (i.e., B is regulated by A, and B does not regulate A). Expression of known clock genes should also be assayed in the new mutant—obviously this can be done even before the affected gene is cloned. Effects on expression can range from altered levels and disrupted cycling to changes in the subcellular distribution. Interpretations of these findings must take the cyclic nature of the circadian process into account. For instance, low levels of a clock protein that negatively regulates its own transcription could result from a regulatory mutation that affects gene expression, but they could also arise from a mutation that reduces stability of the clock protein. In the latter event, RNA levels would be expected to be high due to reduced negative feedback.

Transgenic Clock Gene Expression

Expression of transgenic clock genes in organisms allows analysis of circadian clock mechanisms at the molecular and whole organism level. In all the model organisms employed by chronobiologists, techniques for the introduction of exogenous genes (i.e., transgenes) have been developed. These techniques allow the insertion of "designer" genes created by the experimenter into the genomes of the organism. The types of experiments that transgenic approaches allow are too numerous to completely describe here, but a few general approaches are worth mentioning. Overexpression of a clock protein can be accomplished by fusing the gene to a promoter that is highly expressed in a particular tissue. This produces a genotype that is complementary to the null genotype. For instance, if clock gene A represses the transcription of clock gene B, then a null mutant of A will express high levels of the mRNA for B, while overexpression of A from a transgene will produce low levels of the mRNA for B. Overexpression of a gene that encodes cycling RNA and/or protein will sometimes eliminate cyclic expression of the gene product(s). This approach can thus be used to test the importance of the cycling of that gene product to the overt rhythm. If output rhythms are disrupted under conditions that eliminate RNA/protein cycling of a specific gene, then most likely cyclic expression of the gene in question is required for a rhythmic output.

Tissue specific expression of a transgene can be achieved by fusing the gene to a promoter that is expressed exclusively in specific cell types in an organism that is null for the endogenous copy of the gene. This allows one to assesses the effect of the clock gene in that specific tissue, without the confounding effects of expression in other tissues. For instance, expression of the Drosophila *per* gene exclusively in several central brain neurons produces circadian activity rhythms in a *per* null mutant fly, which has arrhythmic activity without expression of the transgene. This experiment supports the view that these cells are the site of the clock controlling circadian activity. In several organisms (e.g., Neurospora and mice), it is possible to replace the endogenous gene with a nonfunctional transgenic copy, thereby producing a null mutant. This is an important accomplishment if classical forward genetics has not produced a null mutation for the gene.

Another important approach is the analysis of transgenes with genetically engineered mutations in particular regions of the protein coding region or the promoter. These mutations may abolish some particular aspect of the gene's proper function, such as circadian regulation of its expression or its protein–protein interaction with another clock protein. Such a result establishes the importance of different protein and promoter domains for specific biochemical functions. In the case of promoter analysis, reporter genes are usually employed, for the reasons described above. Some reporter genes (such as luciferase) can be imaged in living organisms and allow circadian regulation of promoters to be followed for many days in a single organism. While analysis of mutant promoters and genes can be pursued in vitro or cell culture, transgenic organisms allow the effects on whole organism circadian physiology to be assessed.

■ OTHER METHODS OF ISOLATING CIRCADIAN GENES

Homology to Known Clock Genes in Other Organisms

The mammalian *period* (*per*) genes were isolated or identified on the basis of their homology to Drosophila *per*. Conversely, the isolation of the Drosophila *Clock* gene was facilitated by the previous existence of the mammalian *Clock* gene. Thus, this cross-species approach is clearly a powerful method to identify circadian-relevant genes. Until recently, polymerase chain reaction (PCR) technology was used to amplify homologous sequences and thereby isolate a gene from another species. However, as the genomes of an increasing number of organisms are being sequenced, the most efficient method is to scan computer databases of genomic sequence for the sequence of interest. This area of biology, termed *bioinformatics*, is now as important to molecular biologists as are standard "bench" experiments.

Interaction with Known Clock Proteins

The most common method for this purpose is the yeast two-hybrid assay (Fig. 2.5). Using this assay, one can either test for an interaction between two known proteins or else conduct a random screen to identify new proteins that interact with a protein of interest. Typically, a cDNA encoding the protein of interest ("bait") is fused to sequences encoding the DNA binding domain of a transcription factor and expressed in yeast. The protein with which interaction is being tested ("prey") is fused to an activation domain that cannot by itself associate with the DNA binding protein fused to the bait. A functional transcription factor is generated only when the two protein domains (the bait and the prey) interact and bring the DNA binding and activation domains together. The functional transcription factor is detected on the basis of its ability to bind and activate its target promoter sequence, which is fused to a reporter gene. The reporter typically produces a colored product or a product necessary for survival of the yeast. For a random screen, a library is generated in which each cDNA is fused to the activation domain of the transcription factor. Identification of interacting proteins is based on the color assay or the ability of the yeast to grow on a particular nutrient-deficient medium (the nutrient is produced only when a functional transcription factor is generated).

It is also possible to analyze protein–protein interactions more directly with biochemical assays that can be applied to proteins expressed in vitro, in bacteria, in cell culture or in vivo. For instance, an antibody which recognizes one clock protein can be used to "immunoprecipitate" that clock protein by linking the antibody to a bead, which settles out of solution; the bead brings with it the clock protein recognized by the antibody, and (importantly) any other proteins that form stable protein/protein interactions with the clock protein. The recovery of these other proteins with the beads shows that these proteins interact with the clock protein. Another approach requires the fusion of a short ligand binding domain to one clock protein. For instance, the protein glu-

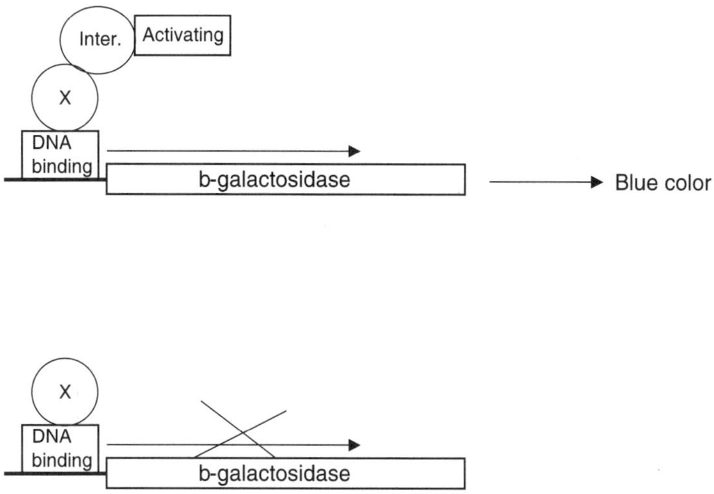

Figure 2.5. Identification of interacting proteins through the yeast two-hybrid assay. The protein of interest (protein X) is fused to the DNA binding domain of a transcription factor. An entire library of cDNA sequences is fused to a domain that is capable of activating transcription ("Activating"). Transcription of the reporter occurs only when the DNA binding and activating domains are brought together through the interaction of protein X with a protein in the library ("Inter."). Transcription of the reporter produces a protein that either allows growth of the yeast on specific media or renders a blue color to the yeast colony. In the example shown here, the reporter is β-galactosidase, which confers a blue color.

tathione-*S*-transferase (GST), which binds glutathione tightly, can be fused to a clock protein. The clock protein will now be retained on beads which are coated with glutathione, thereby allowing the recovery of the clock protein along with the beads. If another protein that is not fused to GST is also recovered on the beads, a protein–protein interaction with the clock protein is demonstrated.

Assays for Circadian Regulation

Differential Display/Subtractive Hybridization Experiments

The finding that expression of many circadian genes cycles with a ~24-hour period has prompted researchers to use this property to identify new circadian genes. In principle, the approach consists of searching for RNAs whose levels vary at different times of day. Prior to the development of genomics technology (described below), the most popular methods for such screens were subtractive hybdridization and differential display. Both methods involve the isolation of RNA at different times of day and the synthesis of the corresponding cDNAs from these RNAs. In subtractive hybridization, the cDNAs from one time of day are hybridized to the cDNAs from another time of day and those that fail to hybridize, because they are not present at the other time of day, are cloned. These represent RNAs that are expressed predominantly at a specific time(s) of day. Differential display experiments are based on the same principle, but consist of "displaying" all the RNA products expressed at a specific time of day and comparing those with ones expressed at a different time of day. In practice, radiolabeled cDNAs are synthesized from each RNA population using random primers, and these are then separated on a denaturing gel. Radiolabeled "bands" that are found in RNA samples extracted at some times of day, but not at others are cut out of the gel and cloned. This technology has been refined in different ways to reduce the frequency of false positives and has been quite successful in identifying cycling genes.

Presence of Circadian Elements in the Promoter

As noted in Chapter 1 and elsewhere in this book, some of the known clock proteins are transcription factors that bind to specific DNA sequences. These sequences, commonly termed "consensus sites," can be used to indicate whether the gene is likely to be clock-controlled. Now that the genomes of many different organisms have been sequenced, one can, in theory, scan databases of genomic sequence to identify putative clock-controlled genes. The caveat is these consensus sites are usually quite frequent and, in many cases, nonfunctional. Sometimes they are functional, but not under control of the clock, as there are non-clock transcription factors that recognize the same sequences. It has become clear that regions flanking the consensus sites can play a role in determining specificity and functionality. However, until specific sequences or structures are known for these flanking regions, it is difficult to distinguish a clock-controlled consensus site from a non-clock-controlled one.

Microarrays

This is yet another example of the applications of genomics technology. As mentioned above, there has been considerable emphasis on sequencing entire genomes as well as large numbers of randomly selected cDNAs from several organisms. As a result, there is a wealth of sequence information that can be exploited for high-throughput analysis. The goal of this type of analysis is to identify all the genes whose regulation is altered in the course of a biological/pathological process. Thus, it can be used to identify genes that are up- or downregulated when cells become cancerous, when a particular behavioral state is induced or when a specific stimulus is introduced. From the circadian point of view, it is the variation of gene expression over the course of a day–night cycle that is of interest, as well as the effects of various clock mutations on gene expression. Consequently, this technology has been used to identify cycling genes in several systems. The example described below (see also Fig. 2.6) describes the isolation of circadian regulated genes from Drosophila heads.

On the basis of the known genome sequence, it is possible to photochemically

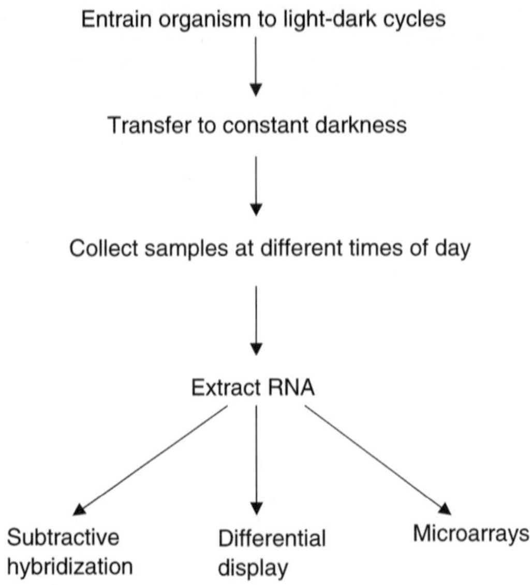

Figure 2.6. Isolation of circadian genes through molecular screens for time-of-day-specific expression. Three different approaches that have been successfully used to isolate genes whose expression changes over the course of a day are shown. With the advent of microarrays, the other two methods have largely been phased out, although they may still be useful in organisms where microarrays have not been generated because of lack of sufficient genomic information.

synthesize short oligonucleotides complementary to unique regions of all the predicted 13,600 genes in the Drosophila genome. The DNA for all of these genes can be fixed as spots on glass slides; the small size of the spots and the precision of the arraying process allow spots for all of the predicted Drosophila genes to be arrayed on a small glass slide (a "microarray" or "gene chip"). These slides are hybridized with fluorescently labeled nucleic acids synthesized from mRNAs collected at different timepoints. The amount of fluorescently labeled nucleic acid that hybridizes to a particular spot is proportional to the amount of mRNA for that gene in the total mRNA, and is determined by measuring the fluorescent intensity of the spot after hybridization. Hybridizations of fluorescently labeled nucleic acids prepared from different circadian timepoints track the changing patterns of gene expression on a genomic scale.

Its is clear that the molecular analysis of circadian rhythms has advanced to a point where it lends itself to the most sophisticated, cutting-edge technology. In addition to the approaches described here, which are common to most organisms, other elegant methods have been used in specific organisms. For instance, in both Drosophila and mammals, chimeric animals (mosaics) have been made in which some cells of the animal are mutant for a particular clock gene and others are wild type. In Drosophila, this approach helped to identify the central clock cells in the Drosophila brain, because circadian activity rhythms are only altered by the mutation if it is present in these central clock cells. In mammals, the study of chimeric animals showed that properties of circadian rhythms such as periodicity, rhythm strength, and phase are conferred by different, overlapping subsets of cells in the SCN. Thus, a combination of approaches is being successfully used to further our understanding of clock function.

FURTHER READING

Ewer, J, Frisch, B, Hamblen-Coyle, MJ, Rosbash, M, Hall, JC (1992): Expression of the period clock gene within different cell types in the brain of Drosophila adults and mosaic analysis of these cells' influence on circadian behavioral rhythms. *J Neurosci* **12**: 3321–3349.

Frisch, B, Hardin, PE, Hamblen-Coyle, MJ, Rosbash, M, Hall, JC (1994): A promoterless period gene mediates behavioral rhythmicity and cyclical per expression in a restricted subset of the Drosophila nervous system. *Neuron* **12**: 555–570.

Gekakis, N, Saez, L, Delahaye-Brown, AM, Myers, MP, Sehgal, A, Young, MW, Weitz, CJ (1995): Isolation of timeless by PER protein interaction: Defective interaction between timeless protein and long-period mutant PERL. *Science* **270**: 811–815.

Harmer, SL, Hogenesch, JB, Straume, M, Chang, HS, Han, B, Zhu, T, Wang, X, Kreps, JA, Kay, SA (2000): Orchestrated transcription of key pathways in Arabidopsis by the circadian clock. *Science* **290**: 2110–2113.

Konopka, RJ, Benzer, S (1971): Clock mutants of *Drosophila melanogaster*. *Proc Natl Acad Sci (USA)* **68**: 2112–2116.

Kornmann, B, Preitner, N, Rifat, D, Fleury-Olela, F, Schibler, U (2001): Analysis of circadian liver gene expression by ADDER, a highly sensitive method for the display of differentially expressed mRNAs. *Nucleic Acids Rese.* **29**: E51–51.

Loros, JJ, Denome, SA, Dunlap, JC (1989): Molecular cloning of genes under control of the circadian clock in Neurospora. *Science* **243**: 385–388.

Low-Zeddies, SS, Takahashi, JS (2001): Chimera analysis of the Clock mutation in mice shows that complex cellular integration determines circadian behavior. *Cell* **105**: 25–42.

Siwicki, KK, Eastman, C, Petersen, G, Rosbash, M, Hall, JC (1988): Antibodies to the period gene product of Drosophila reveal diverse tissue distribution and rhythmic changes in the visual system. *Neuron* **1**: 141–150.

Welsh, S, Kay, SA (1997): Reporter gene expression for monitoring gene transfer. *Curr Opin Biotechnol* **8**: 617–622.

Part II

MOLECULAR CONTROL OF CIRCADIAN RHYTHMS: ANIMAL MODELS

3

DROSOPHILA MELANOGASTER: A MODEL SYSTEM FOR MOLECULAR CHRONOBIOLOGY

Jeffrey L. Price

AN INTRODUCTION AND HISTORICAL PERSPECTIVE

The demonstration of a circadian rhythm in Drosophila eclosion by Kalmus and Bunning in the 1930s initiated the development of Drosophila into one of the primary model systems for chronobiologists. *Drosophila pseudoobscura* was employed by Colin Pittendrigh to explore the formal properties of phase resetting and to formulate the general circadian principle of temperature compensation in the 1950s (see Chapter 1). In 1971, Ronald Konopka and Seymour Benzer isolated the first single gene "clock" mutants in *Drosophila melanogaster*, thereby initiating genetic analysis of the clock mechanism. With the advent of Drosophila molecular biology in the 1970s, it became possible to isolate the genes affected by Drosophila clock mutations. This combination of phenomenological, genetic, and molecular genetic approaches in Drosophila has led to a virtual explosion in our knowledge of clock mechanisms. Fortunately, the mammalian clock employs many of the same genes as the Drosophila clock, so the models derived for the Drosophila clock have been readily applied to the mammalian clock. Studies of the mammalian clock have become a potent force in their own right, allowing a synergy that has facilitated work in both Drosophila and mammals.

Circadian Rhythms of Drosophila

The circadian rhythm demonstrated by Kalmus and Bunning in the 1930s is called

Molecular Biology of Circadian Rhythms. Edited by Amita Sehgal
ISBN 0-471-41824-2

the "eclosion rhythm." The term "eclosion" refers to the emergence of the fly from the pupal case at the end of its metamorphosis from the larval to adult stage. Eclosion is said to be "gated" by the clock, because a fly that is developmentally ready to eclose during the late day or night nevertheless refrains from eclosing until the next circadian "gate," which occurs during an 8–10-hour interval beginning 1–2 hr before dawn. While an individual fly ecloses only once in its lifetime, a population of flies with a broad distribution of developmental stages produces recurrent eclosion activity around the time of dawn in a light–dark cycle (LD), or at the time of subjective dawn in constant darkness (DD). Eclosion activity is defined as the number of flies emerging in fixed intervals (e.g., hourly intervals), and the numbers for successive intervals are plotted as a function of collection time. Such plots show a circadian rhythm, in which the peaks of eclosion are approximately 24 hours apart in wild-type flies entrained to light–dark cycles (LD) and then transferred to constant darkness (DD) (Fig. 3.1a). Circadian control of eclosion is mediated by at least four hormones (ecdysone, ecdysis triggering hormone, eclosion hormone, and crustacean cardioactive peptide).

With the advent of electronic data collection systems, locomotor behavior became the most widely assayed circadian behavior in Drosophila. In this assay, individual flies are placed in glass tubes, and their activity is detected as deflections of an infrared light beam that is passed through the tube. *Actograms*, in which activity is plotted as a function of time, show a circadian rhythm. The flies are active at the times of lights on and off in LD, and during the subjective day in DD (Fig. 3.1b). Superficially, these records resemble the sleep/wake records of mammals.

It has been shown that the Drosophila activity rhythm has more than just a superficial resemblance to human sleep–wake cycles. Drosophila rest is associated with prolonged immobility, a characteristic posture, and unresponsiveness to sensory stimuli, as is mammalian sleep. Another similarity is that flies respond homeostatically to disrupted rest with an increased tendency to rest. This response is affected by several clock mutations. Drosophila rest is suppressed by exposure to caffeine and enhanced by cyclohexyladenosine and hyrdoxyzine, which have similar effects on sleep in mammals. Finally, wakefulness in Drosophila is associated with elevated levels of several mRNAs that are also elevated in the rat cerebral cortex during wakefulness in the rat, and the capacity to tolerate sleep deprivation is increased by expression of heat shock genes, which also confer resistance to other stresses such as heat shock. Hence, a genetic and molecular analysis of "sleep–wake" in Drosophila may yield insights into the mechanisms and function of human sleep, in addition to the role of circadian rhythms in the regulation of this process.

Drosophila Clock Proteins

These proteins are expressed in the central brain and in peripheral oscillators. A group of adult brain neurons contains a circadian oscillator that controls locomotor activity and eclosion. However, there is evidence for circadian clock control of olfaction, retinal physiology, response to starvation, responsiveness to dopamine, sensitization to cocaine, sensitivity to ethanol, ovarian diapause, Malpighian tubules (the fruitfly kidney), and the gut in *Drosophila melanogaster.* Initially, the broad tissue distribution of clock gene expression was surprising, as the head had been shown to be the relevant site for regulation of locomotor activity, and it was thought that clocks would be found exclusively in the nervous systems of higher organisms. We now know that the circadian oscillator is largely cell autonomous, and that many of

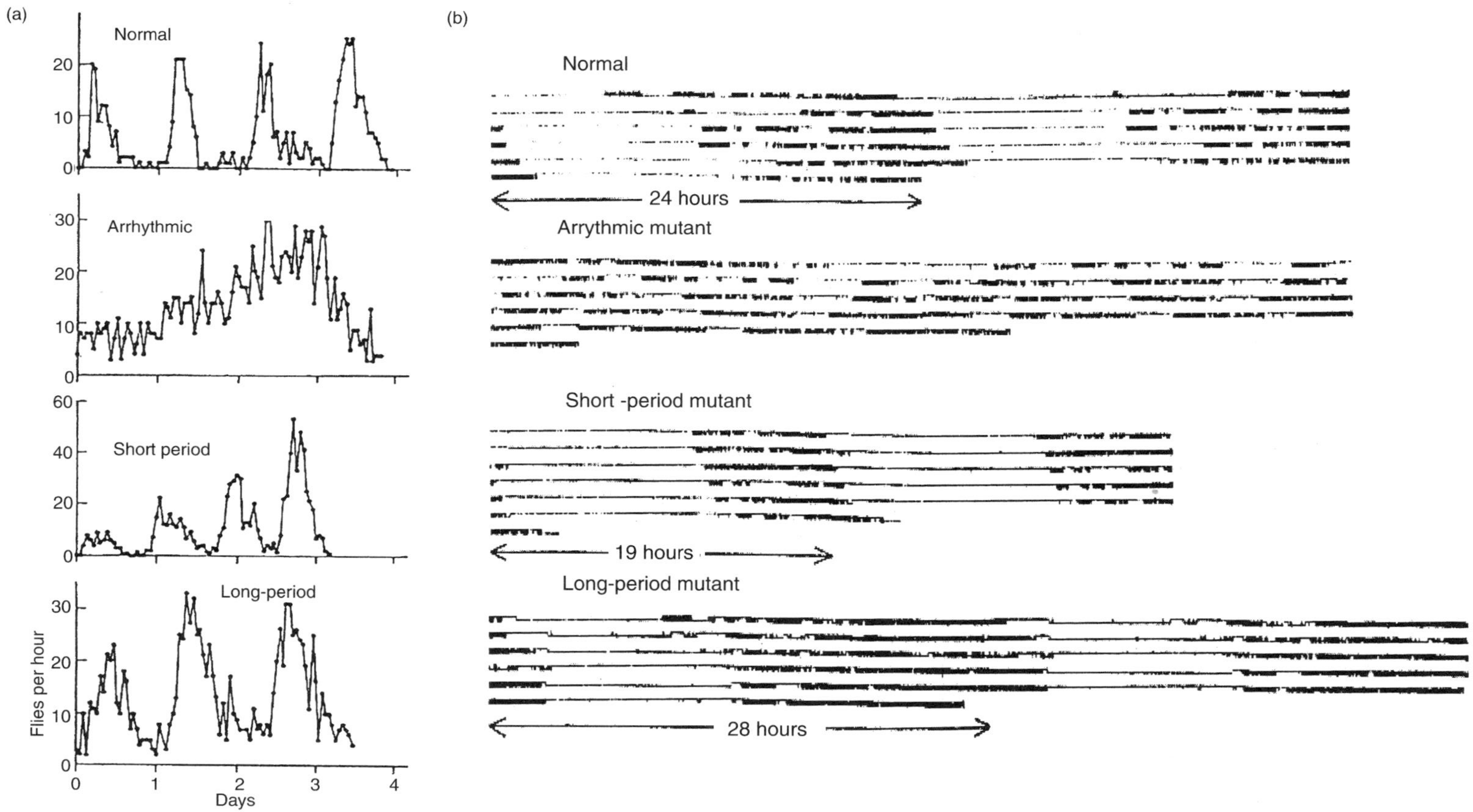

Figure 3.1. Examples of eclosion (a) and locomotor activity (b) in wild-type and *per* mutant flies. The flies had previously been entrained to L:D 12 hours:12 hours but are monitored here in DD. For A, the number of flies emerging from pupae every hour is plotted as a function of time, while in B an activity event for an individual fly is recorded as a deflection of a chart recorder pen. The activity in B is double-plotted on a timescale that approximates the period of the endogenous clock, so that active periods occur at approximately the same time in successive circadian days. In the arrhythmic mutant fly, there are no organized bouts of activity. [Parts (a) and (b) both reprinted with permission from Konopka and Benzer (1971).]

the physiological processes listed above are controlled by peripheral and nonneural clocks, which are dispersed throughout the body. These peripheral clocks are discussed in more detail in Chapter 9.

There are no documented circadian behaviors or physiological responses in Drosophila larvae, but circadian rhythms in gene expression do occur in several small groups of neurons, and are affected by the various clock mutations, as are the circadian rhythms of adults. One of these groups of neurons is thought to survive metamorphosis and ultimately to form some of the adult neurons that control circadian rhythms of locomotor activity and eclosion. This group is likely to mediate larval "time memory," in which locomotor and eclosion rhythms of adults kept in DD can be entrained by prior exposure to LD during the larval stage.

GENETIC ANALYSIS OF CIRCADIAN RHYTHMS IN DROSOPHILA

We begin our study of the Drosophila clock mechanism by considering the classical genetic analysis of the *period* (*per*) gene. This work illuminated features of the tissue and cellular locus of the circadian clock controlling locomotor and eclosion behavior. Then, we consider the molecular genetic studies of *per* and other genes constituting the core oscillator mechanism. Next, we consider the mechanism of the input or entrainment pathways, and finally we consider the mechanisms for coupling the clock to outputs such as locomotor activity.

Single-Gene Clock Mutations

In 1971 Konopka and Benzer were the first to isolate single-gene clock mutations in any organism. The original three mutations were isolated by testing fly lines, each of which contained a mutagenized sex (X) chromosome, for aberrant eclosion. Three lines eclosed abnormally. When these three lines were entrained to LD and then subjected to hourly collections in DD, one exhibited a short-period profile, one a long-period profile, and one an arrhythmic eclosion profile (Fig. 3.1a). Their locomotor activity rhythms were affected in the same way (Fig. 3.1b). Genetic tests showed that all of these mutations affected the same gene, which was dubbed *period* (or *per*, for brevity). In accordance with the considerations outlined in Chapter 2, the centrality of the *per* gene to the clock oscillator mechanism was suggested by the effects of the mutants on both behavioral outputs of the clock (eclosion and locomotor activity) and the isolation of long-period, short-period, and arrhythmic alleles. The short-period allele was dubbed per^S, the long-period allele per^L, and the arrhythmic allele per^0.

Locomotor Activity Rhythms Produced in the Head

The *per* gene functions primarily in a group of neurons within the head to produce locomotor activity rhythms. In the first experiment, the head of a per^S fly was transplanted into the abdomen of an arrhythmic per^0 fly (Fig. 3.2a). The transplantation resulted in a short-period circadian rhythm of locomotor activity, corresponding with the genotype of the transplanted head, in a fly

Figure 3.2. Experiments that established that *per* expression in the lateral neurons is necessary and sufficient for locomotor activity rhythms: (a) transplantation of a per^S head into the abdomen of a per^0 fly produces per^S locomotor activity; (b) gynandromorph flies with per^S male heads(s) exhibit a fully mutant (19-hour) rhythm, while flies with per^S/per^+ (s/+) female heads and per^S male bodies exhibit a per^S/per^+ rhythm (21.5 hours); (c) in mutant and transgenic flies with altered *per* expression in the head, rhythmic locomotor activity results only if *per* is expressed in the lateral neurons.

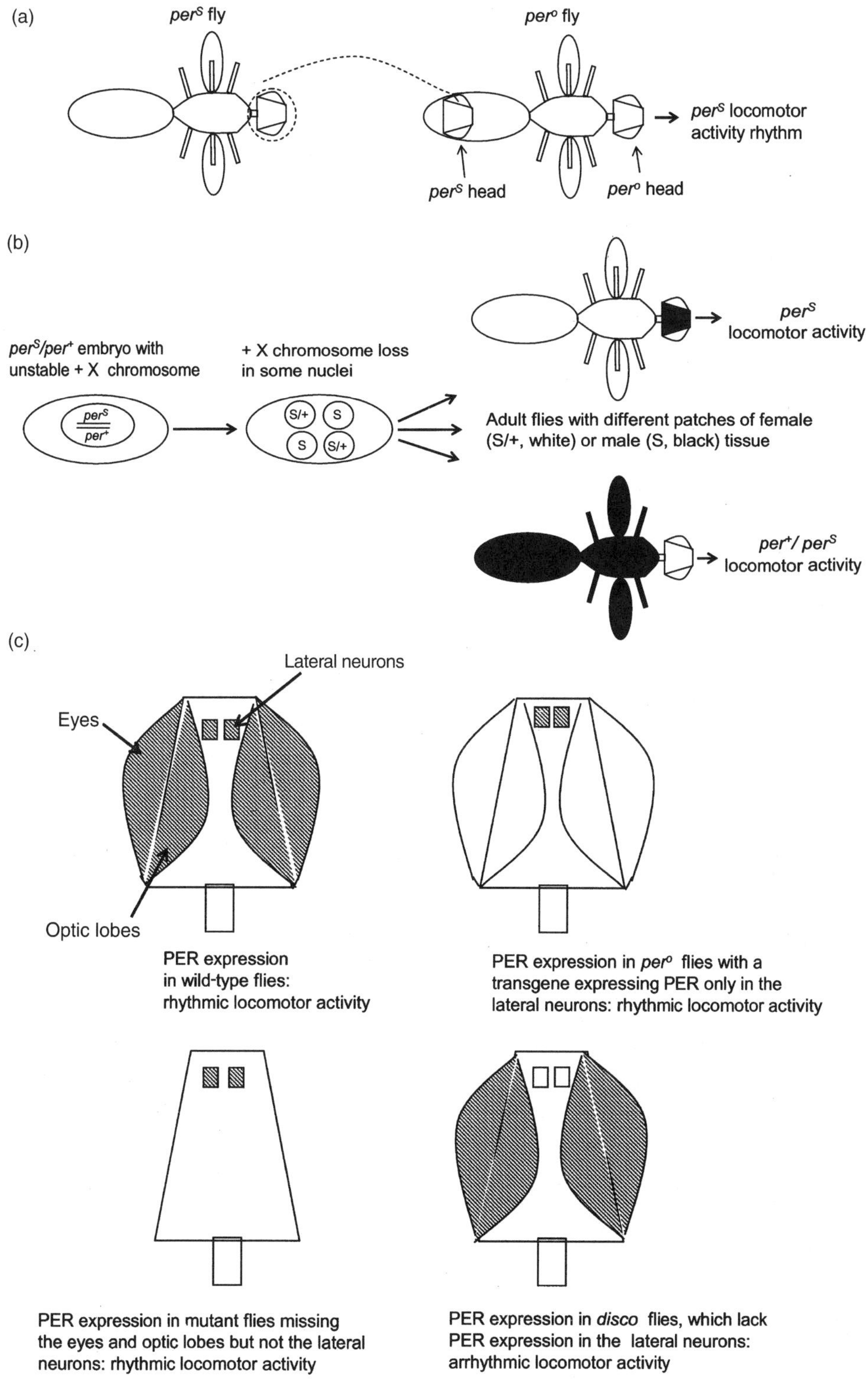
(a)
per^S fly
per^o fly
per^S locomotor activity rhythm
per^S head
per^o head
(b)
per^S/per^+ embryo with unstable + X chromosome
+ X chromosome loss in some nuclei
per^S
per^+
S/+
S
per^S locomotor activity
Adult flies with different patches of female (S/+, white) or male (S, black) tissue
per^+/per^S locomotor activity
(c)
Lateral neurons
Eyes
Optic lobes
PER expression in wild-type flies: rhythmic locomotor activity
PER expression in per^o flies with a transgene expressing PER only in the lateral neurons: rhythmic locomotor activity
PER expression in mutant flies missing the eyes and optic lobes but not the lateral neurons: rhythmic locomotor activity
PER expression in *disco* flies, which lack PER expression in the lateral neurons: arrhythmic locomotor activity

that would otherwise have been arrhythmic. These results established that expression of the *per* mutant gene in the head was sufficient to produce the phenotype. In addition, they suggested that the head produced a diffusible factor that could drive rhythmic locomotor activity, since the transplanted head was unlikely to generate axonal projections to other neurons or the thoracic muscles controlling the legs.

Complementary experiments demonstrated the expression of the *per* mutation in the head was necessary and sufficient for the mutant phenotype (Fig. 3.2b). In females that were heterozygous for per^S or per^0 (per^+/per^0 or per^+/per^S), loss of the X chromosome carrying the per^+ allele was induced in some of the embryonic cells, which then carried only the mutant allele (per^S or per^0) and became male cells (X0 is male in Drosophila). A cell that had lost the per^+ X divided to produce a patch of male tissue, while patches in which the per^+ X had not been lost remained female. Flies generated in this way are part male and part female and are termed *gynandromorphs*. In these cases, the female patches were per^+/per^S or per^+/per^0, while the male patches were per^S or per^0. Since the *per* mutant chromosome also carried another mutation (e.g., a recessive body color mutation), male patches could be identified in the adult fly because they exhibited this mutant phenotype (e.g., mutant body color). Tests of locomotor activity revealed whether a particular gynandromorph had a fully mutant circadian rhythm (19 hours or arrhythmic) or one consistent with a heterozygous genotype (21.5 hours for per^S/per^+ or 24 hours for per^0/per^+). The vast majority of flies with fully mutant rhythms had male patches in the head, demonstrating that the *per* mutant allele must be expressed in the head to affect period length or rhythmicity. Furthermore, it was shown that small groups of cells in the brain were sufficient to confer the fully mutant phenotype, suggesting a discrete central brain locus that did not require *per* expression in the eyes or optic lobes.

Later, it became possible to produce transgenic flies in which genetically engineered *per* genes were placed in the genome of flies. Some of these allowed the expression of *per* only in cells that have been termed the "lateral neurons," because they are found at the side of the central brain, where it is joined to the optic lobe medulla. Expression of *per* in these cells alone rescued the locomotor arrhythmicity of flies that were otherwise per^0, because their endogenous gene was per^0 (Fig. 3.2c). Therefore, expression of *per* in the lateral neurons is sufficient for circadian rhythms of locomotor activity, and expression of *per* is not required in the eyes and optic lobes.

In addition, analyses of various mutations affecting adult eye development showed that the lateral neurons are necessary for circadian rhythms of locomotor activity, while the eyes and optic lobes are dispensable (Fig. 3.2c). Most such mutations have no effects or only minor effects on circadian locomotor activity. Even mutants that are missing the compound eyes entirely can still be entrained by light and have wild-type periods. However, the *disconnected* mutation (*disco*), which disrupts the connection of the compound eye to the optic lobes, also produces a high frequency of locomotor arrhythmia. In these flies, identifiable lateral neurons are almost always absent, but *per* expression persists in the eyes and other sites of expression. Therefore, *per* must be expressed in the lateral neurons for circadian locomotor activity.

Molecular Cloning

Classical genetic analysis of *per* localized the gene for molecular cloning. With the advent of molecular genetic techniques, it became possible to clone genes affected by mutations, thereby leading to a quantum leap in our understanding of many genes. The intense genetic analysis of Drosophila during the twentieth century had produced high-resolution genetic maps of all the Drosophila chromosomes, and *per* was located in particularly well mapped region between the *y* and *w* genes on the X chromosome. Two groups successfully cloned the *per* gene. The laboratories of Jeffrey Hall and Michael Rosbash accomplished this by cloning microdissected DNA from the region of the X chromosome that contained *per*, while the laboratory of Michael Young "walked" to *per* from a previously cloned region by successive isolation of overlapping clones.

Ultimately, introduction of *per* transgenes into per^0 flies identified the *per* transcription unit. In Drosophila, genes that are flanked by sequences from a class of transposable elements known as "P elements" can integrate into the genomic DNA of the fly, thereby producing a transgenic fly carrying a gene chosen by the experimenter. It was shown that introduction of one gene from the genetic interval known to contain *per* could produce wild-type circadian behavior in a fly that contained only the per^0 mutation at the endogenous gene locus. Therefore, this gene was in fact *per*. Transgenes with different amounts of DNA upstream of the coding region of *per* expressed different amounts of *per* mRNA. Those transgenes expressing higher levels of *per* had shorter periods, while longer periods resulted from lower levels of *per* expression.

The mutant per^S, per^L, and per^0 genes were sequenced and found to differ at single nucleotides from the wild-type gene. In the per^S and per^L mutant genes, the nucleotide changes produced a single amino acid change, while the per^0 mutation produced a translational stop codon in the amino-terminal part of the reading frame. At best, the per^0 gene could encode only a truncated *per* protein, but even a truncated *per* protein has never been seen to accumulate in per^0 flies. Presumably, the per^0 translation product is not stable enough to accumulate, and the per^0 mutation is described as a null mutation, meaning that it lacks all gene activity. By contrast, the per^S and per^L genes are still functional enough to support rhythmicity, albeit with altered period length.

■ CIRCADIAN OSCILLATIONS OF *period* AND *timeless* GENE PRODUCTS

Circadian oscillations of these genes are generated by feedback of *period* and *timeless* proteins on their gene products.

PER Protein Versus *per* mRNA

The circadian oscillation of *per* protein (PER) lags the oscillation of *per* mRNA by about 6 hours. Since *per* was the first clock gene to be cloned, it was the first clock gene for which circadian oscillations were demonstrated. *per* mRNA and protein were shown to oscillate with a circadian rhythm (Fig. 3.3) in all tissues except the ovary. These oscillations continued in DD and were affected by the various *per* mutations in the same way that circadian behavior was affected. per^S flies had short-period molecular oscillations, per^L flies had long-period molecular oscillations, and per^0 flies did not produce the molecular oscillations. A transgene expressing wild-type *per* protein (PER) could rescue rhythmic oscillations of the endogenous per^0 mRNA, since per^0 mRNA, which is constitutively expressed in the per^0 mutant without a transgene, exhibited circadian oscillations in a per^0 fly that also contained a wild-type *per* transgene. Thus, PER protein was required for a feedback regulation of its own gene products. Because of their persistence in DD and the effects of the mutants on the period of both behavior and the molecular oscillations, the resulting molecular oscillations of the *per* gene products were thought likely to be part of the long-sought circadian oscillator mechanism.

Intriguingly, the phases of the *per* mRNA and protein rhythms were quite distinct (Fig. 3.3); PER protein reaches its peak about 6 hours after the peak in *per* mRNA, and *per* mRNA levels are already decreasing at this time. PER protein levels are lowest when *per* mRNA levels are starting to accumulate. The interesting phase relationship posed two immediate questions:

1. What posttranscriptional control mechanism results in the long lag between *per* mRNA accumulation and PER protein accumulation?
2. Why does *per* mRNA begin to fall when high levels of PER protein have accumulated?

We will address the second question first, and return to the first question in a later section.

PER Transcription

PER negatively regulates its own transcription. In all tissues except the ovary, in which *per* expression is not rhythmic, PER accumulates in nuclei with a circadian rhythm. *per* mRNA levels begin to fall when PER becomes strongly nuclear, and begin to rise after nuclear PER levels have declined during the day. This phase relationship of

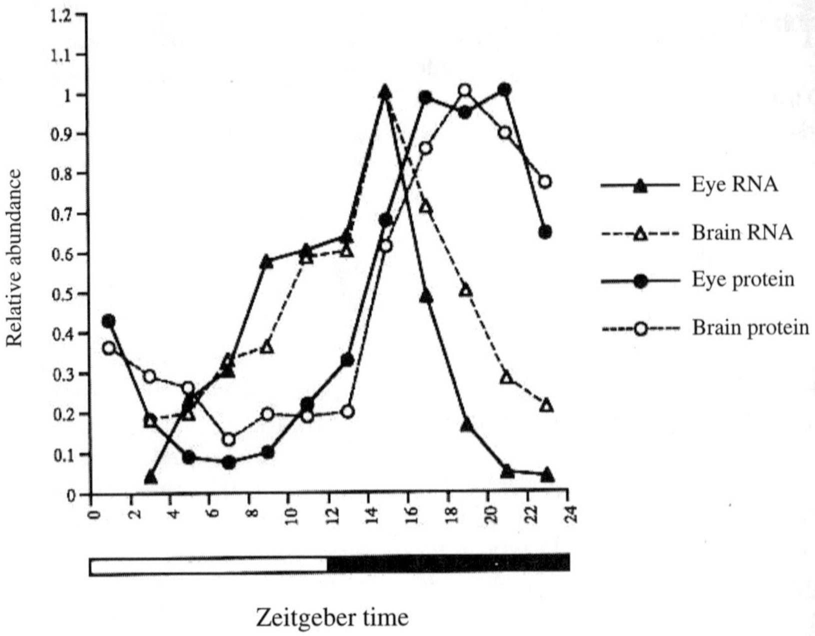

Figure 3.3. *per* mRNA and protein levels exhibit circadian oscillations in both the eyes and brain. The mRNA oscillation was detected with an RNAse protection assay, in which a radioactively labeled *per* probe is hybridized to Drosophila RNA. The protein oscillation was detected with an immunoblot analysis. [Reprinted from Zeng et al. (1994): Constitutive overexpression of the Drosophila *period* protein inhibits *period* mRNA cycling. *EMBO J* **13**(15): 3590–3598 by permission of Oxford University Press.]

per mRNA and nuclear PER protein suggested that PER protein might negatively regulate transcription of its own mRNA. It was possible to test the hypothesis that PER negatively regulates its own expression by generating a line of flies in which PER was constitutively expressed at high levels in the eye (Fig. 3.4a). The coding region for the *per* gene was placed downstream of the rhodopsin promoter, which produced constitutively high levels of eye-specific *per* mRNA and protein. As a result, the oscillation of *per* mRNA expressed from the endogenous gene in the eye was suppressed (Fig. 3.4a, lower panel), and the level of expression was similar to the levels expressed at the trough of the wild-type cycle. These results were consistent with an intracellular negative feedback mechanism, since the endogenous *per* gene was repressed only in the eyes. Molecular oscillations persisted in other tissues (e.g., the brain; Fig. 3.4a, top panel), as did circadian rhythms of locomotor activity, which do not derive from the eye (see discussion above).

The effects of PER on its mRNA were shown to be mostly transcriptional because transgenes containing the *per* promoter fused with the bacterial β-galactosidase, chloramphenicol acetyl transferase (CAT), or luciferase (luc) produced circadian oscillation of the reporter gene, which lacked any part of *per* except the regions that could bind transcription factors (Fig. 3.4b, top panel). Rhythmic transcription of the reporter gene required PER, because the oscillation of the reporter gene did not occur in a per^0 fly. The phase of the oscillation produced by the *per* promoter did not precisely match the phase of endogenous *per* (Fig. 3.4b, top), while a reporter gene that contained enough *per* mRNA sequence as well as the reporter sequence did oscillate in phase with endogenous *per* mRNA (Fig. 3.4b, bottom). An extensive analysis of this difference has shown that regulatory sequences within the transcribed part of *per* and a circadian regulation of *per* mRNA stability both contribute to *per* mRNA oscillations, in addition to the transcriptional effects that are mediated by the upstream promoter.

The PAS Domain

PER contains a protein–protein interaction region called the *PAS domain*. Initially, the conceptual translation of *per's* nucleotide sequence did not provide much information about the function of PER protein. Subsequent to the cloning of *per*, the conceptual translations of several genes cloned from Drosophila and mammals showed weak homology to a region of *per* that was termed the "PAS" region, after the three proteins in which the homology was first observed (*P*ER, the *a*ryl hydrocarbon receptor of mammals, and the *s*ingle-minded gene in Drosophila). Most of these other genes with a PAS region also have a basic helix–loop-helix (bHLH) domain, which identifies them as members of a bHLH/PAS transcription factor family, several of whose members are molecular components of the circadian oscillator in Drosophila and mammals (Fig. 3.5a). PER does not contain a bHLH domain, which is necessary to bind DNA (Fig. 3.5b). Therefore, it cannot bind DNA directly. However, biochemical experiments have established the PAS region as a protein–protein interaction domain, suggesting that PER may bind to other proteins. Further genetic analysis in Drosophila identified other factors that do bind PER to produce post-translational control of PER protein and the transcriptional control of *per* mRNA.

tim Gene Products

The *timeless* (*tim*) gene products also oscillate. Like the *per* gene, *tim*, the second Drosophila clock gene shown to be part of the circadian oscillator was identified in a screen for recessive mutants that affected eclosion. The first *tim* mutant produced

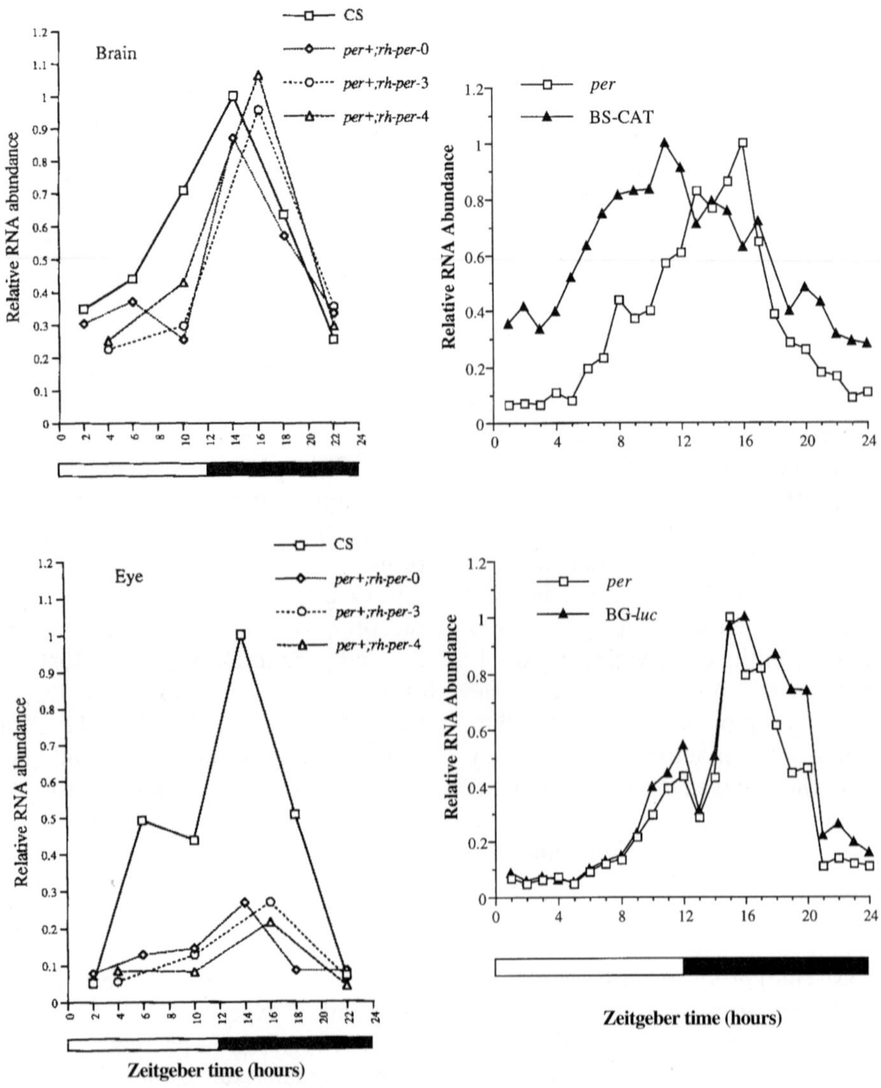

(a) (b)

Figure 3.4. The production of *per* mRNA oscillations involves intracellular negative feedback by PER protein, transcriptional regulation of the *per* promoter, and posttranscriptional effects on *per* mRNA. (a) *per* mRNA oscillations are observed in both the eyes and brains of wild-type flies (CS), while flies in which PER is overexpressed from the eye-specific rhodopsin promoter exhibit *per* mRNA oscillations in the brain but not the eye. In the eyes of these overexpressing flies, the endogenous *per* mRNA, which is quantitated here instead of the transgenic mRNA, is expressed at constitutively low levels. [Reprinted from Zeng et al. (1994): Constitutive overexpression of the *Drosophila period* protein inhibits *period* mRNA cycling. *EMBO J* **13**(15): 3590–3598 by permission of Oxford University Press]. (b) Transgenic reporter genes driven by the *per* promoter were introduced into flies expressing wild-type PER. The levels of the BS-CAT reporter mRNA, which does not contain any *per* mRNA sequences but does contain the promoter regions that are responsive to transcriptional regulation, oscillate with a circadian rhythm that differs from the oscillation of *per* mRNA, while a BG-luc reporter mRNA containing the 5′ part of the *per* mRNA oscillates in phase with *per* mRNA, presumably because it is regulated both transcriptionally and posttranscriptionally. [Reprinted from Stanewsky et al. (1997): Multiple circadian-regulated elements contribute to cycling *period* gene expression in *Drosophila*. *EMBO J* **16**(16): 5006–5018 by permission of Oxford University Press.]

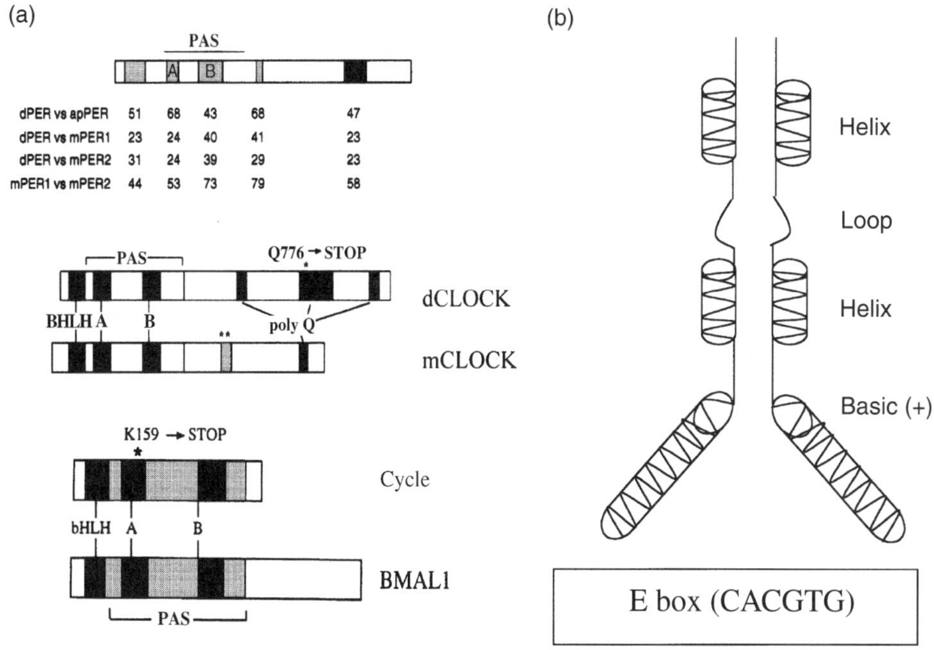

Figure 3.5. Several clock proteins found in both Drosophila and mammals contain PAS domains. (a) Alignment of similar PAS proteins from mammals, fruitflies, and silkmoths. The PAS region, which can mediate protein/protein interactions, contains short regions that are almost direct copies of each other [(a) and (b)]. PER also contains other regions that are conserved in mouse (m) and silkmoth (ap) PER (% identities shown). CLOCK and CYCLE contain basic helix–loop-helix (bHLH) regions that bind to DNA. CLOCK also contains a polyglutamine-rich (polyQ) C terminus, which is likely to activate RNA polymerase at the promoter to which the CLOCK/CYC heterodimer binds. The mutations in the Drosophila and mouse *Clk* genes are found in a polyglutamine region, while the Drosophila cyc^0 mutation occurs in the PAS region. [Top, reprinted from Shearman et al. (1997): Two *period* homologs: circadian expression and photic regulation in the suprachiasmatic nuclei. *Neuron* **19**: 1261–1269, copyright (1997) with permission from Elsevier Science. Middle, reprinted from Allada et al. (1998), copyright (1998) with permission from Elsevier Science. Bottom, reprinted from Rutila et al. (1998): CYCLE is a second bHLH-PAS clock protein essential for circadian rhythmicity and transcription of *Drosophila period* and *timeless*. *Cell* **93**: 805–813, copyright (1998) with permission from Elsevier Science.] (b) bHLH transcription factors generally bind as dimers consisting of two different peptides (e.g., CLK and CYC). The dimers are formed by protein–protein interactions involving the two pairs of alpha helices, while E-box elements in the DNA are bound by the basic regions of the heterodimer.

arrhythmic locomotor activity and eclosion. This mutant (tim^0) is the result of a deletion in the middle of the *tim* coding region. The deletion apparently produces a "null" mutant with no evidence of residual function. In addition, *tim* mutants that shorten circadian period, lengthen circadian period, and suppress some of the per^L phenotype have been identified. *tim* mRNA and protein undergo circadian oscillations with phases quite similar to those of *per* gene products (Fig. 3.3). In particular, there is also the same lag between the *tim* mRNA and protein oscillations. *tim* and *per* mutants affect these oscillations just as they affect behavior.

PER/TIM Collaboration

The *timeless* protein (TIM) associates with PER to produce post-translational control of PER levels and nuclear accumulation of a PER/TIM complex. With the isolation of the *tim*^{0} mutant, it was now possible to investigate the effects of one mutation (*tim*^{0} or *per*^{0}) on the expression of the other gene products (*per* or *tim*, respectively). The results were quite striking. In the *tim*^{0} mutant (Fig. 3.6a), *per* mRNA levels did not oscillate in a circadian manner and were expressed at levels approximating the median of the wild-type cycle. Even more striking were the effects on PER protein. The levels of PER detected by immunoblot analysis were constitutively low—equivalent to the levels found at the trough of the wild-type cycle. So the presence of TIM somehow stabilized PER in the wild-type flies.

Additional insight was gained by examining the effects of the *tim*^{0} mutation on reporter transgenes, in which bacterial β-galactosidase was fused to the first half of *per* and expressed under the control of the *per* promoter. This protein was expressed at high and equivalent levels in both wild type and *tim*^{0} flies, because the fusion of β-galactosidase to PER somehow stabilized the remaining parts of PER to the proteolysis that acted on PER itself in the *tim*^{0} mutant. More importantly, the subcellular localization of the reporter protein (PER-βGAL) was altered in the *tim*^{0} mutants. PER-βGAL accumulated in the nuclei of wild-type flies but was exclusively cytoplasmic in the *tim*^{0} mutant. The effect of *tim*^{0} was not a general one for all nuclear proteins, because nuclear accumulation of other nuclear proteins was normal in *tim*^{0} flies. In fact, the deletion of a region of PER containing the PAS region of homology and another interaction sequence [the cytoplasmic localization domain (CLD); see text below] produced a β-galactosidase reporter fusion protein that localized to nuclei even in *tim*^{0} flies. In summary, TIM was shown to be necessary for stabilizing PER in the cytoplasm and producing its nuclear accumulation and oscillations.

A complementary analysis of the effects of *per*^{0} mutants on the *tim* gene products (Fig. 3.6b) showed that *per* was important for the circadian regulation of *tim*. *tim* RNA does not oscillate in a *per*^{0} mutant in either LD or DD. While TIM protein levels oscillate in LD in the *per*^{0} mutant, TIM is expressed at constitutively high levels in DD in the *per*^{0} mutant, in contrast to the persistent oscillation observed in wild-type flies. These high levels of TIM accumulate only in the cytoplasm, so PER is required for nuclear accumulation and circadian oscillations of TIM, just as TIM is required for these features of PER's regulation. However, since TIM can accumulate to high levels in the dark in the *per*^{0} mutant, PER is not required to stabilize TIM, while TIM is required to stabilize PER.

The findings suggested that both the stability and nuclear translocation of PER and TIM might be regulated by the circadian clock. Because the oscillations of TIM levels could be driven by LD in the absence of a *tim* mRNA oscillations, the stability of TIM was also suggested to be regulated by light (as we shall see, the regulation of TIM by light is part of the entrainment pathway). Consistent with a regulation of nuclear translocation, PER and TIM have been seen to accumulate in the cytoplasm of the lateral neurons prior to nuclear accumulation during the circadian cycle.

PER and TIM interact directly through domains that include nuclear and cytoplasmic localization signals. In addition to its isolation through positional cloning, the *tim* gene was identified on the basis of its interaction with PER. This was in the course of a yeast two-hybrid screen designed to look for PER-interacting proteins. Consistent with the idea of a direct intracellular interaction of the two proteins, the *tim* gene products were expressed in the head in the same places where *per* was seen. In fact, two

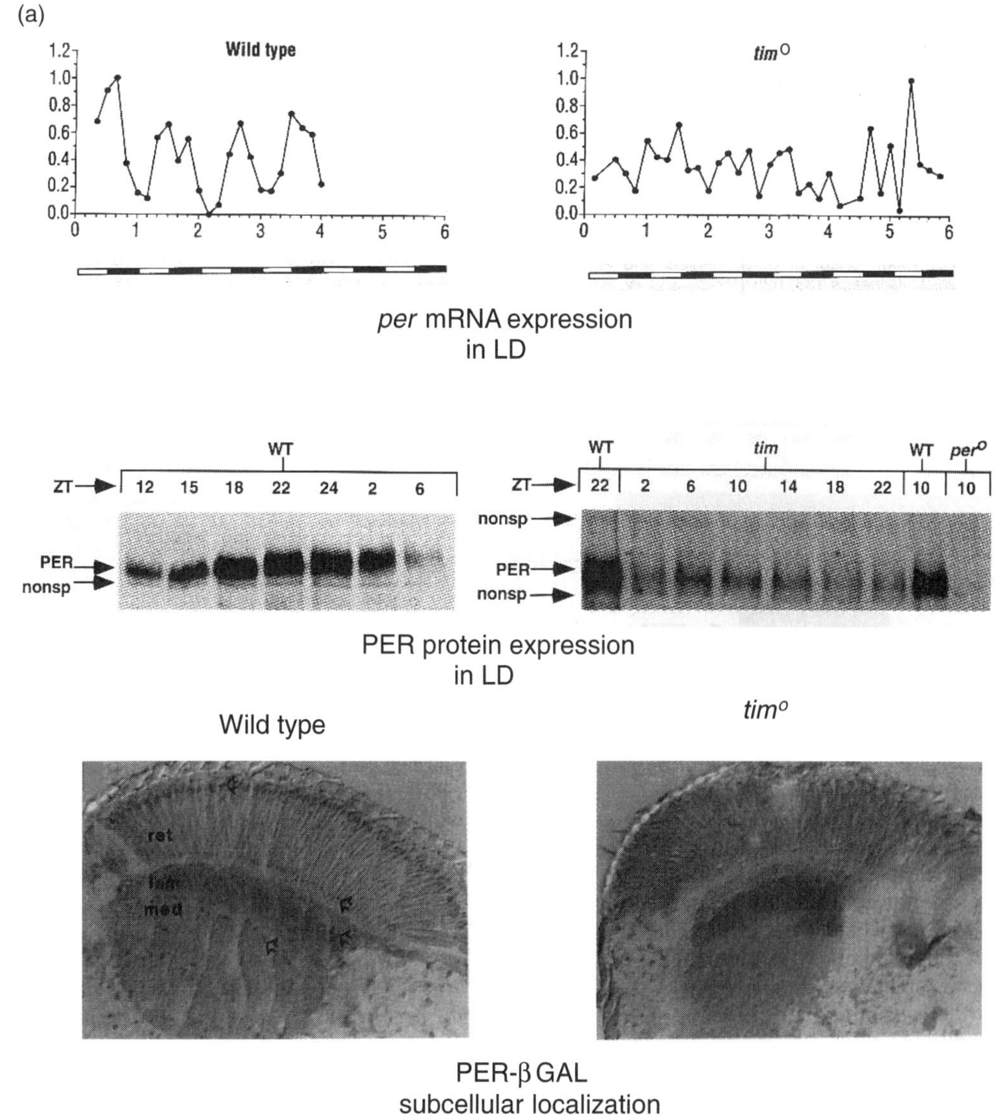

Figure 3.6. PER and TIM regulate each other transcriptionally and posttranscriptionally. (a) *per* expression in wild-type (left) and *tim*^{0} mutant (right) heads. *per* mRNA (analyzed by RNAse protection assay) did not exhibit circadian oscillations in *tim*^{0} flies. PER proten (analyzed by immunoblot analysis, in which the bound anti-PER antibody results in the emission of a luminescent signal) was expressed at constitutively low levels in *tim*^{0} flies (nonsp—nonspecific crossreacting band). The low levels of PER preclude its detection in situ, so a PER-βGAL reporter, which accumulates to high levels in *tim*^{0} flies, was detected instead. It is expressed in the same tissues as in wild-type flies but fails to localize to nuclei (arrowheads). The nuclei are best visialized in the retina where they appear as a layer. [Top, reprinted with permission from Sehgal et al. (1994): Loss of circadian behavioral rhythms and *per* RNA oscillations in the Drosophila mutant *timeless*. *Science* **263**: 1603–1606, copyright (1994) American Association for the Advancement of Science. Middle, reprinted from Price et al. (1995) with permission of Oxford University Press. Bottom, reprinted with permission from Vosshall et al. (1994), copyright (1994) American Association for the Advancement of Science.]

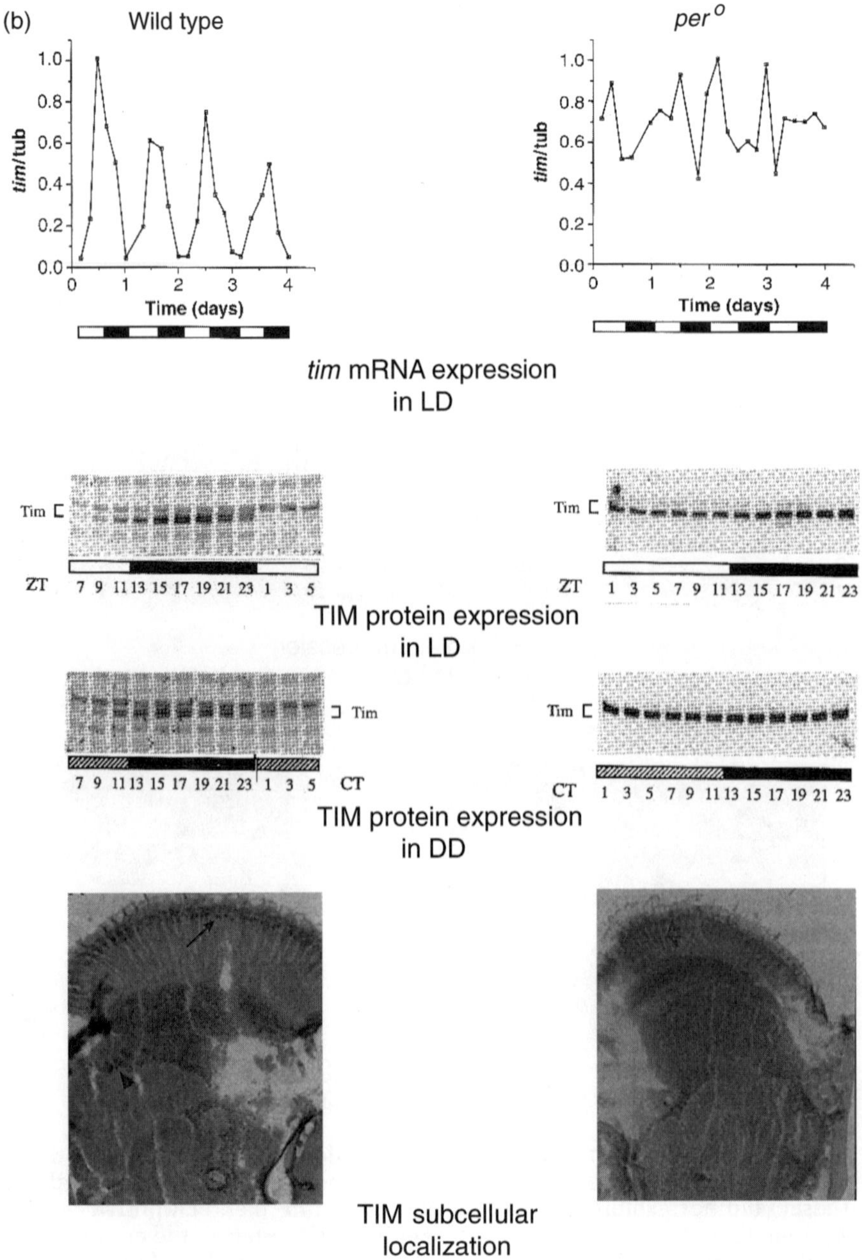

Figure 3.6. *Continued.* (b) *tim* expression in wild-type (left) and *per*^{0} mutant (right) heads. Detection was accomplished as in (a). The *per*^{0} mutation eliminates the circadian oscillations of *tim* gene products and the nuclear accumulation of TIM, but it does not eliminate LD-driven cycles of TIM protein or generally depress TIM protein levels. [Top, reprinted with permission from Sehgal et al. (1995): Rhythmic expression of *timeless*: A basis for promoting circadian cycles in *period* gene autoregulation. *Science* **270**: 808–810, copyright (1995) American Association for the Advancement of Science. Middle, reprinted by permission from Zeng et al. (1996): A light-entrainment mechanism for the Drosophila circadian clock. *Nature* **380**: 129–135, copyright (1996) Macmillan Publishers Ltd. Bottom, reprinted from Myers et al. (1996): Light-induced degradation of TIMELESS and entrainment of the Drosophila circadian clock. *Science* **271**: 1736–1740, copyright (1996) American Association for the Advancement of Science.]

(c)

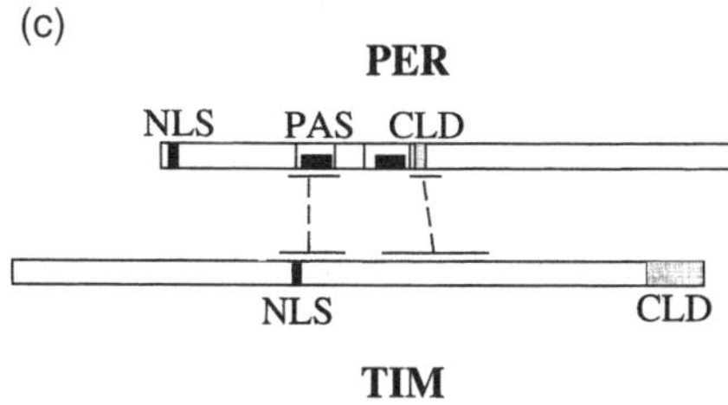

Figure 3.6. *Continued.* (c) The mutual regulation of *per* and *tim* involves protein/protein interactions between PER and TIM. Both proteins contain putative nuclear localization signals (NLSs) thought to mediate nuclear import. In addition, both proteins contain cytoplasmic localization domains (CLDs), which retain PER (or TIM) in the cytoplasm unless the other is coexpressed. PER also contains the previously described PAS region [black bars denote (a) and (b) repeats]. Interactions between PER and TIM occur as shown. [Reprinted from Saez and Young (1996), copyright (1996) with permission from Elsevier Science.]

distinct interaction domains exist within both PER and TIM (Fig. 3.6c). Within PER, one of these was the PAS region, and the other was termed a *cytoplasmic localization domain* (CLD) because it also functions to retain PER in the cytoplasm in the absence of TIM. Deletion of this CLD causes PER to move to the nucleus of Drosophila tissue culture cells even without TIM, while PER which includes the CLD will move into the nucleus only if TIM is coexpressed. Furthermore, the per^L mutation, which lies in the PAS region of *per*, reduces the interaction of PER and TIM and delays the nuclear localization of PER^L at high temperature, thereby further establishing the link between PER/TIM association and nuclear localization.

Within TIM, one of the interaction domains includes its nuclear localization signal, which interacts with the PER PAS domain. The other interaction domain in TIM interacts with the CLD of PER. TIM also has a cytoplasmic localization domain, which retains TIM in the cytoplasm unless PER binds to TIM. All these data argue that TIM and PER coregulate their subcellular localization by direct binding of regions that regulate subcellular localization, thereby masking the cytoplasmic localization domains and unmasking nuclear localization signals in both proteins. This interaction is also likely to stabilize PER by protecting it against proteolytic degradation in the cytoplasm. Once in the nucleus, PER and TIM negatively regulate their own promoters, in a manner that we shall examine in the next section.

dCLOCK AND CYCLE

These circadian oscillator elements form a transcription factor that positively regulates the *per* and *tim* promoters, negatively regulates *dClock*, and is inhibited by binding of PER and TIM.

The E Box

dClock (*dClk*) and *cycle* (*Cyc*) encode basic helix-loop helix (bHLH)/PAS transcription factors that bind to an "E" box within the *per* and *tim* promoters. Several research directions converged to elucidate a role for these two genes in the Drosophila circadian clock. The *Clk* gene was identified first in mice from studies of the first circadian mouse mutant to be isolated (also called *Clk*; see Chapter 4). The conceptual translation of the mouse *Clk* gene showed that it encoded a basic helix–loop helix transcription (bHLH) factor that also contained a PAS region of homology (Fig. 3.5). As discussed above, the PAS region also is found in PER and has been shown to mediate protein–protein associations. The bHLH domain exists in a large number of transcription factors, in which the basic (b) region binds to a short six-nucleotide sequence (CACGTG; the E box) within the promoters of certain genes. Binding of the basic regions usually requires bHLH factors to dimerize with each other. The dimerization is mediated through a region containing two alpha helices connected by a loop (the HLH region; Fig. 3.5). Because mouse *Clk* protein (CLK) also contained a PAS region, it seemed possible that it might interact with a mouse PER via the PAS region, and that the Drosophila circadian clock might also utilize a dCLK protein. Both predictions would be validated, but here we will concentrate on the Drosophila *Clk* protein (dCLK; see Chapter 4 for more on the mammalian clock genes).

Attempts to clone a Drosophila *Clk* gene on the basis of homology to the mouse gene coincided with attempts to clone the genes affected by two clock mutations in Drosophila. Both of these mutations cause arrhythmic circadian behavior when homozygous. The *Jrk* mutation turned out to affect the Drosophila version (ortholog) of the mouse *Clk* gene, and so it was renamed $dClk^{Jrk}$. The other mutation (cyc^0) also affected a bHLH/PAS transcription factor protein (CYC; Fig. 3.5a). CYC and dCLK proteins were shown to associate with each other both in vitro and in vivo, thereby forming a heterodimer (Fig. 3.7a). Although all bHLH transcription factors bind to an E-box-containing DNA sequence, a particular bHLH heterodimer will recognize only the subset of E-box sequences that have a particular DNA sequence in the flanking region. Both the *per* and *tim* promoters have E-box sequences in them, and an E-box sequence in the *per* promoter has been shown to be necessary for normal circadian oscillations of *per* mRNA.

So an obvious hypothesis to test was whether this particular E box was a target for the dCLK/CYC heterodimer. This was tested by introducing a luciferase reporter gene under the control of the *per* or *tim* E box into a Drosophila cell line, along with DNA that expressed dCLK. Luciferase is the enzyme that produces flashes of light in fireflies, and it can produce a glow in transgenic fruit fly cells, too. Since CYC was

Figure 3.7. dCLOCK and CYCLE associate to form a transcription factor that positively regulates the *per* and *tim* promoters by binding to them, and this binding is negatively regulated by PER and TIM. (a) A yeast two-hybrid assay, in which dCLOCK and CYC (or dBMAL1) are shown to associate. dCLOCK is fused to a DNA binding domain (LEXA), while dBMAL1 is fused to a domain that activates RNA polymerase (VP16). Transcription of a β-galactosidase (β-Gal) gene can only occur if a protein/protein association between dCLK and dBMAL occurs. Because β-GAL cleaves a substance in the food to produce a blue color, transcription of the β-Gal gene (and therefore a protein–protein interaction) is demonstrated in yeast colonies that turn blue (seen as darker colored colonies here). (b) dCLK/CYC activates transcription of *per* through the *per* E box, and PER and TIM inhibit this activation. Introduction of a dCLK expressing transgene into a cell line that already expresses CYC activates expression of a reporter gene driven by the *per* E box. Mutation of the E box or introduction of both PER and TIM expressing transgenes reduces the activation. Activation is measured by the light emission that is catalyzed by luciferase. [Parts (a) and (b) reprinted with permission from Darlington et al. (1998), copyright (1998) American Association for the Advancement of Science.] (c) dCLK is required for activation of the *per* promoter in vivo. A luciferase reporter gene is rhythmically expressed from the *per* promoter in wild-type or $+/dClk^{Jrk}$ flies but it is constitutively expressed at low levels in a homozygous $dClk^{Jrk}$ mutant. [Reprinted permission from Allada et al. (1998), copyright (1998) with permission from Elsevier Science.] (d) dCLK/CYC binds to the E box in vitro, and this binding is reduced by PER, TIM, or PER/TIM. The indicated proteins are added to the radioactively labeled E-box sequence. The addition of dCLK and CYC reduces the mobility of the radioactively labeled DNA on a gel, thereby producing a band. Although PER and TIM do not produce novel mobility shifts (E1 and E2 are produced by contaminating proteins in the assay), they do reduce the amount of dCLK/CYC-dependent complex. [Reprinted from Lee et al. (1999) with permission from the American Society for Microbiology.]

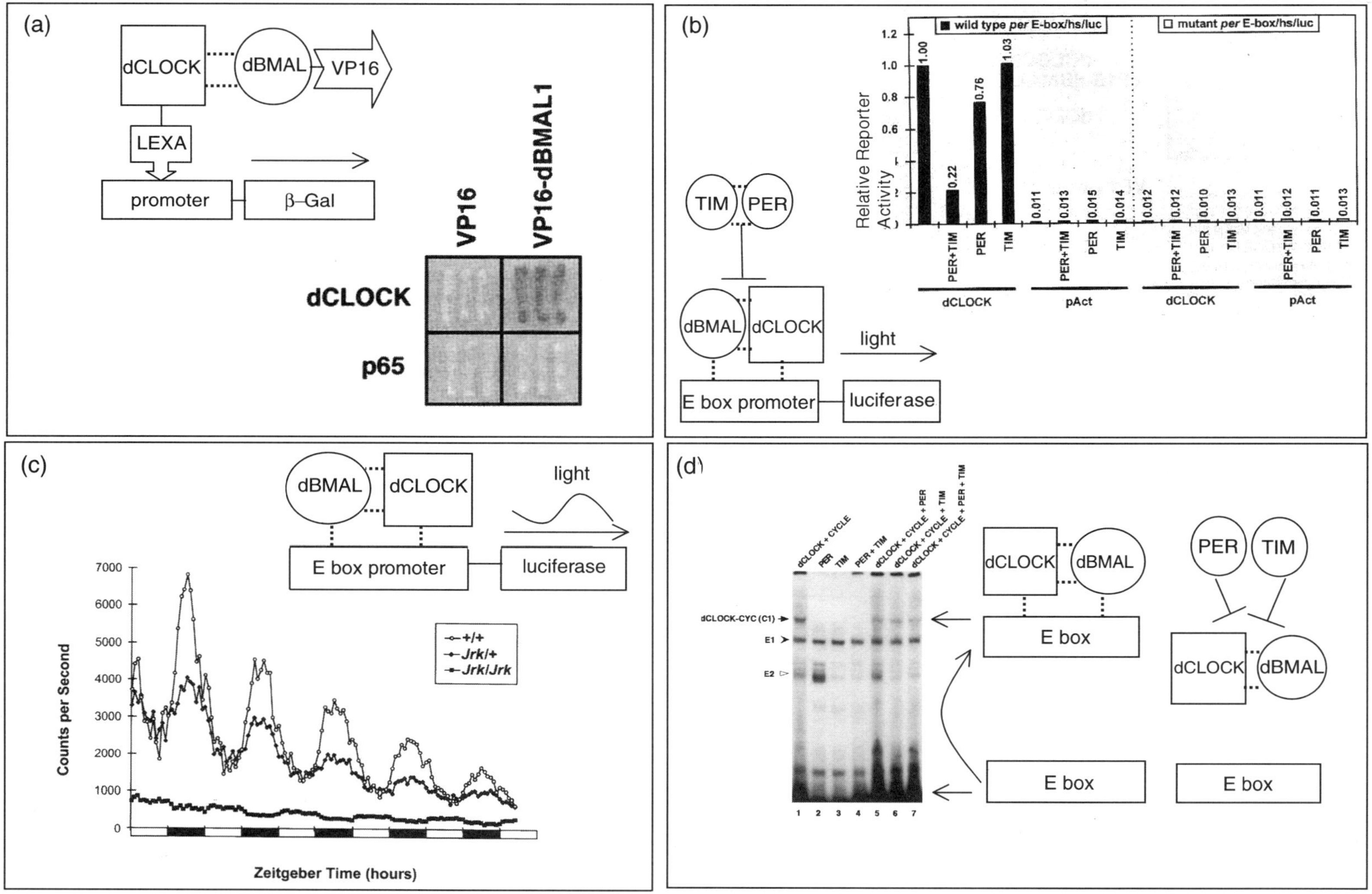
(a)
dCLOCK
dBMAL
VP16
LEXA
promoter
β–Gal
VP16
VP16-dBMAL1
dCLOCK
p65
(b)
TIM
PER
dBMAL
dCLOCK
E box promoter
luciferase
light
wild type per E-box/hs/luc
mutant per E-box/hs/luc
Relative Reporter Activity
1.00
0.22
0.76
1.03
0.011
0.013
0.015
0.014
0.012
0.012
0.010
0.013
0.011
0.012
0.011
0.013
PER+TIM
PER
TIM
dCLOCK
pAct
(c)
dBMAL
dCLOCK
E box promoter
luciferase
light
+/+
Jrk/+
Jrk/Jrk
Counts per Second
Zeitgeber Time (hours)
(d)
dCLOCK + CYCLE
PER
TIM
PER + TIM
dCLOCK + CYCLE + PER
dCLOCK + CYCLE + TIM
dCLOCK + CYCLE + PER + TIM
dCLOCK-CYC (C1)
E1
E2
1 2 3 4 5 6 7
dCLOCK
dBMAL
E box
PER
TIM

expressed from the endogenous *cyc* gene of the cells, the *dClk* transgene resulted in a dCLK/CYC heterodimer, which caused a significant increase in the luciferase "glow" of the cell line (Fig. 3.7b). Since single nucleotide changes in the *per* E-box sequence eliminated the stimulation of luciferase by dCLK, the effect of dCLK was shown to be mediated by the E box. Another line of evidence showing that dCLK and CYC regulated *per* and *tim* transcription comes from assays of endogenous *per* and *tim* levels and transcription rates in the $dClk^{Jrk}$ and cyc^0 mutants. These are constitutively low in the mutants (e.g., Fig. 3.7c), as expected for mutations that eliminate positively acting transcription factors for the *per* and *tim* promoters. Finally, when purified dCLK and CYC proteins are mixed together and added to E-box DNA in vitro, they can be shown to bind to the DNA (Fig. 3.7d).

PER/TIM–dCLK/CYC Binding

PER and TIM bind to the dCLK/CYC heterodimer and abrogate its E-box binding. If PER and TIM are expressed in the Drosophila cell line along with dCLK, endogenously expressed CYC and the *per* E-box-driven luciferase, the levels of luciferase are reduced relative to its expression in the absence of PER and TIM (Fig. 3.7b). Coexpression of both PER and TIM is required for this negative feedback. One reason for this requirement is that neither PER nor TIM is nuclear unless both are expressed. In vivo, the binding of PER and TIM with dCLK occurs during the night but not during the latter part of the day—consistent with the times at which *per* and *tim* transcription are inhibited. In vitro, this binding abrogates the binding of dCLK/CYC to the *per* promoter (Fig. 3.7d) without disrupting the dCLK/CYC interaction. Therefore, these data argue that PER and TIM exert negative feedback by binding to dCLK/CYC and preventing them from binding to the *per* and *tim* genes. However, other data do not support this mechanism in its entirety, as incubation of Drosophila head extracts with DNA containing a *tim* E box produces comparable amounts of a bound dCLK/CYC at all times of day. The high levels of PER and TIM bound with dCLK/CYC in head extracts isolated from night timepoints would be predicted to reduce the amount of complex formed during the night.

Since both PER and TIM bind to dCLK/CYC in vivo, it is not clear what the relative contributions of PER, TIM, and PER/TIM are to negative feedback in vivo. In vitro, either PER or TIM alone can inhibit binding of dCLK/CYC to the E box (Fig. 3.7d). In vivo, nuclear PER persists after TIM is largely eliminated by light, and negative feedback also persists for much of this time. Moreover, in the tim^{UL} mutant, elimination of nuclear TIM^{UL} protein by light enhances repression of *per* and *tim* mRNA, demonstrating that PER alone is a more effective negative regulator than is PER/TIM^{UL}. In tissue culture cells, mutant PER missing a CLD can transport to nuclei and repress the *per* promoter in the absence of TIM.

Interlocked Feedback Loops

The Drosophila clock consists of interlocked feedback loops. The levels of *dClk* mRNA oscillate with a circadian rhythm that has a phase opposite that of *per* and *tim* mRNA (Fig. 3.8a). There is almost no phase lag between *dClk* mRNA and protein, both of which peak in abundance around the time of dawn, while *per* and *tim* mRNA levels peak at the time of dusk. The time at which *dClk* mRNA peaks is also the time at which PER and TIM are most strongly localized to nuclei, suggesting that PER and TIM might positively regulate d*Clk* mRNA. Consistent with this notion, *dClk*

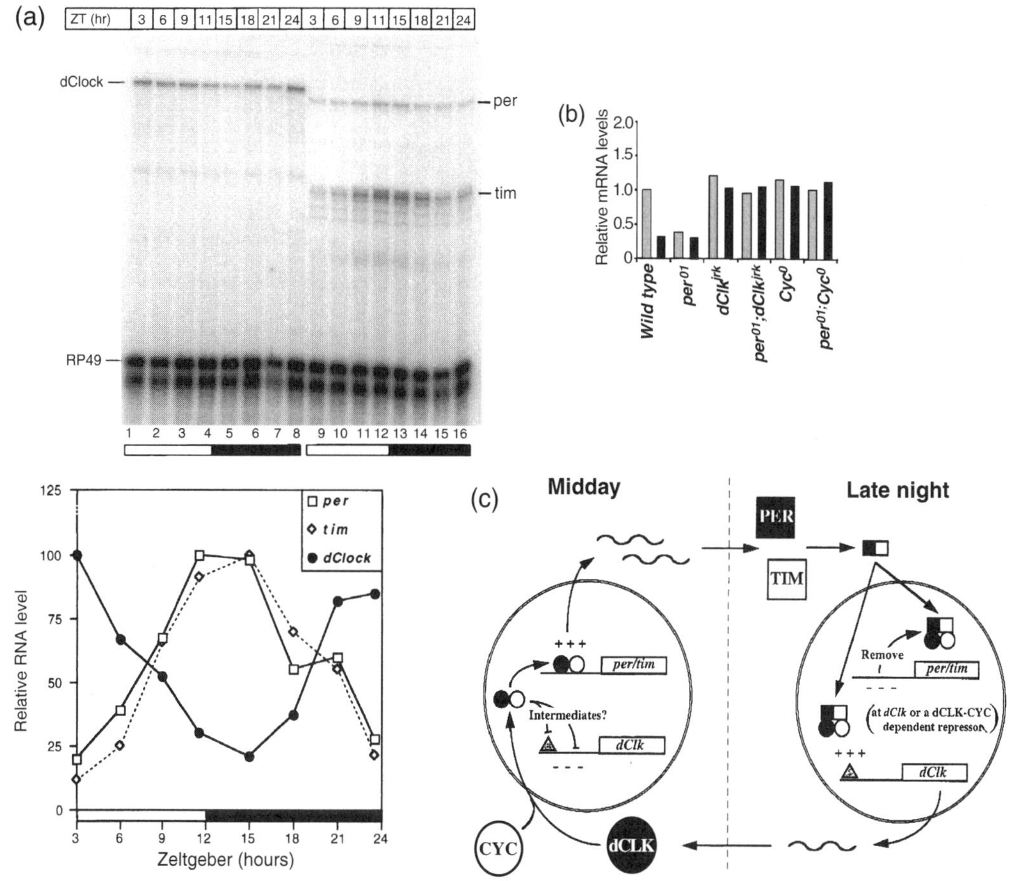

Figure 3.8. *per* and *tim* positively regulate *dClk* mRNA levels and negatively regulate *per* and *tim* mRNAs—most likely in both cases through a negative regulation of dCLK/CYC. (a) The *dClock* mRNA oscillates with a phase opposite to the *per* and *tim* mRNA oscillations. The upper panel is an autoradiograph of a gel demonstrating these oscillations. The radioactive band is produced by hybridization of the indicated mRNA to a complementary radioactively labeled RNA produced from the cloned gene. The hybridization event protects the radioactive "probe" from digestion by RNAse, and the amount of protected probe is therefore a measure of the complementary mRNA present. In the lower panel, the amount of radioactivity for each band has been normalized to the amount of radioactivity for a constitutively expressed mRNA (RP49, also visualized on the gel). [Reprinted from Bae et al. (1998): Circadian regulation of a *Drosophila* homolog of the mammalian *Clock* gene: PER and TIM function as positive regulators. *Mol Cell Biol* **18**: 6142–6151 with permission from the American Society for Microbiology.] (b) *dClk* mRNA is constitutively expressed at low levels in the per^0 mutant and at high levels in the $dClk^{Jrk}$, cyc^0. per^0; $dClk^{Jrk}$, and per^0; cyc^0 mutants. *dClk* mRNA levels were quantitated as in (a) from early evening time points (black bars) and early subjective day timepoints (light bars). (c) Model suggesting how negative regulation of dCLK/CYC by PER/TIM can produce oppositely phased oscillations by interlocked feedback loops. The large circles represent the nucleus. [Parts (b) and (c) reprinted with permission from Glossop et al. (1999), copyright (1999) American Association for the Advancement of Science.]

mRNA and protein are expressed at constitutively low levels in per^0 and tim^0 flies, which do not express any functional PER or TIM (respectively; e.g., Fig. 3.8b).

How might PER and TIM produce such "positive feedback?" An important insight comes from the observation that *dClk* mRNA levels are constitutively high in $dClk^{jrk}$ or cyc^0 single mutants, and also in double mutants such as per^0; $dClk^{Jrk}$ and per^0; cyc^0 (Fig. 3.8b). Taken together with the opposite effects of per^0 or tim^0 mutants on *dClk* mRNA, these results argue that dCLK/CYC negatively regulates the mRNA levels of d*Clk*, and that PER and TIM operate upstream of dCLK/CYC to eliminate this negative feedback. The elimination of the two sequentially acting negative feedbacks produces a positive effect on d*Clk* in the double mutants (Fig. 3.8b). It is not known how PER and TIM abrogate negative feedback by dCLK/CYC, but they are likely to do so in the same way they abrogate positive feedback by CLK/CYC—by binding to the CLK/CYC heterodimer (Fig. 3.8c). It is now known how dCLK/CYC negatively regulates *dClk* mRNA (Fig. 3.8c). It does so indirectly by activating a repressor of the *dClk* promoter—*vrille* (to be discussed in a later section).

■ POSTTRANSLATIONAL REGULATION OF CIRCADIAN OSCILLATOR PROTEINS BY PHOSPHORYLATION

The function of several circadian oscillator proteins is regulated posttranslationally by phosphorylation. In addition to rhythms of level and nuclear accumulation, PER and TIM are phosphorylated rhythmically. This rhythm is observed as a change in the mobility of PER and TIM on SDS (sodium dodecyl sulfate)–polyacrylamide gels (e.g., Figs. 3.6a and 3.6b, middle panels). In addition, the dCLK protein is phosphorylated, but the function of this phosphorylation is still unknown. In the case of PER and TIM, maximal levels of phosphorylation are observed before the proteins disappear, suggesting that phosphorylation targets them for degradation. Genetic analysis has identified kinases that are involved in the phosphorylation of both these proteins, and has elucidated some of the roles for this phosphorylation.

The *dbt* Gene Product

The *double-time* (*dbt*) gene product regulates the stability and nuclear accumulation of PER protein. PER becomes progressively phosphorylated after its synthesis during late night, and it becomes maximally phosphorylated during morning, when high levels of nuclear PER are also detected. Then, levels of PER begin to decline. However, the role of this phosphorylation was not known until the isolation of the *dbt* mutants allowed the Drosophila clock to be studied under conditions in which the phosphorylation of PER is altered. *dbt* mutations that shorten, lengthen, or eliminate circadian rhythmicity have all been identified. The mutation which shortens circadian period expedites both the phosphorylation of PER and its subsequent disappearance. The mutations that lengthen the circadian period delay or prolong PER phosphorylation, and result in persistently high levels of PER after TIM disappears in the morning. Two *dbt* mutations produce molecular and/or behavioral arrhythmicity. One of these is also a pupal lethal. Both produce constitutively high levels of PER whose phosphorylation state is constitutive (either constitutively low or constitutively heterogeneous). So there is a correlation between the timing or amount of PER phosphorylation and the turnover of PER in the *dbt* mutants. This correlation is consistent with a role for phosphorylation in triggering the degradation of PER.

Casein Kinase I

dbt encodes an ortholog of the vertebrate ε/δ casein kinase I isoforms. This type of protein kinase is thought to phosphorylate threonines and serines. While many protein kinases have to be activated by the binding of a ligand to a receptor or by a second messenger [e.g., Ca^{2+}, diacylglycerol or cyclic adenosine monophosphate (cAMP)], casein kinase I has not been shown to be directly activated by this type of mechanism. Instead, it prefers to phosphorylate substrates that have already been phosphorylated on a threonine, serine, or tyrosine one to two amino acids upstream from the target serine or threonine. So *dbt* protein (DBT) may be activated to phosphorylate PER by its own prior kinase activity on PER, or by the activity of other, messenger-dependent kinases on PER. It has been shown that DBT can bind with PER in vitro and in cell culture, and this interaction is likely to enhance the targeting of PER as a substrate. *dbt* mRNA and protein are expressed at the same sites as *per* mRNA in the adult head, so DBT is expressed in a manner that allows it to undergo direct interactions with PER.

It is certain that the DBT kinase has many other substrates in addition to PER, since many mutations in this gene are lethal, with effects on cell survival and growth in imaginal discs (these have been dubbed "discs overgrown" or "dco" mutations). In vertebrates and yeast, casein kinase I has been shown to be involved in processes as diverse as DNA repair, cell cycle progression, vesicle trafficking, Wnt signaling (a pathway important for development), and circadian rhythms.

So far, it has not been possible to produce enzymatically active recombinant DBT protein, so a demonstration that DBT directly phosphorylates PER has not been possible in flies. However, casein kinase I has been shown to bind and phosphorylate mammalian PER, and to decrease the stability of mammalian PER in tissue culture cells (see Chapter 4). Moreover, there are many potential casein kinase I phosphorylation sites in both mammalian and Drosophila PER, and sequential phosphorylation of these may explain the progressive phosphorylation of PER that is observed during the circadian cycle.

In the case of two of the three long-period *dbt* mutants and the arrhythmic mutant that also produces pupal lethality, it is clear that the mutations reduce the kinase activity of DBT. The two long-period *dbt* mutations, which change two amino acids in DBT, have been introduced into the yeast homologue, where they reduce the activity of the kinase in vitro. The lethal *dbt* allele (*dbt*^{P}) arises from insertion of a P transposable element, which strongly reduces the level of *dbt* mRNA and produces hypophosphorylated PER in vivo. Therefore, lowered levels of DBT activity can reduce or delay phosphorylation of PER and lead to elevated or persistent expression of PER.

PER in Cytoplasm

Wild-type DBT acts to destabilize PER in the cytoplasm. The effects of altering PER phosphorylation are seen most clearly in the P-element-induced mutant of *dbt*, which strongly depresses DBT expression and causes pupal lethality. As mentioned above, PER is constitutively hypophosphorylated in this mutant (Fig. 3.9a). Since this mutant also results in pupal lethality, adult circadian behaviors cannot be assayed. However, it is possible to assay the rhythms of *per* in the larval lateral neurons, which mediate the larval time memory discussed earlier. In the lateral neurons of this mutant, PER protein is expressed at constitutively high levels in both DD and LD, rather than with the oscillations that are found in wild-type larvae (Fig. 3.9a). Moreover, PER is seen in parts of the larval central nervous system where it is not normally detected (Fig. 3.9b).

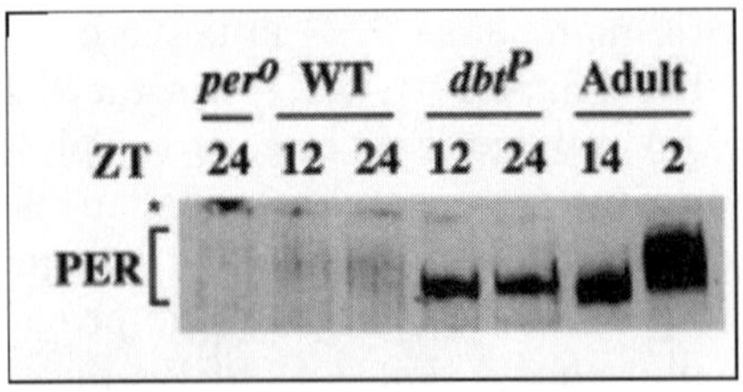

PER protein levels are higher in *dbt*^{P} larvae than in wild type larvae, and PER is not highly phosphorylated in *dbt*^{P} larvae

(a)

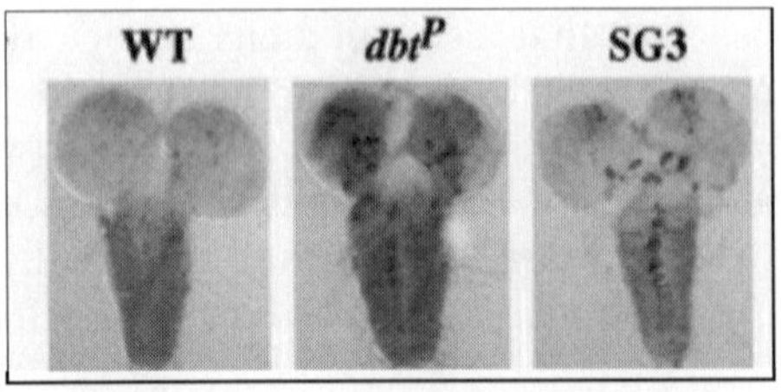

Like PER-βGAL (SG3) but unlike PER (WT) in wild type larvae, PER in *dbt*^{P} larvae is expressed in parts of the central nervous system where TIM and PER do not accumulate in wild type larvae

(b)

Figure 3.9. Low levels of *dbt* expression lead to high levels of constitutively expressed PER that is hypophosphorylated. (a) Extracts of larvae or adults collected at the indicated timepoints were immunoblotted with an antibody for PER, which generated a luminescent band. PER is much more highly expressed in *dbt*^{P} larvae than in wild-type larvae; *per*^{0} larvae do not express PER and therefore serve as a negative control. The mobility of PER in *dbt*^{P} is comparable to that of PER in adults at ZT14, when PER is not extensively phosphorylated. (b) Immunolocalization of PER in the central nervous systems of wild-type and *dbt*^{P} larvae, and of a PER-βGAL fusion protein (SG3) expressed from a *per* promoter in wild-type larvae. The extensive sites of PER expression in *dbt*^{P} larvae represent bonafide sites of *per* transcription. The absence of most of these PER sites in wild type most likely results from post-transcriptional destabilization. In *dbt*^{P} this regulation is lost. [Reprinted from Price et al. (1998), copyright (1998) with permission from Elsevier Science.]

The overexpression is not the result of high levels or otherwise aberrant *per* mRNA expression, because the levels and sites of *per* mRNA expression appear to be normal in the mutant. For instance, a reporter gene under the control of the *per* promoter expresses its mRNA in tissues where *per* mRNA is also expressed. The resulting PER-βGAL fusion protein, which is more stable than PER itself, accumulates in many sites where PER is not observed in wild type, as does PER in *dbt*^{P} (Fig. 3.9b). Therefore, the dramatic accumulation of PER in *dbt*^{P} larvae is most likely to arise posttranscriptionally.

In fact, it seems most likely that the observed hypophosphorylation of PER produces the high levels of PER in this *dbt* mutant, in sites where TIM expression is not readily detected. In the *tim*^{0} mutant, recall that PER is cytoplasmic and expressed at very low levels. Apparently, PER does not require high levels of TIM for its stability in the absence of DBT and in the absence of wild-type phosphorylation of PER. This hypothesis was confirmed by subjecting larvae to constant light (LL), which eliminates PER in wild-type larvae by destabilizing TIM. In the *dbt* mutant, PER levels were unaffected by LL, thereby establishing that PER stability is largely TIM-independent. It seems that wild-type DBT protein is likely to destabilize PER in the cytoplasm unless high levels of TIM have accumulated. Thus, the cytoplasmic destabilization of PER seems to be one result of DBT-dependent phosphorylation of PER.

PER in Nuclei

DBT also regulates nuclear accumulation of PER. In the *dbt*^{S} mutant, the accumulation of total PER levels is slightly advanced relative to wild type in LD, and yet the accumulation of nuclear PER is delayed by 3 hours. In addition, the time course of *per* mRNA is altered in the *dbt*^{S} mutant, suggesting an alteration in the timing of nuclear feedback. The delay in accumulation of nuclear PER does not occur in the *per*^{S} mutant, which has an endogenous period quite similar to the *dbt*^{S} period (19 and 18 hours, respectively). These results suggest that DBT regulates the nuclear accumulation of PER independently of its regulation of cytoplasmic stability. It may do so by preferentially altering the stability of nuclear PER relative to cytoplasmic PER. Since PER declines to lower levels in the nuclei of *dbt*^{S} flies, it is likely that the stability of nuclear PER is reduced by the *dbt*^{S} mutation during this time. As well, analysis of the long-period *dbt* mutants and a *per*^{S}; *dbt* double mutant have argued that these mutants alter the stability of PER during the time of its nuclear decline.

However, the effect of the *dbt*^{S} mutation on delayed accumulation of nuclear PER may also result from an effect on the nuclear transport of PER. Mammalian casein kinase I has this effect on mammalian PERs expressed in cultured cells. It binds to PER and phosphorylates it, thereby masking or unmasking a nuclear localization signal, and either antagonizing or promoting nuclear transport of PER (depending on the isoform of PER). Likewise, Drosophila DBT could antagonize or promote the nuclear transport of PER, and DBT^S would then be a stronger antagonist or weaker promoter than wild-type DBT. The combined analysis of casein kinase I and PER in mammals and Drosophila has shown than multiple aspects of PER's temporal program are regulated by its phosphorylation.

DBT/PER Association In Vivo and Nuclear Localization

While the DBT protein and mRNA levels do not oscillate, the amount of nuclear DBT does oscillate (Fig. 3.10). High levels of nuclear DBT are found during the late night and morning, at times when PER is nuclear, while mostly cytoplasmic localization of DBT is found during the early evening, when PER is cytoplasmic. Therefore, DBT may localize to the nucleus or cytoplasm at different times in part because of its association with the PER/TIM complex. Accordingly, its associations with PER could mediate both effects on cytoplasmic and nuclear regulation of PER (Fig. 3.10).

DBT Regulation of PER-Driven Feedback in Nucleus

DBT regulates both the initiation and termination of feedback exerted by PER in the nucleus. Destabilization of cytoplasmic PER and delay of its nuclear accumulation are predicted to delay initiation of the negative and positive feedback that PER and TIM exert in the nucleus (Fig. 3.10). Delays in negative feedback are thought to be essential for circadian rhythmicity, because mathematical modeling of the negative feedback loop has argued that immediate negative feedback would produce a constitutive equilibrium; the level of PER protein that accumulates would maintain the equilibrium level of *per* mRNA by repression, while in turn that level of *per* mRNA would maintain a constant level of PER. This is exactly the type of phenotype observed in the larval lateral neurons of the P-element-induced *dbt* mutant, in which the high levels of PER that accumulate are thought to exert immediate feedback in the nucleus (Fig. 3.9). In wild-type flies, the delay in negative feedback allows high levels of *per* and *tim* mRNA to be expressed before they are repressed, thereby producing molecular oscillations. Presumably, associations between PER and TIM may eventually stabilize PER and promote nuclear localization of the PER/TIM complex, thereby terminating the delay in initiation of negative feedback. Alternatively, extensive phosphorylation of PER (or TIM; see discussion below), or some other signal,

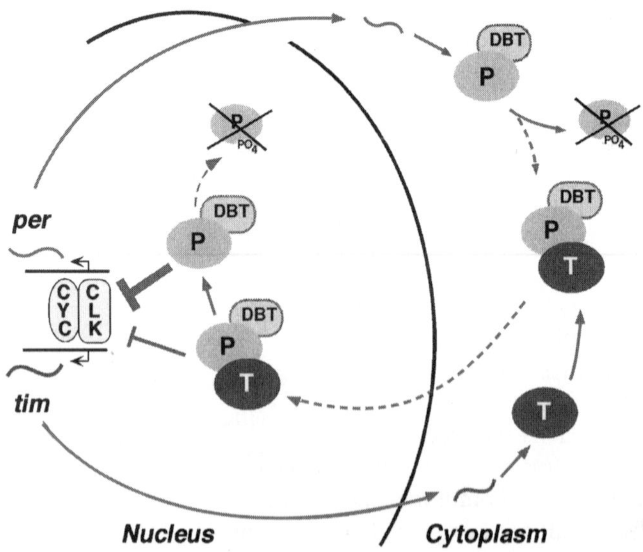

Figure 3.10. Model for regulation of the timing of nuclear feedback processes by DBT. DBT binds with PER (P) and phosphorylates it in the cytoplasm, thereby causing it to be degraded and possibly antagonizing the nuclear import of PER. TIM stabilizes PER and causes it to be translocated into the nucleus. When light or the circadian clock eliminates TIM, DBT, which is transported into the nucleus with PER, is able to phosphorylate nuclear PER and target it for degradation, thereby terminating the repression of CLK/CYC by PER. [Reprinted from Kloss et al. (2001): Phosphorylation of PERIOD is influenced by cycling physical associations of DOUBLE-TIME, PERIOD, and TIMELESS in the *Drosophila* clock. *Neuron* **30**: 699–706, copyright (2001) with permission from Elsevier Science.]

ultimately triggers nuclear accumulation of PER and TIM.

Since TIM is light-sensitive, light reduces the levels of nuclear TIM before nuclear PER levels are reduced, and the termination of feedback in the nucleus is likely to result when PER levels are subsequently reduced to some threshold level. Indeed, *per* and *tim* mRNA do not begin to accumulate until several hours after nuclear TIM is no longer detectable. Since DBT regulates the time course of the decline in nuclear PER, it also regulates the time at which both positive and negative nuclear feedback are terminated. However, the final stages of PER phosphorylation may require the elimination of TIM from the PER/TIM complex (Fig. 3.10), because maximal phosphorylation of PER is delayed in the tim^{UL} mutant, which produces a TIM^{UL} protein that persists much longer than does wild-type TIM in DD. Light eliminates the TIM^{UL} protein and produces more rapid phosphorylation of PER. So TIM may antagonize DBT-dependent phosphorylation and turnover of PER, while DBT antagonizes the stabilization of PER (and possibly its nuclear translocation) effected by TIM in both the cytoplasm and the nucleus (Fig. 3.10).

It is predicted that a *dbt* mutation that increases the cytoplasmic and nuclear stability of PER should tend to expedite the initiation of nuclear feedback, but delay the termination of the same. Depending on whether the effects on initiation or termination are larger, the mutant could either produce a shorter- or longer-period clock. Perhaps this is why both short- and long-period *dbt* mutants can be isolated. In the case of the *dbt*^{S} mutant, analysis of resetting of the clock by light shows a shortening of the subjective day, when PER is disappearing from the nucleus and negative feedback is terminated. A shortening of this part of the circadian program evidently leads to a shorter overall period for the circadian program, despite the delayed accumulation of PER in the nucleus that is observed in LD.

The *sgg* Protein Kinase

The *shaggy* (*sgg*) protein kinase regulates the timing of nuclear feedback by phosphorylating TIM. *sgg*, which has a role in the determination of segment polarity, was originally identified in screens for mutations that affect the segmentation of the Drosophila embryo. Loss of function mutations lead to a loss of segment polarity (i.e., there is no difference between the anterior and posterior parts of the segment). It is now known that *sgg* functions as part of the *wingless* (*wg*) pathway (Fig. 3.11a). WG protein binds to the *Dfrizzled2* receptor on the cell surface, and the signal transduction pathway that is activated by this interaction downregulates the activity of the *sgg* kinase (also known as glycogen synthase kinase 3) on armadillo protein (a β-catenin). When active, *sgg* protein (SGG) causes the phosphorylation of *armadillo*, which leads to its degradation. When SGG activity is downregulated in response to the WG signal, *armadillo* protein is able to accumulate and move to the nucleus, where it activates transcription of a battery of genes that confer the anterior/posterior fate of the segment.

sgg was identified as a clock gene in a screen for genes whose overexpression caused aberrant circadian locomotor activity rhythms (Fig. 3.11b). A promoter with a UAS sequence, which binds the yeast transcription factor GAL4, was moved around the Drosophila genome, so that it landed at many different random places. Then, overexpression of genes in the regions where this promoter landed was achieved by crossing many of these random insertion lines to a *tim*-GAL4 driver, which expresses GAL4 from the *tim* promoter and therefore specifically causes the UAS-driven genes to be expressed in clock tissues. In one of these lines, the UAS sequence had inserted within the *sgg* gene, thereby causing its overexpression in *tim*-expressing cells. This overexpression causes a shortening of circadian period, an accelerated accumulation of PER/TIM in nuclei, and an accumulation of a hyperphosphorylated form of TIM. Conversely, genotypes in which SGG activity can be specifically lowered in the adult, so that embryonic lethality is bypassed, lengthen circadian period and lead to hypophosphorylation of TIM and overexpression of PER and TIM. In vitro, a vertebrate ortholog of SGG (glycogen synthase kinase-3β) has been shown to phosphorylate TIM, so the effects on TIM phosphorylation and nuclear accumulation appear to be direct. TIM is apparently a target for SGG just as is β-catenin in the *wg* pathway, except that SGG facilitates the nuclear accumulation of TIM instead of antagonizing it, as it does for β-catenin.

■ ENTRAINMENT OF DROSOPHILA CIRCADIAN RHYTHM BY LIGHT

Light-Induced TIM Degradation

Intriguingly, the onset of light leads to a rapid reduction in the levels of TIM, so TIM is unstable in the presence of light (Fig. 3.12a). This instability is present in both wild-type flies and in *per*^{0} flies. The response of TIM to light in both *per*^{0} and wild-type flies demonstrates that the effect of light on TIM requires neither PER nor a functional clock. Therefore, the response of TIM to light can be considered a part of an input pathway that mediates entrainment to light. In wild-type flies, constant light leads to arrhythmicity, which is correlated with the

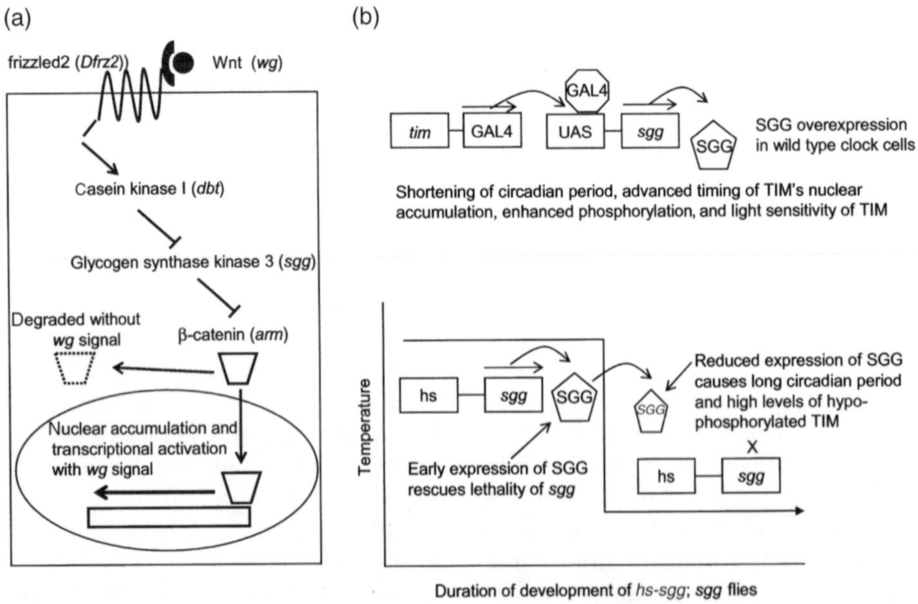

Figure 3.11. The *dbt* and *sgg* protein kinases have roles in both the *Wnt* (*wg*) signaling pathway and the circadian clock. (a) Model for the role of casein kinase I and *sgg* in the *wg* pathway. The role for casein kinase I has been shown in Xenopus, while the role for *sgg* has been shown in many systems (including Drosophila). Casein kinase I activates the *wg* pathway by repressing SGG, which otherwise causes β-catenin to be degraded. (b) A role for *sgg* in the Drosophila circadian clock has been demonstrated by overexpressing SGG in *tim* expressing cells (top), or by lowering SGG protein levels in the adult (bottom). The UAS-GAL4 system was used to overexpress SGG specifically in *tim* cells. *tim* is the promoter of the *timeless* gene, GAL4 is a yeast transcription factor, and UAS is the promoter that GAL4 binds and activates. Bottom, the lethality associated with the *sgg* mutant was rescued by expressing a *sgg* transgene, under control of a heat shock promoter, during development. By decreasing the temperature in adults, the effects of reduced *sgg* expression on circadian rhythms were determined. The heat shock promoter (hs) activates *sgg* at elevated temperatures and is turned off at lower temperatures.

rapid degradation of TIM in response to light and a more gradual decline of PER levels as well. Recall that PER requires TIM for its stability, so constant light produces low levels of PER as a consequence of the low levels of TIM.

It is possible to explain the wild-type Drosophila PRC on the basis of the observed *per* and *tim* oscillations, the sensitivity of TIM to light, and the model for the molecular mechanism of *per* and *tim* oscillations (Fig. 3.12b). During the first half of the night, when phase delays are obtained in response to short light pulses, light degrades mostly cytoplasmic TIM, which is not contributing to nuclear feedback. The presence of high levels of *tim* mRNA allows the resynthesis of TIM, stabilization of PER by TIM and transport of PER/TIM complex to the nucleus; phase delays are produced in the process. By contrast, during the second half of the night, most PER and TIM are nuclear and their mRNA levels are low, so degradation of TIM in response to light expedites the termination of nuclear feedback, under conditions in which resynthesis of TIM and PER at high rates is not possible; early termination of nuclear feedback leads to a phase advance. During the subjective day, the circadian clock has

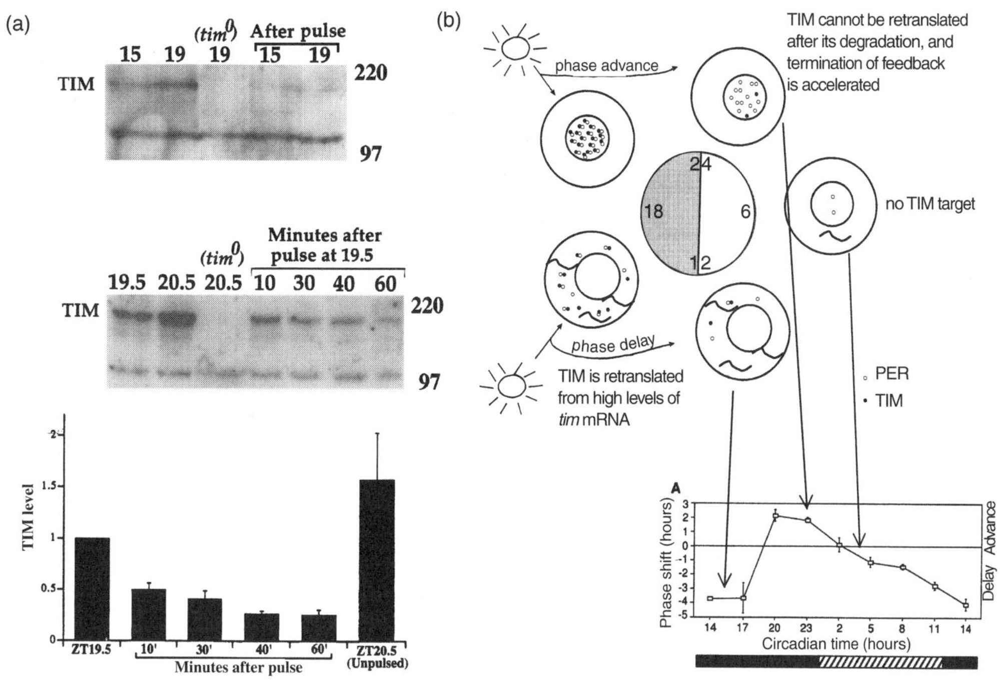

Figure 3.12. The resetting of circadian phase in response to timed light pulses entails rapid degradation of TIM. (a) TIM protein was assayed by immunoblot analysis in the heads of wild-type or *tim*^{0} flies collected at the indicated circadian time in the dark, or after a light pulse. In the top panel, flies were collected after a light pulse at ZT15 (produces a phase delay) or at ZT19 (produces a phase advance), whereas in the middle panel flies were collected at the indicated times after initiation of a light pulse at ZT19.5. In the bottom panel, the signal intensity for TIM was quantitated for several experiments like the one in the middle panel. [Reprinted from Hunter-Ensor et al. (1996), copyright (1996) with permission from Elsevier Science.] (b) The phase response curve for wild-type flies in response to 10-minute light pulses is explained by the response of TIM and *per*/*tim* cycling. Flies were entrained to LD: 12 hours: 12 hours and then released into DD. The average phase of flies that received a light pulse at the indicated circadian time was subtracted from the average phase of flies that had not received the pulse to calculate phase shifts. The hatched bars denote the times when lights would have been illuminated if the LD cycle had continued. [Reprinted from Hunter-Ensor et al. (1996), copyright (1996) with permission from Elsevier Science. Also, reprinted with permission from Myers et al. (1996): Light-induced degradation of TIMELESS and entrainment of the *Drosophila* circadian clock. *Science* **271**: 1736–1740, copyright (1996) American Association for the Advancement of Science.] Delays are thought to be produced when PER/TIM are cytoplasmic and RNA levels are high. Advances result when the proteins are nuclear and RNA levels are low.

lowered TIM levels, so there can be little or no further effect of light. The amount of phase resetting is therefore little or none.

The degradation of TIM in response to light has been investigated under various conditions that address its relevance to resetting of the behavioral rhythms by light. As the amount of light delivered in the light pulse is reduced by shortening the light pulse or reducing the intensity of light, the magnitude of phase resetting and TIM degradation are reduced in parallel. In addition, the sensitivity of the phase-resetting response to the wavelength of the light pulse parallels the sensitivity of TIM degradation to the wavelength. These parallels are consistent with a role for the degradation of TIM in the resetting of circadian phase.

Cryptochrome: A Dedicated Circadian Photoreceptor

Since TIM is not homologous to any known photoreceptor and is not directly sensitive to light, it was thought to be the downstream target of a clock photoreceptor pathway. Mutant adults lacking eyes or visual transduction pathways have rhythms that can be entrained to light and can degrade TIM in response to light, so the photoreceptors and biochemistry used for vision are not necessary for entrainment of circadian rhythms by light (Fig. 3.2). Instead, it seemed likely that at least one nonvisual photoreceptor pigment and pathway were involved in entrainment of the Drosophila clock by light.

Genetic screens and genome-scale analysis identified a circadian cryptochrome photoreceptor. Identification of this photoreceptor benefited from revolutionary advances that have been occurring in molecular genetic analyses of many model organisms. The genetic screen made use of a clock-controlled reporter gene, consisting of firefly luciferase fused to the *N*-terminal two-thirds of *per* and transcribed from a *per* promoter (Fig. 3.13). Firefly luciferase metabolizes the organic molecule luciferin to produce light emission. When expressed under the control of the *per* promoter, luciferase mRNA and protein oscillate with a circadian rhythm. Fruitflies fed luciferin produce a rhythmic glow resulting from the oscillating levels of luciferase, which thereby "reports" the timing of the circadian clock. Mutations in other, endogenous genes of the fruitfly are expected to alter the rhythmicity and period of this glow if the affected genes contribute to the circadian oscillator or

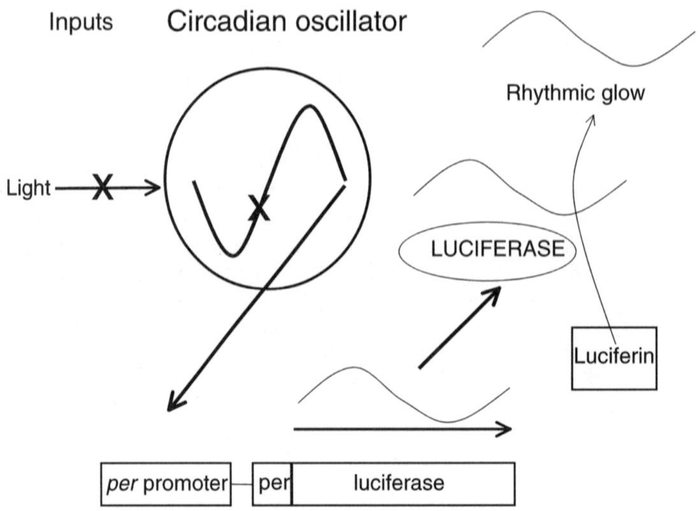

Figure 3.13. Design of a genetic screen for mutants that affect the circadian oscillation of *per* transcription. Circadian oscillations of luciferase transcription are driven by the *per* promoter, which responds to clock-driven alterations of dCLK/CYC-dependent activation. The resulting circadian oscillations of luciferase activity produce a rhythmic glow in flies fed luciferin (the substrate for luciferase). The glow can be measured and analyzed in single flies, just as locomotor activity can be (Fig. 3.1). A mutation (X) that affects the input pathway or the oscillator mechanism is predicted to alter the circadian glow rhythm.

input pathways that regulate the *per* promoter (Fig. 3.13). In such a screen, a mutation affecting a clock photoreceptor was identified because it produced an arrhythmic glow of luciferase.

The affected gene also turned up as an expressed sequence in the Berkeley *Drosophila* Genome Project, which was actively engaged in sequencing the entire *Drosophila* genome at the time (likewise, the Human Genome Project was undertaking the same task for the human genome). The sequence attracted the interest of clock researchers because it was homologous to a class of blue-light photoreceptors in plants, and to a class of enzymes called *photolyases*, which are activated by blue light to repair DNA damaged by ultraviolet light (Fig. 3.14). Resetting of the Drosophila circadian rhythm is maximally responsive to blue light, so a blue-light photoreceptor had long been implicated in circadian photoreception. The Drosophila gene was dubbed *cry* in recognition of its homology to the plant blue-light photoreceptors, which were called *cryptochromes*. Plant cryptochromes and photolyases bind to two chromophores: pterins (MTHF or 8-HDF) and flavins. The conceptual translation of the Drosophila *cry* gene contains domains with sequence homology to the domains that bind these chromophores in photolyases and plant cryptochromes (Fig. 3.14).

The circadian phenotype of the mutant (cry^b) was consistent with a role for the gene in circadian photoreception (Fig. 3.14). The cry^b mutation is predicted to change an amino acid in the domain that binds the flavin chromophore, and so it is likely to disrupt the detection of light that

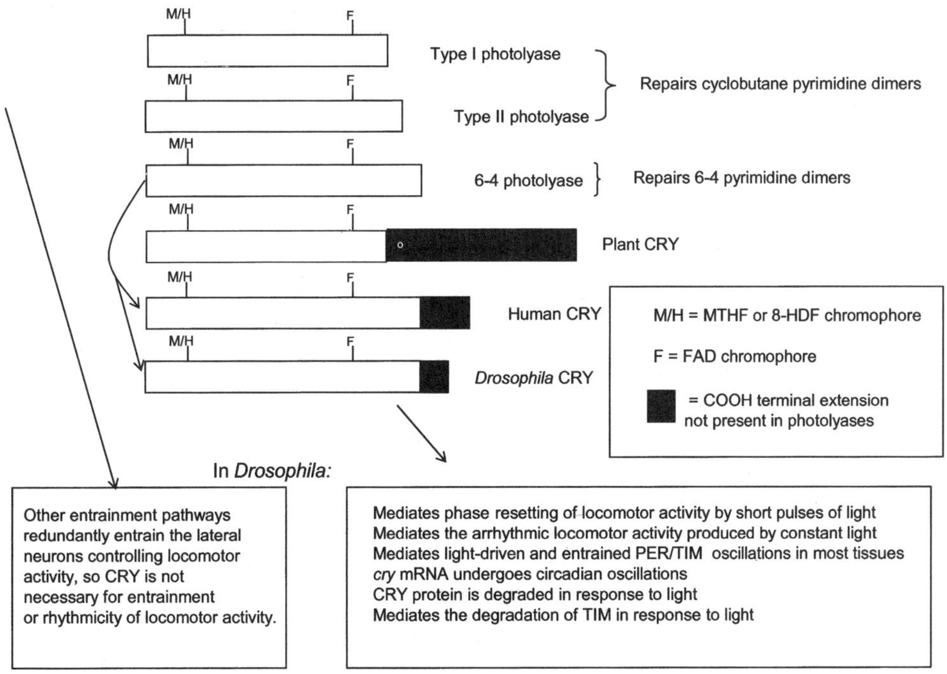

Figure 3.14. Drosophila *cry* is homologous to photolyases and plant cryptochromes and has the properties of a circadian photoreceptor. All of these proteins have binding sites for two kinds of chromophores (FAD and MTHF/8-HDF). Human and Drosophila CRYs are more homologous to the 6–4 photolyases than to the other family members. Note that not all photoreceptor functions are subserved by Drosophila *cry*.

is mediated by this chromophore. In the *cry^b* mutant, TIM is no longer degraded in response to light in most tissues, and both PER and TIM are constitutively expressed at moderately high levels in most expression sites. One exceptional site is a subset of the lateral neurons, which, as discussed above, are the cells that control rhythmic locomotor activity. Here, PER and TIM continue to oscillate and are entrained by light. Likewise, locomotor activity continues to be rhythmic in the *cry^b* mutant, and it is entrained by LD cycles. However, the capacity of short light pulses to reset the clock is virtually eliminated, so the *cry^b* PRC is quite flat. In complementary experiments, CRY overexpression in clock cells alters the sensitivity of the phase response to pulses of light relative to wild-type flies. Finally, the arrhythmic locomotor activity that normally results from LL is not obtained in *cry^b* mutants (i.e., the rhythms resemble those obtained in DD).

These results argue that *cry* is not an essential component of the molecular oscillator in lateral neurons, but rather that it is a component of a somewhat redundant input pathway that responds to light. While some effects of light on lateral neurons are entirely dependent on *cry*, others are also responsive to other photoreceptors. The constant levels of TIM that are observed in most clock tissues could result from an absence of photic entrainment and light-mediated TIM degradation in these tissues. However, evidence presented in a later chapter will argue that *cry* is likely to be a component of the oscillator mechanism in some peripheral oscillator tissues outside the lateral neurons (see Chapter 9).

CRY/TIM Association and Degradation

In response to light, CRY associates with TIM and both are degraded. In wild-type flies, CRY protein is light-sensitive, such that it degrades in response to light and is expressed at high levels only during the dark period. Moreover, it associates directly with TIM in yeast only in the presence of light (Fig. 3.15a). The CRY^b protein, which is predicted to be unresponsive to light because it cannot bind the flavin chromophore, does not associate with TIM in either light or dark. Hence, CRY is capable of a light-influenced association with TIM in the absence of any other Drosophila-specific component besides TIM. In a Drosophila cell line, coexpression of CRY with dCLK, PER, and TIM abrogates the repression of a dCLK-activated reporter gene, but only if the cells are exposed to light. In the dark, PER/TIM can repress dCLK-activated transcription even if CRY is coexpressed. Moreover, coexpression of the CRY^b protein does not eliminate repression by PER/TIM even in the presence of light (Fig. 3.15b).

In addition, a light-dependent association between PER and CRY has been detected in yeast and a Drosophila cell line. Deletion of the *C*-terminal end of CRY allows this association with PER, as well as the one with TIM, to occur in both the light and the dark. These results have been interpreted to show that the *C*-terminal end of CRY mediates the light-dependent interactions with PER and TIM. It has been proposed that the *C* terminus binds a nuclear repressor of the PER and TIM interactions in the dark, while light eliminates the CRY/repressor interaction and produces an interaction with PER and TIM instead. Deletion of the *C*-terminal region would also eliminate the interaction with this repressor and produce interactions with PER or TIM in both light and dark. The role of the PER/CRY interaction in the entrainment or function of the clock is not yet known.

Phosphorylation and Ubiquitination

The response of TIM to light involves phosphorylation and degradation by the ubiquitin proteasome pathway. The ubiquitin proteasome pathway is one of the major

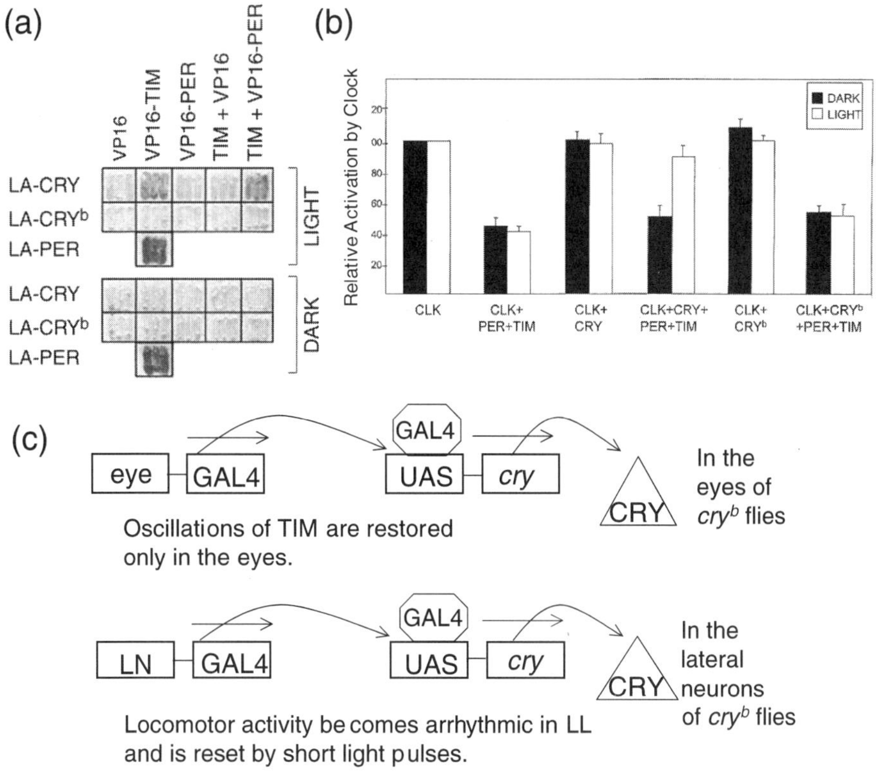

Figure 3.15. CRY undergoes a light-influenced association with TIM and acts intracellularly. (a) The yeast two-hybrid assay (described in Fig. 3.7a) shows that CRY associates with TIM in the light but not the dark. Proteins designated "LA-" are expressed as fusions with the LEXA DNA-binding domain, while those designated "VP16-" are expressed as fusions with the VP16 RNA polymerase-activating domain. CRY could also form a three-component (ternary complex) with PER, if TIM was present to complex with both PER and CRY (hence the interaction detected with LA-CRY, TIM, and VP16-PER; however, as described in the text, subsequent studies also found direct interactions between PER and CRY). CRY^b cannot associate with TIM in either light or dark, and PER associates with TIM in either light or dark. (b) CRY disrupts the repression of dCLK by PER/TIM, but only in the presence of light. CRY^b does not disrupt PER/TIM function. The cell culture assay is described in Figure 3.7b. [Reprinted with permission from Ceriani et al. (1999), copyright (1999) American Association for the Advancement of Science.] (c) Expression of CRY only in the eyes or only in the lateral neurons rescues the *cry* mutant phenotype only in the cells where CRY is expressed. The GAL4/UAS expression system is described in Figure 3.11b. The eye-specific ("eye") promoter is the rhodopsin promoter, and the lateral neuron-specific ("LN") promoter is the pigment dispersing factor promoter.

protein degradation pathways in the cell. In this pathway, lysines on the protein targeted for degradation are covalently modified with chains of a small peptide monomer called *ubiquitin*. These polyubiquitin chains mediate the binding of the modified protein to the 26*S* proteasome—a large, cylindrical macromolecule thought to unfold ubiquitinated substrates and force the polypeptide backbone through a proteolytic core, which degrades the protein. Inhibitors of the proteasome pathway inhibit the degradation of TIM in response to light (Fig. 3.16). In addition to ubiquitination, the degradation

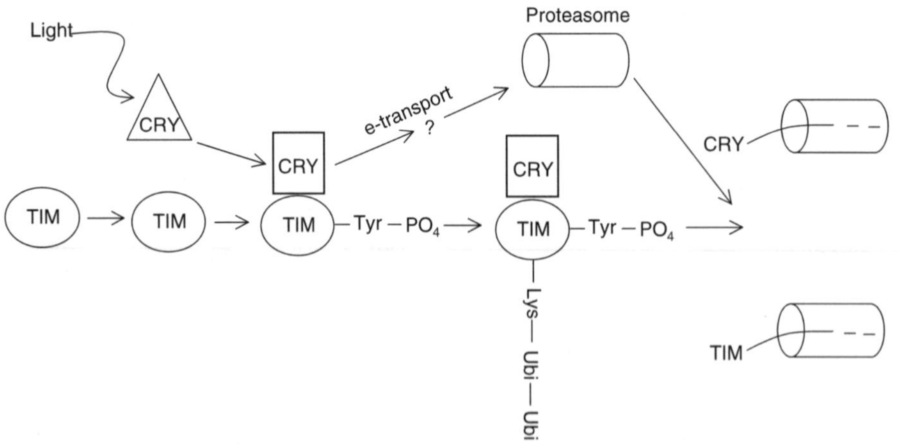

Figure 3.16. Model for the light-dependent degradation of TIM and CRY by the proteasome. Phosphorylation of TIM on tyrosine (Tyr-PO_4) is thought to be necessary, as is the addition of ubiquitin peptides (Ubi) to lysines in TIM.

of TIM in response to light requires tyrosine kinase activity. Elevated levels of phosphotryrosine can be detected in TIM after a light pulse, and inhibitors of tyrosine kinases prevent the degradation of TIM in response to light.

The degradation of CRY in response to light is also inhibited by proteasome inhibitors, by inhibitors of electron transport, and by several mutations in *cry*. The model is that light induces a conformational change in CRY, which then donates electrons from reduced flavin to downstream components that signal degradation of CRY (Fig. 3.16). In a cell line expressing CRY and exposed to light, inhibition of electron transport elevates the levels of ubiquitinated TIM (in this cell line, TIM is ubiquitinated, but not degraded in response to light). Thus, electron transport is required to turn off CRY-mediated ubiquitination of TIM in response to light; presumably, light converts CRY into an active form that promotes ubiquitination and degradation of TIM, while electron transport terminates this activity by converting CRY to an inactive form or triggering the degradation of CRY.

CRY Mediation of Photoreception in Clock Cells

If CRY mediates entrainment by directly binding to TIM, then CRY should mediate photoreception within the clock cells, rather than in a separate photoreceptor cell type that communicates synaptically with the clock cell. Recall that the compound eye and visual transduction pathway are not required for clock photoreception. To test the possibility that CRY mediates cell-autonomous photoreception, experimental control of CRY's tissue specific expression is required. As described for SGG above, it is possible to express genes in different, specific cell types by using the yeast GAL4 transcription factor as an intermediary (Fig. 3.15c). A promoter that is activated by GAL4 because it has been genetically engineered to have binding sites for GAL4 (termed “UAS”) was linked to a coding sequence for *cry* and introduced as a transgene into flies. In addition, promoters that were active in diverse cell types were linked to the coding region for GAL4, producing a number of transgenes, each of which expresses GAL4 in a distinct cell type.

For instance, by hooking GAL4 to the rhodopsin promoter, the transgene will be expressed only in certain photoreceptors of the eye. By hooking it to the *pdf* promoter, which is expressed only in the lateral neurons (and that is described in more detail later), the transgene will express GAL4 only in the lateral neurons. If a fly carrying a tissue-specific GAL4 transgene is crossed to a fly carrying a UAS-*cry* transgene, any progeny inheriting both transgenes will now express *cry* in the same tissue-specific pattern as GAL4, since GAL4 is driving expression of *cry* (Fig. 3.15c).

Using this methodology, it is possible to show that expression of wild-type *cry* in a cry^b mutant produces a wild-type phenotype (a "rescue") only in those tissues where *cry* is expressed (Fig. 3.15c). If rhodopsin-GAL4 is the driver, oscillations of TIM are rescued only in the eye but not in the body. If the lateral neuron-specific *pdf*-GAL4 is the driver, the defective PRC and the lack of arrhythmicity in response to LL are rescued in the cry^b mutant (recall that entrainment of lateral neurons and rhythmic locomotor activity is still obtained in the cry^b mutant, so there is nothing to rescue with respect to these). The results are consistent with a role for *cry* that is cell autonomous, and with the role of the lateral neurons in circadian locomotor activity.

Non-cell-Autonomous Entrainment

Since circadian rhythms of locomotor activity can still be entrained to 12-hour : 12-hour light : dark cycles in the cry^b mutant, CRY cannot be the only circadian photoreceptor for the lateral neurons controlling locomotor behavior. In support of this conclusion, locomotor activity in a $gl;cry^b$ double mutant does not entrain to LD cycles. The *gl* mutation eliminates all known "visual" photoreceptors in the adult fly: the two compound eyes, the three ocelli (light-sensitive organs at the top of the head), and the two internal, rhodopsin-positive cell groups that lie underneath the retina and project to the lateral neurons (the Hofbauer–Buchner eyelet; Fig. 3.17). Therefore, in the $gl;cry^b$ double mutant, there is an absence of both functional CRY protein and any input from visual photoreceptors external to the lateral neurons. With both types of input gone, the rhythms of PER and TIM are not entrained in the lateral neurons, and behavioral rhythms cannot be entrained. Evidently, there is functional redundancy in the input pathways to the lateral neurons, such that the lateral neurons are still photically entrainable as long as either a CRY pathway or visual pathway is available.

■ REGULATION OF CIRCADIAN-CLOCK-CONTROLLED OUTPUTS BY THE OSCILLATOR MECHANISM

Function of the Molecular Oscillator

The molecular oscillator drives the oscillations of numerous clock-controlled genes. In theory, any gene which has an E-box-containing promoter that is responsive to the dCLK/CYC heterodimer can be regulated by the molecular oscillator described in this chapter (Fig. 3.18). These clock-controlled genes need not be part of the mechanism that generates the molecular oscillations, but they may encode factors that translate the molecular oscillator into actual changes in cellular biochemistry and physiology: the clock outputs. Screens for these genes have involved molecular approaches that identify mRNAs whose levels change over the course of the day (see Chapter 2). Only a handful of such genes have been described in Drosophila by single-gene-based approaches, but as this chapter goes to press, many more candidates are emerging from whole-genome-based approaches ("genomics").

In a screen for genes regulated in response to LD cycles, 20 genes termed *dregs* (for *D*rosophila *r*hythmically *ex*-

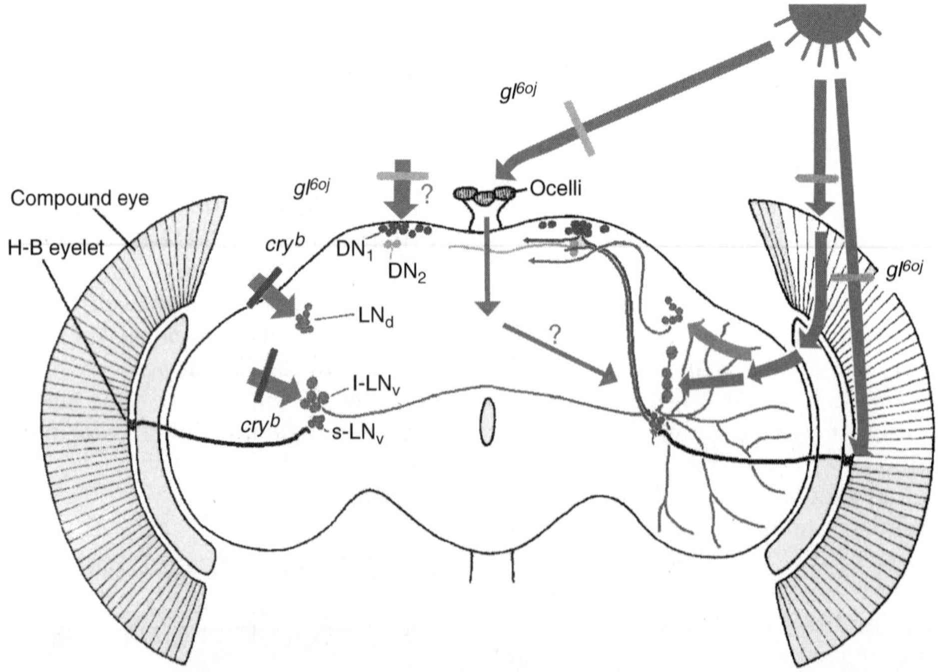

Figure 3.17. Redundant entrainment pathways for the lateral neurons as revealed by the effects of various mutations on entrainment. The *cry^b* mutation disrupts the Drosophila cryptochrome photoreceptor. The *gl* mutation prevents development of all the cells that express PER in the adult head except the lateral neurons. Therefore, a *gl;cry* double mutant has no entrainment pathways. [Reprinted from Helfrich-Forster et al. (2001): The circadian clock of fruit flies is blind after elimination of all known photoreceptors. *Neuron* **30**: 249–261, copyright (2001) with permission from Elsevier Science.] The cells that are potentially circadian photoreceptors include the dorsal and ventral (large and small) lateral neurons (LN_d, l LN_v, s LN_v), the compound eye, the ocelli (three light-sensing organs on top of the head), the H-B eyelet (a cluster of photoreceptors between the retina and the optic lobe), and a cluster of dorsal neurons (DN). The pathways blocked by each of the two mutations, *gl^{60j}*; and *cry^b*, are indicated by the bars on the arrows.

pressed genes) were identified in a collection of cDNAs expressed in the head but not the early embryo. All but three of these have evening peaks of mRNA expression, like those of *per* and *tim*, while the expression of the other three peaked in the morning. Some of these genes show regulation by LD cycles and food, in addition to (or instead of) regulation by the endogenous clock, while others show oscillations in DD and require *per* function. Only one of the genes (alcohol dehydrogenase) has a known function. In addition, two transcription factors (*vrille* and *crg-1*) and a ligand binding factor (*takeout*) have been identified in separate screens for genes whose mRNAs exhibit circadian oscillations or regulation.

vrille

vrille is a clock-controlled gene that also affects clock function. *vrille* is a member of the basic leucine zipper class of transcription factors, and it has a dCLK/CYC responsive site in its promoter. As is predicted by this feature of its promoter, *vrille* mRNA exhibits oscillations that are in phase with those of *per* and *tim* in LD and

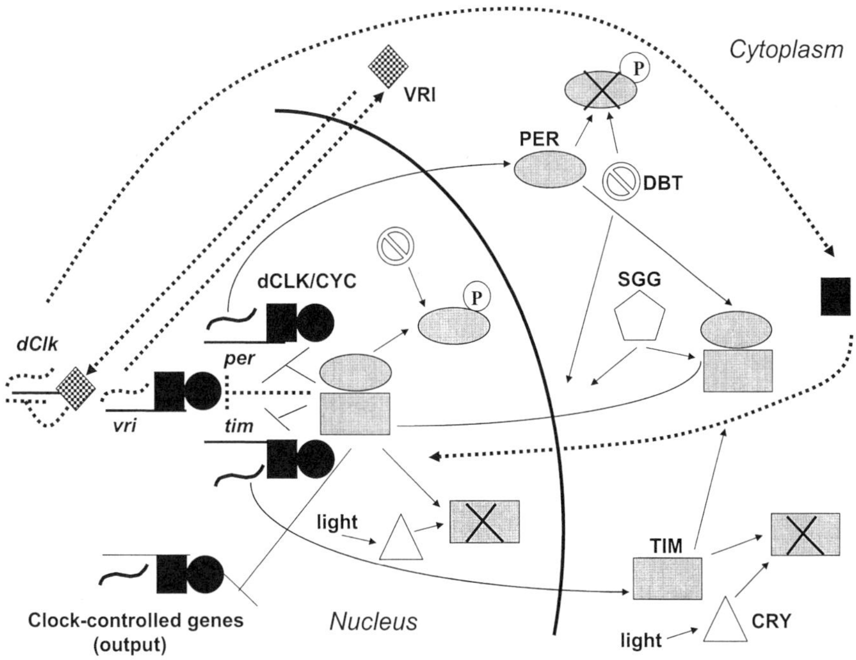

Figure 3.18. A molecular model for the circadian clock in Drosophila. The rhythmic accumulation of PER and TIM proteins results in the formation of a PER/TIM complex, which negatively regulates two feedback loops involving dCLK/CYC. In one of these loops (solid lines) dCLK/CYC positively regulate *per* and *tim* transcription, and in the other loop (dotted lines) dCLK/CYC negatively regulates *dClk* mRNA by activating *vri* expression. Light, CRY, and SGG regulate a lag in nuclear feedback by PER/TIM through their effects on TIM, and light entrains the oscillator through CRY-mediated degradation of TIM. DBT regulates the initiation and termination of nuclear feedback through its effects on PER. In addition to driving transcription of *per* and *tim*, dCLK/CYC is thought to activate transcription of numerous clock-controlled genes, which translate the rhythm of the circadian oscillator into circadian outputs of biochemistry and cell physiology.

DD, and the oscillations are eliminated by the *per*^{0} mutation. In the adult head, *vrille* is expressed in photoreceptor cells and the lateral neurons, in a pattern that is identical to the expression pattern of *tim*. While mutations in *vrille* are homozygous lethal, heterozygotes are viable and have slightly shortened circadian periods. Overexpression of *vrille* in clock cells produces long periods and arrhythmicity and depresses expression of *per*, *tim*, *dClk* and pigment dispersing factor (PDF, a neuropeptide released by the lateral neurons; more on PDF below). The *dClk* promoter is negatively regulated by *vrille*, and all these effects on clock gene products may therefore derive from effects on *dClk*. *vrille* is therefore part of the central oscillator mechanism, although it was isolated in a screen for output genes.

takeout

The *takeout* (*to*) gene mediates resistance to starvation. Its mRNA oscillates with a phase that is several hours delayed relative to *per* and *tim* mRNA. While the 5′ upstream region of *to* has a potential E box, this region of the *to* promoter does not confer circadian oscillations to reporter genes and does not confer inducibility by dCLK/CYC in the cell culture assay. Therefore, the mechanism for circadian regulation of *to* is not well understood, although it is clear that known clock genes are involved, because *to* mRNA is downregulated in several null mutations of known Drosophila clock genes. In addition to circadian oscillations, *to* mRNA exhibits induction in response to starvation. For reasons that are not clear, this induction requires the presence of a functional circadian clock, as it does not occur in clock-null mutants (per^0, tim^0, and cyc^0). The *to* gene confers resistance to starvation, because a mutation in *to* leads to more rapid lethality in response to starvation. It is possible that clock regulation of *to* contributes to metabolic and food-related oscillations. This possibility is intriguing, in light of the regulation of mammalian PAS-bHLH factors by redox state (see Chapter 4).

Genomic Methods

Genomic approaches have yielded several hundred possible clock-controlled genes. The Berkeley *Drosophila* Genome Project resulted in the nucleotide sequence of the majority of the genes in Drosophila. This allowed microarray analysis of the type described in Chapter 2 to identify circadian changes in gene expression. Initial attempts to analyze circadian changes in gene expression by several groups have produced somewhat different tabulations of clock-controlled genes. With refinement of the technology and confirmation of candidate genes, a comprehensive list should emerge. Already it seems that many of the output genes differ in phase from *per*, *tim*, and *dCLk*. Many of them are involved in common processes, such as learning and memory/synapse function, vision, olfaction, locomotion, detoxification, proteolysis, immunity, and metabolism. None of the clock-controlled genes exhibit circadian oscillations in the Clk^{Jrk} mutant, so there is no evidence for any circadian oscillator independent of the core oscillator mechanism described in this chapter. Many genes that do not exhibit circadian oscillations nevertheless exhibit altered levels of constitutive expression in $dClk^{Jrk}$ or per^0 mutant flies. The basis for these altered expression levels is not known.

Posttranscriptional Regulation

On the basis of our current model for this oscillator, promoters with E boxes seem likely to be the most direct target in output pathways. However, rhythmically expressed transcripts will produce rhythmically expressed proteins. These proteins can produce downstream rhythms of other proteins, whose transcription is not necessarily rhythmically controlled.

Cyclic Activity of cAMP Response Element Binding (CREB) Protein

The capacity of CREB to activate transcription from a cAMP response element (CRE)-driven luciferase reporter oscillates with a circadian rhythm that is affected by *per* mutations just as circadian behavior is affected. Since CRE sites are targets of CREB, the cycling of this reporter is indicative of cyclic CREB activity. The level of CREB does not oscillate, so it must be regulated posttranslationally—perhaps by phosphorylation. Mutations in the *Creb* gene shorten the period of circadian behavior and blunt *per* mRNA and protein oscillations. In fact, the *per* promoter has several putative CRE elements, so CREB activity may feed back to regulate the *per* promoter. Alternatively or additionally, CREB activity may be part of an output pathway regulating locomotor activity, as circadian regulation of the CREB activity is blunted in mutants that specifically affect output pathways rather than the central clock mechanisms. For instance, the *NF1* mutation, which upregulates a mitogen activated protein kinase pathway involved in output to locomotor rhythms, and the *dfmr1* mutation, which disrupts synapse formation, both blunt the rhythm of CREB activity (see text below). Since CREB is involved in consolidation of memories, circadian regulation of CREB may couple the circadian clock with learning and memory, which is thought to be consolidated during sleep in vertebrates.

pdf

The neuromodulator, pigment dispersing factor (*pdf*), is an important circadian output for circadian control of locomotor activity and eclosion. The work that led to this knowledge entailed an elegant combination of comparative endocrinology and molecular genetic analysis. Pigment dispersing hormones were isolated as substances that mediate dispersion and translocation of light-adaptational pigments in crustaceans. Antibodies to these crustacean hormones detect immunoreactivity in the Drosophila lateral neurons, which are the site of the clock that controls rest–activity rhythms. Therefore, a molecular analysis of Drosophila pigment dispersing factor (*pdf*) was undertaken, in order to determine what role this factor might have in Drosophila circadian rhythms. Several clock mutants affect the expression of *pdf*; mutations in *dClk* and *cyc* reduce *pdf* mRNA, and overexpression of *vrille* reduces the level of *pdf* peptide (PDF) without affecting the levels of *pdf* mRNA. While levels of *pdf* mRNA and protein do not oscillate, PDF rhythmically accumulates at the terminals of the lateral neurons, and this rhythm is absent in the per^0 and tim^0 mutants. It is likely that this rhythm is effected by rhythmic transport and/or release from the lateral neuron terminals.

A *pdf* mutant that does not express this peptide has a high frequency of arrhythmic locomotor activity. In addition, the dorsal central brain, to which the lateral neuron axons project, is the only site of overexpression that significantly alters the rhythmicity of locomotor activity and eclosion. This result indicates the importance of this region to the clock output pathway. Because neither elimination nor overexpression of PDF completely eliminates rhythms of locomotor activity, it is likely that other locomotor outputs exist.

The importance of proper synaptic connections for the PDF-releasing lateral neurons is suggested by studies of Drosophila *dfmr1* mutants. These mutants do not express a homolog of a human gene that is mutated in one of the most common forms of mental retardation: the fragile X syndrome. The Drosophila mutants display a high frequency of arrhythmic circadian locomotor activity. While oscillations of PER and TIM are normal in the lateral neurons, the projections of the lateral neurons are abnormal. Hence, while the lateral neuron circadian oscillator is apparently normal in these flies, it is not able to communicate its temporal information to its outputs, because of abnormal axonal pathfinding or synapse formation.

lark

The *lark* mutation was isolated in a screen for mutations affecting the timing of eclosion (Table 3.1). Heterozygous *lark* mutant flies eclose early in an LD cycle, but do not exhibit an altered circadian period of eclosion in DD, nor any alterations in circadian control of locomotor activity. The muta-0tion produces embryonic lethality when homozygous. *lark* protein (LARK), which is predicted to be an RNA binding protein, exhibits an oscillation at the protein level, while *lark* mRNA does not oscillate. High levels of protein are expressed during the day, and the oscillation persists in DD and requires functional PER. The high levels of LARK during the day, when eclosion behavior is suppressed, are consistent with a function for LARK as a repressor of eclosion. While LARK is localized in the nuclei of most or all pupal neurons, it is also

TABLE 3.1. Circadian Mutants in *Drosophila*

Gene Name	Phenotype of Mutants	Function of Protein	Circadian Targets
period (per)[a,c]	ar, short, or long circadian behavioral and molecular rhythms	Interacts with TIM, dCLK and CYC to negatively regulate dCLK/CYC	Oscillator
timeless (tim)[a,b,c]	ar, short, or long circadian behavioral and molecular rhythms	Interacts with PER, dCLK and CYC to negatively regulate dCLK/CYC	Oscillator
dClock (dClk)[a]	ar circadian behavioral and molecular rhythms	Interacts with CYC to positively regulate *per* and *tim* mRNA and to negatively regulate *dClk* mRNA; both types of regulation repressed by PER/TIM	Oscillator
cycle (cyc)	ar circadian behavioral and molecular rhythms	Interacts with dCLK to positively regulate *per* and *tim* mRNA and to negatively regulate *dClk* mRNA; both types of regulation repressed by PER/TIM	Oscillator
doubletime (dbt)[c]	ar, short, or long circadian behavioral and molecular rhythms	Casein kinase I ortholog which interacts with PER to produce phosphorylation of PER; regulates stability and timing of nuclear feedback by PER	Oscillator
shaggy (sgg)	Short or long circadian behavioral and molecular rhythms	Glycogen synthase kinase III ortholog which phosphorylates TIM to regulate the timing of TIM's light sensitivity and nuclear feedback	Oscillator
vrille (vri)[a]	Short, long, and damped circadian behavioral and molecular rhythms	Transcription factor which is positively regulated by dCLK/CYC and which negatively regulates *per*, *tim* and *dClk* mRNA	Oscillator
cryptochrome (cry)[b,d]	Arrhythmic and high levels of PER and TIM in most tissues; rhythmic behavior and molecular oscillations in lateral neurons.	Blue-light photoreceptor which dimerizes with TIM to effect degradation of both TIM and CRY	Input; peripheral oscillators
lark[e]	Early eclosion in heterozygotes; homozygous lethal	RNA-binding protein that may regulate the mRNA of a hormone for eclosion	Eclosion (output)
takeout (to)[d]	Hypersensitive to starvation; no effect on molecular oscillations	Most likely a ligand-binding protein	Starvation (output)

Creb[f]	*per* oscillation is damped, circadian period of locomotor activity shortened	cAMP response element binding protein: transcription factor regulated by phosphorylation	Possibly PDF response (output)
Neurofibromatosis-1 (NF-1)	Arrhythmic locomotor activity; no effect on molecular oscillations	Inactivates signaling through the RAS protein kinase pathway	Possibly PDF response (output)
protein kinase A (PKA)	Loss of some locomotor activity rhythmicity; no effect on molecular oscillations	A protein kinase activated by cAMP	Locomotor activity (output)
dunce (dnc)	Altered phase response curve and slight shortening of period	cAMP phosphodiesterase	?
pigment dispersing factor (pdf)	Loss of locomotor activity and eclosion rhythmicity	Neuropeptide which accumulates rhythmically in lateral neuron terminals	Locomotor, eclosion outputs
fragile X syndrome gene (dfmr1)	Loss of some locomotor rhythmicity; abnormal synapse formation; no effect on molecular oscillations	mRNA binding protein which alters connections of lateral neurons with postsynaptic neurons	Locomotor, eclosion outputs
disconnected (disco)	Loss of locomotor and eclosion rhythmicity	Required for proper pathfinding of the larval and adult optic nerve and lateral neuron determination	Clock cells which control behavior

[a] mRNA and protein oscillate with a circadian rhythm.
[b] Levels of protein acutely depressed by light.
[c] Nuclear localization of protein oscillates with a circadian rhythm.
[d] mRNA oscillates with a circadian rhythm.
[e] Protein but not mRNA oscillates with a circadian rhythm.
[f] Activity oscillates with a circadian rhythm.

expressed in the cytoplasm of neurons that release a peptide involved in triggering eclosion. It is possible that LARK contributes to a translational control mechanism that regulates the synthesis or release of this hormone (crustacean cardioactive peptide). However, mutations in the RNA binding domains of lark do not affect the circadian regulation of eclosion, so lark may not affect eclosion by binding to mRNA.

PKA and Nf1

Protein kinase A (PKA), which is stimulated by cAMP, is specifically involved in the circadian regulation of locomotor activity, because mutations in the PKA catalytic and regulatory subunits produce a high frequency of locomotor arrhythmia without affecting eclosion or the oscillations of *per*. Modulations of cAMP levels may also be involved in some part of the clock mechanism, as *dunce* mutations, which reduce the level of cAMP-specific phosphodiesterase, alter the Drosophila PRC, shorten the circadian period, and enhance a circadian oscillation of cAMP levels.

Null mutations of the *neurofibromatosis-1 (Nf1)* gene also specifically disrupt locomotor activity rhythms. *per* and *tim* oscillations are not affected by these mutations, consistent with a function downstream of the clock. Since the NF1 protein was known to signal through PKA in flies, the obvious explanation was that its effects on rhythms were mediated by PKA. However, it turned out that NF1 acts through the Ras/MAPK (mitogen-activated protein kinase) pathway to effect clock output. MAPK activity is elevated in *NF1* mutants, and Ras/MAPK mutations that reduce MAPK signaling suppress the arrhythmia caused by *Nf1*. It is likely that release of PDF from lateral neuron terminals stimulates MAPK activity in the postsynaptic neurons, because a circadian oscillation in phospho-MAPK (the activated kinase) is observed in the vicinity of the dorsal PDF projections and levels of phospho-MAPK are reduced in the *pdf* null mutant.

■ PRESENT QUESTIONS AND FUTURE PERSPECTIVES

While the genetic analysis in Drosophila has been remarkably successful in integrating many observations and diverse genes into a coherent model (Fig. 3.18), there are still some phenomena that are not explained by the model put forward in this chapter. One of these phenomena is temperature compensation, which refers to the relative constancy of circadian period at different temperatures and is a defining feature of circadian rhythms (see Chapter 1). Attempts have been made to explain temperature compensation within the framework of the existing Drosophila model, but there has been little progress. Another unexplained phenomenon is that the *per* mutations affect the periodicity of an ultradian rhythm of approximately one minute. This is a rhythm in the courtship song that the male Drosophila sings to the female—a song that ends with mating, if it is sung well! The song consists of pulses of wing beats, and the frequency of these pulses oscillates with a period of about a minute in wild-type flies. The period is shorter in per^S flies, longer in per^L flies, and arrhythmic in per^0 flies. The role proposed for *per* in the circadian molecular oscillator is on a timescale that cannot explain the effects on these ultradian rhythms.

Within the framework of the model proposed in this chapter, the oscillations of *per* and *tim* gene products exhibit complexities that are not explained by the model. For instance, other promoter sequences besides E boxes contribute to the cycling regulation of transcription, and there is also posttranscriptional regulation of *per* mRNA and of some other clock-controlled mRNAs. Additionally, circadian rhythms of behavior occur in some transgenic flies in which *per*, *tim*, or *dClk* are expressed from promoters that do not produce normal circadian oscillations of these mRNAs. Thus, rhythmic expression of *per*, *tim*, or *dClk* mRNA is not necessary for circadian rhythms. Nevertheless, these transgenes support rescue of circadian rhythms of PER, TIM, or dCLK proteins. It is possible that rhythmic feedback on *per/tim* and *dClk* transcription is not an essential function for the oscillations of PER and TIM in the circadian oscillator mechanism. Perhaps oscillating transcription of one or more of the other genes that are regulated by clock-controlled E boxes (e.g., *vrille*) is sufficient to drive the post-transcriptional oscillations of PER and TIM. Alternatively and more profoundly model-altering, PER and TIM may be able to drive their oscillations entirely through feedback loops that are posttranscriptional. If the latter possibility materializes, the major function of PER and TIM in the circadian oscillator has yet to be elucidated.

It has been shown that electrical activity of the lateral neuron pacemaker cells is required for persistent circadian oscillations of PER and TIM, as well as for circadian rhythms of behavior. Loss of electrical activity was produced by expressing K^+ channel mutants in the lateral neurons. The expression of these mutant channels does not kill the lateral neurons, because oscillations of PER and TIM persist if the transgenic flies are maintained in LD cycles instead of DD. Potentially, electrical activity contributes in some way to one of the known feedback loops, or to a novel mechanism that is not yet understood.

One of the successes of the Drosophila model is that many of its features have been extended quite well to the mammalian clock, which is covered in Chapter 4. The mammalian clock has three known *per* genes, two *cry* genes, a family of CLK/CYC transcription factors, and a family of casein kinase I genes similar to *dbt*. Apparently, the mammalian TIM is not involved in the clock, but the other mammalian orthologs of Drosophila clock genes have clock roles. CRY appears to be more of a central component in the mammalian clock; it serves to facilitate nuclear accumulation of PER, and it negatively regulates CLK/CYC. PER is also involved in this negative regulation, in the positive regulation of a CYC ortholog, and is induced in response to light. The differences and similarities between the mammalian and Drosophila clocks are discussed more thoroughly in subsequent chapters.

As novel genetic screens are devised and the genome-scale screens for clock-controlled genes are refined, it is highly likely that additional genes will be shown to be part of the clock mechanism. Some of these may be assigned to input and output pathways, but it is possible that the distinction between these divisions of the circadian system may begin to blur, as many of these pathways may relate to the central mechanism as well. It has already become clear that many of the output pathways are controlled by peripheral oscillators rather than distinct output mechanisms. Understanding the integration of multiple oscillators to produce rhythmicity at the organismal level presents a major challenge (see Chapter 9). Clearly, times that are as exciting as the late 1990s through 2003 still lie ahead for molecular chronobiologists.

FURTHER READING

Allada, R, Emery, P, Takahashi, JS, Rosbash, M (2001): Stopping time: The genetics of fly and mouse circadian clocks. *Annu Rev Neurosci* **24**: 1091–1119.

Allada, R, White, NE, So, WV, Hall, JC, Rosbash, M (1998): A mutant *Drosophila* homolog of mammalian *Clock* disrupts circadian rhythms and transcription of *period* and *timeless*. *Cell* **93**: 791–804.

Blau, J, Young, MW (1999): Cycling *vrille* expression is required for a functional *Drosophila* clock. *Cell* **99**: 661–671.

Ceriani, MF, Darlington, TK, Staknis, D, Mas, P, Petti, AA, Weitz, CJ, Kay, SA (1999): Light-dependent sequestration of TIMELESS by CRYPTOCHROME. *Science* **285**: 553–556.

Darlington, TK, Wager-Smith, K, Ceriani, MF, Staknis, D, Gekakis, N, Steeves, TDL, Weitz, CJ, Takahashi, JS, Kay, SA (1998): Closing the circadian loop: CLOCK-induced transcription of its own inhibitors *per* and *tim*. *Science* **280**: 1599–1603.

Edery, I, Zwiebel, LJ, Dembinska, ME, Rosbash, M (1994): Temporal phosphorylation of the *Drosophila period* protein. *Proc Natl Acad Sci* (*USA*) **91**: 2260–2264.

Gekakis, N, Saez, L, Delahaye-Brown, A-M, Myers, MP, Sehgal, A, Young, MW, Weitz, CJ (1995): Isolation of *timeless* by PER protein interaction: Defective interaction between *timeless* protein and long-period mutant PER^{L}. *Science* **270**: 811–815.

Glossop, NR, Lyons, LC, Hardin, PE (1999): Interlocked feedback loops within the *Drosophila* circadian oscillator. *Science* **286**: 766–768.

Handler, AM, Konpoka, RJ (1979): Transplantation of a circadian pacemaker in *Drosophila*. *Nature* **279**: 236–238.

Hardin, PE, Hall, JC, Rosbash, M (1990): Feedback of the *Drosophila period* gene product on circadian cycling of its messenger RNA levels. *Nature* **343**: 536–540.

Hunter-Ensor, M, Ousley, A, Sehgal, A (1996): Regulation of the *Drosophila* protein *timeless* suggests a mechanism for resetting the circadian clock by light. *Cell* **84**: 677–685.

Konopka, RJ, Benzer, S (1971): Clock mutants of *Drosophila melanogaster*. *Proc Natl Acad Sci* (*USA*) **68**: 2112–2116.

Konopka, RJ, Wells, S, Lee, T (1983): Mosaic analysis of a *Drosophila* clock mutant. *Mol Gen Genet* **190**: 284–288.

Kyriacou, CP, Hall, JC (1980): Circadian rhythm mutations in *Drosophila melanogaster* affect short-term fluctuations in the male's courtship song. *Proc Natl Acad Sci* (*USA*) **77**: 6729–6733.

Lee, C, Bae, K, Edery, I (1999): PER and TIM inhibit the DNA binding activity of a *Drosophila* CLOCK-CYC/dBMAL1 heterodimer without disrupting formation of the heterodimer: A basis for circadian transcription. *Mol Cell Biol* **19**: 5316–5325.

Martinek, S, Inonog, S, Manoukian, AS, Young, MW (2001): A role for the segment polarity gene *shaggy*/GSK-3 in the *Drosophila* circadian clock. *Cell* **105**: 769–779.

McNeil, GP, Zhang, X, Genova, G, Jackson, FR (1998): A molecular rhythm mediating circadian clock output in *Drosophila*. *Neuron* **20**: 297–303.

Price, JL, Dembinska, ME, Young, MW, Rosbash, M (1995): Suppression of PERIOD protein abundance and circadian cycling by the *Drosophila* clock mutation *timeless*. *EMBO J* **14**: 4044–4049.

Price, JL, Blau, J, Rothenfluh, A, Abodeely, M, Kloss, B, Young, MW (1998): *double-time* is a novel *Drosophila* clock gene that regulates PERIOD protein accumulation. *Cell* **94**: 83–95.

Renn, SCP, Park, JH, Rosbash, M, Hall, JC, Taghert, PH (1999): A *pdf* neuropeptide gene mutation and ablation of PDF neurons each cause severe abnormalities of circadian behavioral rhythms in *Drosophila*. *Cell* **99**: 791–802.

Saez, L, Young, MW (1996): Regulation of nuclear entry of the *Drosophila* clock proteins *period* and *timeless*. *Neuron* **17**: 911–920.

Stanewsky, R, Kaneko, M, Emery, P, Beretta, B, Wager-Smith, K, Kay, SA, Rosbash, M, Hall, JC (1998): The cry^{b} mutation identifies cryptochrome as a circadian photoreceptor in *Drosophila*. *Cell* **95**: 681–692.

Vosshall, LB, Price, JL, Sehgal, A, Saez, L, Young, MW (1994): Block in nuclear localization of *period* protein by a second clock mutation, *timeless*. *Science* **263**: 1606–1609.

Williams, JA, Sehgal, A (2001): Molecular components of the circadian system in *Drosophila*. *Ann Rev Physiol* **63**: 729–755.

Williams, JA, Su, HS, Bernards, A, Field, J, Sehgal, A (2001): A circadian output in *Drosophila* mediated by Neurofibromatosis-1 and Ras/MAPK. *Science* **293**: 2251–2256.

Zerr, DM, Hall, JC, Rosbash, M, Siwicki, KK (1990): Circadian fluctuations of *period* protein immunoreactivity in the CNS and the visual system of *Drosophila*. *J Neurosci* **10**: 2749–2762.

4

MOLECULAR ANALYSIS OF CIRCADIAN RHYTHMS: NONMAMMALIAN VERTEBRATES

Julie A. Williams

Nonmammalian vertebrates express homologs of many of the known clock genes, although temporal expression varies across species. Some novel components of the circadian clock have also been described. The organization of the circadian system has diverged in nonmammalian vertebrates such that the anatomical components of the clock are widespread. In some birds, for instance, multiple oscillators contribute to behavioral rhythmicity. In zebrafish, many tissues contain self-sustaining oscillators, although the contribution of each of the tissues to behavioral rhythmicity is unknown. In addition, because of the multioscillatory feature of the circadian system, preparation of the pineal or retina in culture has provided an in vitro model for the circadian clock.

■ INTRODUCTION

In previous chapters, we've seen that the roles of clock genes have diverged between insects and mammals. For example, the blue-light photoreceptor, cryptochrome, mediates responses to light in flies and plants, but appears to have lost this function in mammals. Instead, it functions as a key transcriptional repressor in mammals. Three period genes have been described in mammals. Although they are each regulated in a slightly different manner, none of them alone is necessary for maintaining rhythmic behavior. In contrast, flies express one period gene that is absolutely required for circadian behavior. Thus, studying the role of each of clock component in lower vertebrates may clarify not only how they

Molecular Biology of Circadian Rhythms. Edited by Amita Sehgal
ISBN 0-471-41824-2

have evolved but also their function in the circadian system.

A common feature of the circadian system in mammals and fruitflies is that a central oscillator is localized to one particular brain region: the suprachiasmatic nucleus and the lateral neurons, respectively. This "core" clock in nonmammalian vertebrates, in contrast, appears to consist of multiple tissues that are all integral components for circadian behavior, including the retina, a suprachiasmatic area, and the pineal gland. These have been studied most extensively in birds, and are discussed in the following section. In zebrafish, many organs contain self-sustaining oscillators (discussed below), although their respective contribution to circadian behavior remains unknown. Thus, the molecular components of circadian rhythms are outlined in three types of nonmammalian vertebrates, birds, zebrafish, and the African clawed frog, *Xenopus laevis*. Because so little is known about the molecular basis of circadian clocks in reptiles, they are not covered in this chapter.

ORGANIZATION OF CIRCADIAN SYSTEM IN BIRDS

Multioscillatory System

The circadian system in birds is multioscillatory. Circadian outputs in avian species include rhythms in plasma levels of melatonin, body temperature, and locomotor activity such as perch hopping in sparrows. As mentioned above, the organization of the circadian system is quite different from that in mammals in that it is a multioscillatory system. There are three major components in this system: (1) the *pineal gland*, which is the primary site of melatonin synthesis; (2) a *hypothalamic* area that may correspond to the suprachiasmatic nucleus (SCN) in mammals; and (3) the *retina*, which is also a site of melatonin synthesis. The relative contribution of each of these components to circadian output varies widely across species, and may even vary within an individual depending on environmental factors such as temperature and photoperiod relating to the season.

The Pineal Gland

The pineal gland is required for circadian behavior in most birds. The pineal gland, considered the "seat of the soul" by Descartes, functions as both a biological clock and as a photoreceptor in most nonmammalian vertebrates. It develops from an evagination of the roof of the diencephalon and remains attached to the brain via a pineal stalk. In some poikilotherms, there is a second component that detaches from the brain and is located more superficially beneath the skull. In amphibians, this is known as the *frontal organ*, and in some lizards, the *parietal eye*.

The primary cell type in the pineal gland is similar to photoreceptive cells in the retina such that it has an outer segment with stacked disks containing photopigments. This morphology indicates that the animal is capable of detecting changes in light in the environment; however it does not have the complex visual acuity of the eye. The mammalian pinealocytes have lost this photoreceptive characteristic and are secretory in appearance (Fig. 4.1). Furthermore, in lower vertebrates, the photosensory pineal cells make synaptic contacts with neurons that project to the brain. Once again, this is in contrast to mammals, where innervation of the pineal is largely sympathetic and there are no sensory contacts from the pineal at all.

Thus the pineal gland is an ideal model for studying circadian clocks in birds and other species. It contains all three components of the circadian system: (1) a photoreceptive input, (2) a central oscillator, and (3) a regulated output (melatonin).

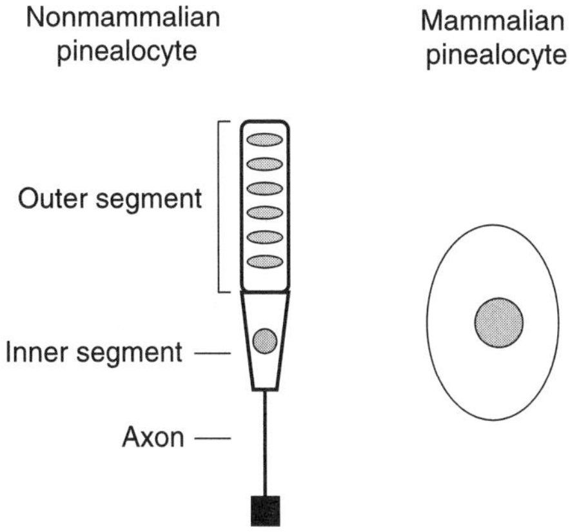

Figure 4.1. Comparison of a mammalian versus a nonmammalian pinealocyte. Mammalian pinealocytes are secretory, and no neural projections have been found from the pineal. Nonmammalian pinealocytes resemble rod photoreceptors; they contain an outer segment with stacked disks, an inner segment, and neural projections.

In nearly all birds, the pineal gland is required for rhythms in locomotor activity, feeding, and body temperature. Pinealectomy abolishes these rhythms, however, with some variability across species. In sparrows, rhythms do not disappear immediately and in quail and pigeon, they persist despite the removal of the pineal gland. These observations indicate that other oscillators can compensate for the loss of the pineal gland. Rhythmicity in pinealectomized birds can be restored with rhythmic doses of melatonin either through injection, through infusion, or in the drinking water. Transplantation of the pineal organ can also restore rhythms. In such cases, the host bird's rhythm is in phase with that of the donor.

The Suprachiasmatic Nucleus

The circadian system in birds includes an SCN. The hypothalamic area in birds is thought to contain a correlate to the mammalian SCN. The avian SCN consists of two nuclei—the medial SCN (mSCN), which is located near the preoptic recess of the third ventricle; and the visual SCN (vSCN), which is slightly more lateral and caudal to the mSCN. It is interesting to note that different anatomical and functional features are associated with each of these nuclei.

The vSCN receives the majority of projections from the retina via the retinohypothalamic tract (RHT) and may thus be important for integrating information about the photoperiod into the circadian system. This is also supported by the observation that the vSCN expresses a cryptochrome (cCRY2; see text below), which is known to be a blue-light photoreceptor in plants and flies. Although its photoreceptive properties are unknown in mammals, CRY is an integral part of the circadian clock in the mammalian system. Communication between the pineal gland and the vSCN is also apparent, since this region expresses melatonin receptors, and lesions of the vSCN abolish rhythmic turnover of norepinepherine in the pineal. Finally, the uptake of 2-deoxy-glucose in the vSCN is rhythmic, which suggests that neural activity in this region oscillates in a circadian manner.

The mSCN, on the other hand, expresses clock genes, including *Clk*, *Per2*, and *Per3*. *Per* RNA in this region is induced by light and oscillates in a circadian manner. Destruction of the mSCN disrupts circadian rhythms in locomotor activity in quail and pigeon, indicating that it is a necessary component of the circadian system in these species.

Taken together, these observations suggest a compartmentalization of the avian SCN. One nucleus, the vSCN, is likely important for mediating light signals from both the retina and the pineal organ, and the other (mSCN) contains a core oscillator and likely relays these signals to other parts of the animal to coordinate behavior or other rhythmic outputs.

The Retina

The retina contains an autonomous oscillator that also contributes to control of circadian behavior. For example, the retina in birds produces melatonin in a rhythmic manner. However, it is not released into the bloodstream from the retina in all species. In sparrow and starling, for instance, plasma melatonin rhythms are abolished after pinealectomy, indicating that retinal melatonin is not released into the bloodstream, nor does it contribute to overt locomotor activity rhythms. In contrast, simultaneous destruction of the eyes and the pineal gland in pigeon and quail abolish melatonin rhythms as well as behavioral rhythms. However, in the quail, severing the optic tract leaves the rhythm in plasma melatonin intact, but abolishes rhythms in behavior.

Molecular Components of the Avian Clock

Most of the known genes that are essential components of the circadian clock have been identified in at least two avian species, including chicken and Japanese quail. The genes isolated include at least two period homologs; *Per2* and *Per3* (each corresponding to the mammalian counterpart); two forms of *Bmal* identified in chicken (*Bmal1* and *Bmal2*); *Clock*, and two homologs of cryptochrome, *cCry1* and *cCry2*.

In both chickens and quail *Per2* RNA oscillates in a circadian manner with a peak in the early morning (ZT0 in quail) and a trough in the early night (Fig. 4.2). In constant conditions, mRNA oscillations are dampened, but still significant. RNA cycling also persists in chick cultured pineal cells, with sustained peak expression from ZT2–6, although it rapidly declines from CT2–6 in DD. Further studies show that *Per2* is rapidly induced by light in both chicken and quail. The mechanism of this induction is discussed in further detail below.

Per3 mRNA has so far been described in detail only in Japanese quail. The RNA cycling is similar to that of *qPer2*, except the rise in mRNA expression is more gradual during the night. *Per2* and *Per3* are expressed throughout the body with highest levels in the retina and pineal gland. Expression was also observed in heart, lung, kidney, and ovary. Low levels of expression were also noted in liver, skeletal muscle, spleen, and testis. *qPer2* and *qPer3* cycle in most tissues examined with peaks at ZT0, except in the ovary. Unlike *qPer2*, *qPer3* is not induced by light.

Clock and two isoforms of *Bmal* (brain and muscle arylhydrocarbon receptor nuclear translocatorlike protein), *Bmal1* and *Bmal2*, were identified in chicken by RT-PCR using degenerate primers and subsequent screening of a chick pineal cDNA library. The *cBmal* genes had 93 and 65% identity with the human *Bmal1* and *Bmal2* genes, respectively. Up to 100% identity was observed in the conserved bHLH and PAS domain repeats. *cBmal1* and *cBmal2* mRNA cycle opposite to the phase of *cPer2*. *cBmal2* is 4 hours delayed relative to *cBMal1*. *cClock* RNA also exhibits circadian cycling, although at reduced amplitude, with peaks occurring at ZT10–18 in LD and CT10–18 in DD (Fig. 4.2).

To determine whether these transcription factors interact, they were subjected to glutathione-*S*-transferase (GST) pulldown assays. This involves expressing at least part of the protein fused to GST in a bacterial vector. The proteins are extracted from bacteria, purified, and added to a mixture containing glutathione-Sepharose beads and an [^{35}S]-methionine-labeled target protein such as *cBmal2*. After an incubation period, the beads are isolated and washed, and the material bound to them is subjected to SDS-polyacrylamide electrophoresis. The GST fusion protein (such as GST-cCLOCK) binds to the beads, and if it interacts with the radiolabeled target protein, then the latter will show up as a single band

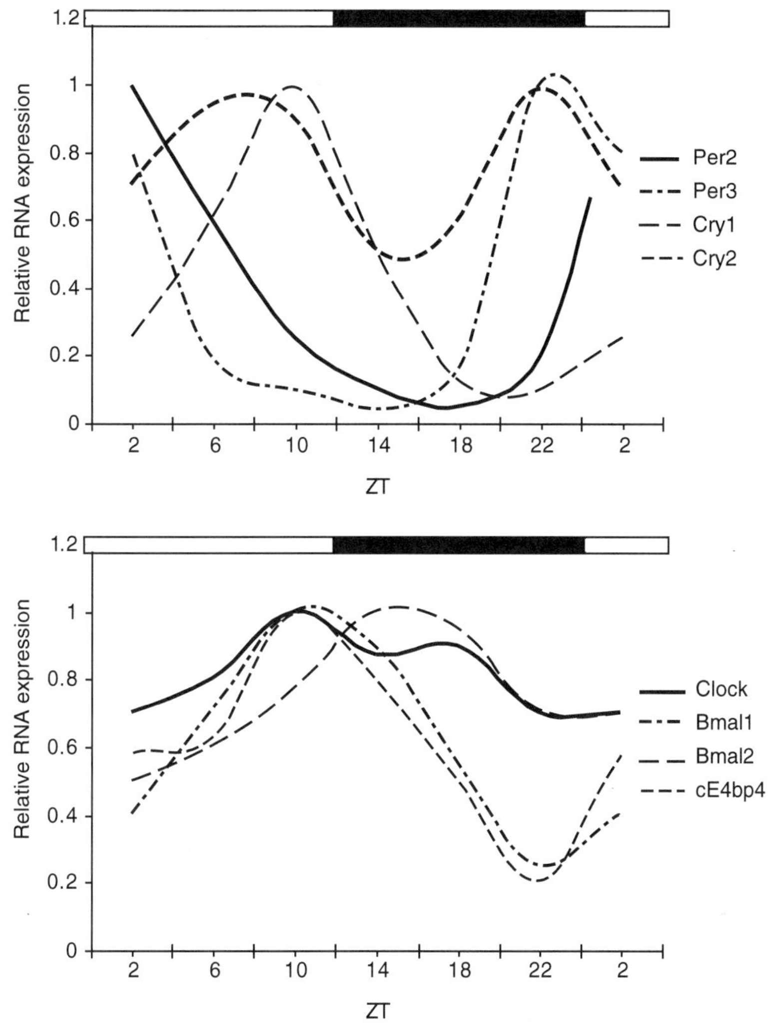

Figure 4.2. Summary of cycling clock genes in the chicken. mRNA levels for each gene are plotted versus time of day [Zeitgeber time (ZT)] in an L:D cycle of 12:12. Open bars at the top represent daytime, starting at ZT0 (lights on); and dark bars represent nighttime, starting at ZT12 (lights off). Levels of RNA do *not* reflect relative expression of the clock genes to each other, but rather how they change during the day.

on the polyacrylamide gel. Indeed, these assays showed interactions between cBMAL1 and cBMAL2, as well as between cCLOCK and each of the two BMAL proteins.

As described in an earlier chapter, the CLOCK:BMAL heterodimer comprises the positive arm of the circadian autoregulatory feedback loop. It binds to an E-box consensus site (CACGTG) in the *per* and *tim* genes (or *per* and *cry* in mammals) to promote their transcription. To determine whether cCLOCK and cBMAL function in a similar manner, Dr. Yoshitaka Fukada and colleagues carried out transcriptional assays using luciferase reporter constructs transfected into human embryonic kidney (HEK) 293 cells. The luciferase reporter genes contained the E-box element and flanking sequences within the promoter regions of *cPer2*, *mPer2*, or mouse vasopressin genes. In all cases, both cCLOCK: cBMAL1 and cCLOCK:cBMAL2 dimers increased luciferase activity in a dose-dependent manner. Interestingly, cBMAL2 inhibits transcription at higher doses.

To test the other arm of the clock, or the negative component, *cPer2* was cotransfected into the cells at varying doses. *cPer2* effectively inhibits cCLOCK: cBMAL-mediated transcription at the E-box. Taken together, these observations indicate that all these proteins have a role in transcriptional regulation that is consistent with both the mammalian and Drosophila models of the circadian clock.

Additional studies by Fukada's group indicated that the chicken clock genes function as such in a relevant system. Overexpression of cBMAL1 or cBMAL2 in cultured pineal cells abolished the rhythm of melatonin synthesis in constant conditions. No effect was observed in LD. Light is known to suppress melatonin synthesis in vivo; thus it is possible that this pathway overrides that driven by cBMAL.

Although cryptochrome (*cry*) gene products across all species examined, including plants, mammals, insects, avians, amphibians, and fish, are structurally similar, their role in the circadian clock has diverged. In flies, for instance, cryptochromes are blue-light photoreceptors important for phase resettting. Absence of *cry* in Drosophila does not affect activity rhythms in constant darkness, but does affect the flies' phase shifting ability in response to light. In mammals, on the other hand, CRY proteins are an integral component of the core clock as they are required for normal behavioral rhythms and on a molecular level, act as transcriptional repressors by regulating CLOCK:BMAL activity. To date, there is no evidence that supports a role of mCRY as a photoreceptor.

Two *cry* genes have been identified in chicken: *cCry1* and *cCry2*. The mRNA of both genes oscillates in LD (Fig. 4.2), but has reduced amplitudes in DD. *cCry1* exhibits peak expression at midday, or mid–subjective day in DD, and *cCry2* peaks during the late night. Northern blot analyses revealed that *cCry2* expression is widespread throughout the animal. High levels of expression were observed in pineal and retina, and also in heart, liver, skeletal muscle, intestine, and brain. Detailed analysis of tissue showed *cCry2* expression in brain regions associated with phototransduction and the visual system such as the vSCN, optic tectum, and the lateral septum. *cCry* mRNA was also observed in the photoreceptive layer of the retina, and in both photoreceptive and interstitial pinealocytes. These data suggest that *cCry* may retain some function as a photoreceptor. However, transcriptional assays revealed that both *cCry1* and *cCry2* repress transcription of *cPer2* by inhibiting CLOCK:BMAL-mediated transcription.

Phase Shifting the Circadian Clock by Photic Pathways

A key feature of a circadian clock is the ability to reset itself in response to a change in the environment. The most drastic change that our clocks are subject to is that which comes with travel over long distances—jet lag. Over several days, we adjust to a new light–dark schedule. What are the mechanisms that mediate this response? In many organisms kept in constant darkness, a single light pulse can reset the phase of activity or the time of day that the organism is active. This is known as a *phase shift*.

Intuitively, one would think the most direct way to transmit signal to the clock is through the retina. However, in birds and other nonmammalian vertebrates, this is not the case. Extraretinal photoreceptors were first described in birds in the 1960s. More recently, a photoreceptive molecule was cloned from chicken pineal gland and named *pinopsin*. Pinopsin localized to regions of pinealocytes that corresponded to the outer photoreceptive segments as described above. Furthermore, pinopsin colocalizes with two GTP binding protein α-subunits known to be involved in mediating responses to light in the pineal, Gt1α and Gq/11α. Interestingly, each G protein is involved in mediating a different kind of light response (Fig. 4.3). One response is pertussis toxin (PTX)–sensitive and involves the immediate inhibition of melatonin synthesis in the presence of light. PTX inhibits some, but not all, G proteins. Gt1α is known to be inhibited by PTX, but Gq/11α is not. Thus, application of PTX to cultured pineal cells results in reduced levels of melatonin, but the phase of the rhythm of melatonin synthesis remains unchanged. The second response to light described in the pineal is the phase shift in the rhythm of melatonin synthesis. It is likely that pinopsin mediates each of these responses through one of these two G proteins.

Several other molecules have been identified as important components of an entrainment pathway. One of these components is mitogen-activated protein kinase (MAPK). The MAPK protein is constitutively expressed in the pineal gland, but the active form of it, phosphorylated MAPK, exhibits a circadian oscillation with peak activity at night. In response to a light pulse, MAPK rapidly dephosphorylates within

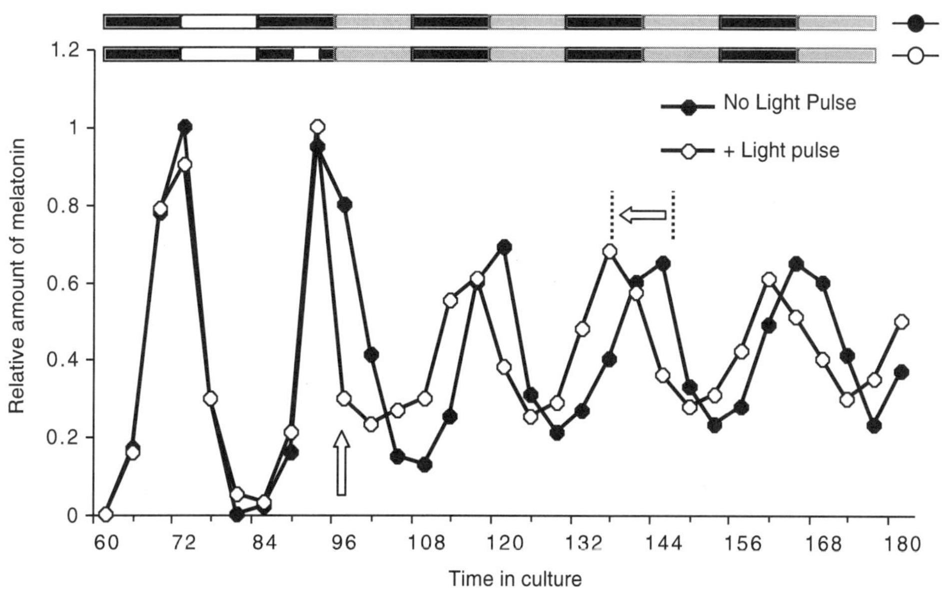

Figure 4.3. Effects of light on melatonin release in cultured chick pineal cells. Pineal cells were maintained in a light–dark cycle in culture for 4 days, and then transferred to constant darkness on day 5 (indicated by the light–dark bars at the top of the chart). On day 5, pineal cells in one of two plates received a four-hour light pulse at CT20 (open circles), while the other plate was kept in the dark (filled circles). Vertical arrow shows the acute affect of light, where melatonin release is suppressed. Horizontal arrow shows a phase-shift in the melatonin rhythm as compared to control. [Adapted from Okano and Fukada (2001): *Microsc Res Tech* **53**: 72–80.]

minutes of the onset of the light pulse. In addition, application of PD98059, an inhibitor of a MAPK kinase (MEK), which is responsible for the phosphorylation or activation of MAPK, phase-shifts the rhythm of melatonin production in cultured chick pineal cells. This effect is both dose-dependent and time-dependent. Together, these observations indicate a role of MAPK in an entrainment mechanism.

In an attempt to identify other genes involved in phase resetting, Dr. Fukada and colleagues used a technique known as differential display PCR (DD-PCR). DD-PCR involves a low-stringency amplification of RNA collected from different conditions. cDNA fragments that are differentially expressed are isolated, reamplified, and cloned for further analysis. In this case, RNA was collected from pineal glands that were isolated from chicks exposed to light pulses at various times throughout the day. Of approximately 5700 PCR products, 99 were affected by light. So far, two of these products have been described in detail.

One product was a chicken homolog of a bZIP transcription factor, *E4bp4* (*cE4bp4*). Among all Drosophila sequences, this gene had highest similarity to *vrille*, one of the Drosophila clock genes (see Chapter 3). *cE4bp4* mRNA exhibited a circadian oscillation, and was rapidly induced by light at all times of day tested. Analysis of the 5′ flanking region of the *cPer2* gene revealed two consensus *cE4bp4* recognition sites. Using this sequence from the *cPer2* gene, transcriptional assays were performed to determine whether cPer expression is affected by *cE4bp4*. Indeed, cotransfection of a *cE4bp4* expression construct repressed *cPer2* expression in a dose-dependent manner. Interestingly, only one of the two *cE4bp4* recognition sites is required for this transcriptional repression.

These observations have lead to filling in another piece of the clock: a light-responsive gene that interacts directly with a core clock component (Fig. 4.4). From the observations discussed above, *cE4bp4* may serve two roles in the clock:

1. Since *cE4bp4* mRNA expression (and likely protein expression) is increasing during the day when *cPer2* mRNA is decreasing, one role of *cE4bp4* may be to down regulate *cPer2* in a circadian fashion.
2. Possibly more importantly, both *cE4bp4* and *cPer2* are induced by light; thus *cE4bp4* may act to curtail the response to light by repressing (and thereby "slowing") *cPer2* induction.

Another interesting product that was isolated from the DD-PCR screen is *cHsp90α*. This gene increases within 5 minutes after light onset, and peaks at 90 minutes to fourfold of baseline levels. *cHsp90α* mRNA cycles during the day with a peak in the early morning. The cycling persists in DD, but at a lower amplitude. The *cHsp90α* protein colocalizes with pinopsin in pinealocytes and was found in both cytoplasm and nucleus, but no cycling in subcellular localization was detected. To date, the role of *cHsp90α* in entrainment is unknown. It is likely that its cycling and light-induced expression are related to circadian and light-induced changes in temperature that are known to occur in the pineal. The human *Hsp90α* interacts with BMAL1 and NPAS2, which indicates some role in the central clock.

■ ORGANIZATION OF CIRCADIAN SYSTEM IN ZEBRAFISH

Locomotor Activity and Melatonin Synthesis

Zebrafish display rhythms of locomotor activity and melatonin synthesis. Activity rhythms have been recorded from both

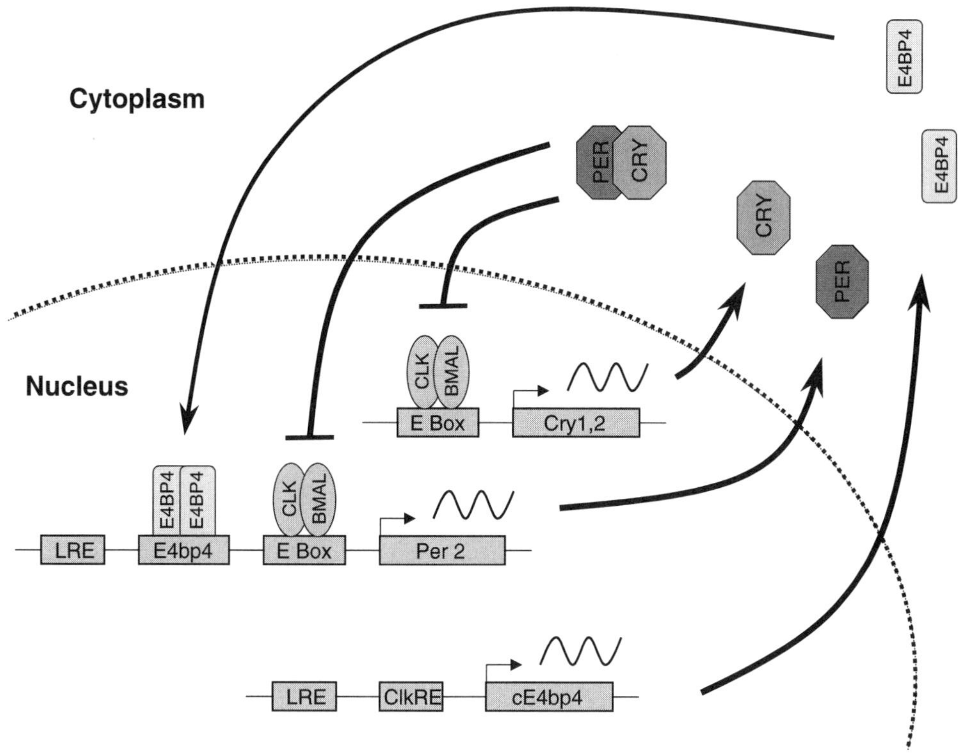

Figure 4.4. Role of the transcription factor *cE4bp4* in the circadian clock. Transcription of chicken clock genes, *per* and *cry*, is mediated by CLOCK and BMAL heterodimers; PER and CRY proteins form heterodimers and translocate back to the nucleus, where they inhibit CLOCK:BMAL transcription; *Per2* transcription is further repressed by cE4BP4 binding to a target sequence in the *per* promoter (LRE = light-responsive element; ClkRE = clock-responsive element). [Adapted from Doi et al. (2001): *Proc Natl Acad Sci* (*USA*) **98**: 8089–8094.]

adult and larval zebrafish. Adults placed in small recording chambers with infrared motion sensors display behavioral rhythms with a large amount of variability. About 70% of the animals show robust rhythms with highest activity during the subjective daytime. Activity rhythms in larval zebrafish measured by video image analysis are more robust; 95% of larvae show rhythmic activity, with highest levels recorded during the day. In both larvae and adults, the average period in constant conditions is approximately 25 hours.

Because of the availability of mutagenesis and transgenic methods in zebrafish, this makes an ideal system for delineating the function of the molecular components of the circadian clock. The behavioral assay allows for the isolation of mutants displaying altered periods or arrhythmic behavior. Other circadian outputs that have been recorded in zebrafish include visual sensitivity and melatonin synthesis.

Rhythms in visual sensitivity have been detected using a couple of different assays. One involves measuring the *electroretinogram* (ERG), which is a simple field potential, measured from the cornea, that represents neural activity in the retina. Another way to quantify visual sensitivity involves the ability of the fish to respond to a threatening visual stimulus. The limit of detection is measured by the extent to which a threatening object is illuminated (and therefore visible to the fish) to evoke an escape response. A rhythmic component has been reported in both assays, with peaks occurring at subjective dawn, and troughs at subjective dusk. Zebrafish expressing a dominant mutation, *night-blindness b* (*nbb*), are deficient in rhythms of visual sensitivity. Several neuronal deficiencies were found in *nbb* mutants, including a reduced number of retinal dopaminergic interplexiform cells. Indeed, chemical destruction of these cells leads to a more profound defect in visual sensitivity rhythms, indicating a role of the dopamine system in the circadian control of visual sensitivity.

Rhythmic melatonin synthesis and release has been recorded from cultured zebrafish pineal gland as well as from retina. Melatonin synthesis in culture is light-sensitive, and rhythms persist in constant conditions for at least a week. This circadian output therefore has the potential for determining how various mutations can affect rhythms at the cellular level. It is interesting to note, however, that the contribution of the pineal gland or of the retina to activity rhythms has not been determined in zebrafish. Preliminary experiments show that activity rhythms in zebrafish are maintained after removal of both the pineal gland and the eye. This is inconsistent with earlier studies in other teleosts in which pinealectomy did abolish behavioral rhythms. Nonetheless, these data indicate that melatonin is likely not required for the circadian control of behavior in zebrafish. As discussed below, tissues throughout the zebrafish contain independent oscillators that are sensitive to light. Perhaps the SCN or any one of these organs is sufficient for driving rhythmic behavior or compensates for the loss of the pineal and/or the eyes. Whatever the case, the question of a central oscillator in zebrafish is an issue that requires further study.

Molecular Components of the Zebrafish Clock

Clock and Bmal

Most of the clock genes that have been identified in zebrafish to date were identified using either PCR or by low-stringency screening of a cDNA library. Shortly after the cloning of the mouse *Clock* (*Clk*) gene, Paolo Sassone-Corsi and colleagues isolated the homologue in zebrafish. The predicted gene product of *zfClk* is an 894 amino acid protein that is highly similar to that in mouse, sharing 80% identity, but it shares only 53% identity with the Drosophila *Clk* gene. *zfClk* contains a bHLH-PAS DNA binding and protein–protein interaction domain, a PAS-B region, and a glutamine-rich region corresponding to the transcriptional activation domain, consistent with the *Clk* proteins found in mouse and fruitfly.

Analysis of *Clk* mRNA expression by RNase protection assay revealed a robust oscillation in both the pineal and retina with peak expression at the onset of the dark period in an L:D 14:10 cycle (Fig. 4.5). *Clk* oscillation persists in DD, indicating that it is likely an integral component of the zebrafish clock. *Clk* mRNA also exhibits circadian oscillations in other tissues, including brain, heart, kidney, and spleen. No cycling is observed in testes. Interestingly, when various organs were isolated in culture, *Clk* cycling persisted for several days in constant darkness. Further studies of heart and kidney organ cultures showed that *Clk* expression in these tissues entrains to LD cycles. These observations indicate the existence of photosensitive peripheral pacemakers in zebrafish, similar to those observed in Drosophila (see Chapter 9 for further discussion of multiple oscillators).

In order to identify other components of the zebrafish clock, Sassone-Corsi's group proceeded to isolate partners of the *zfClk*

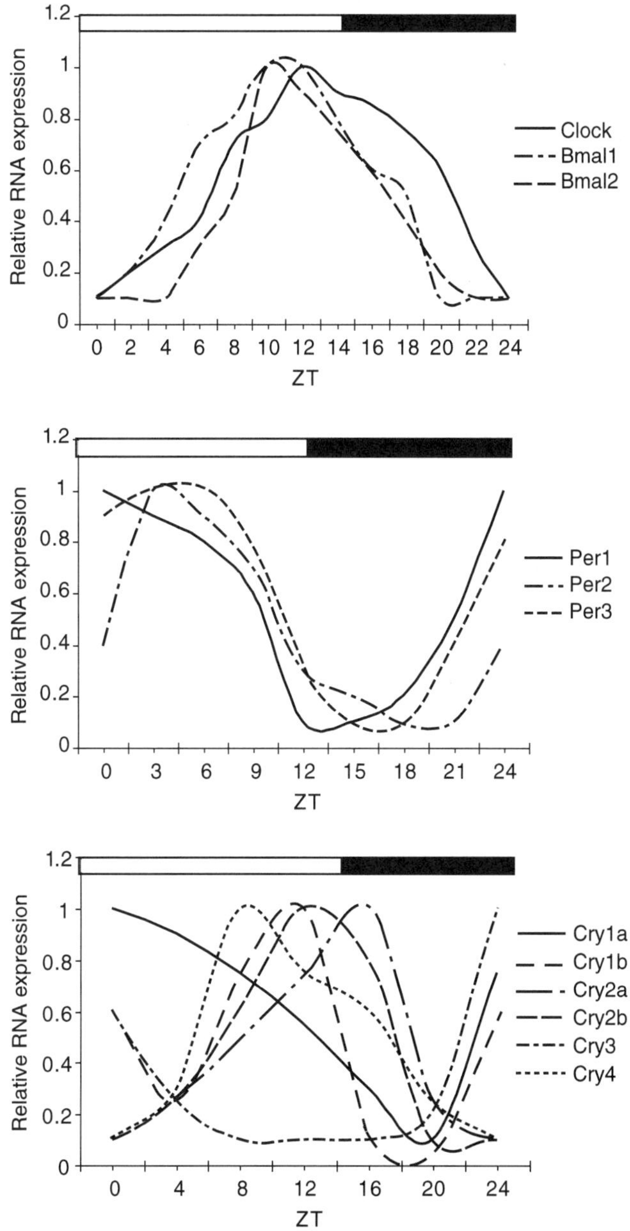

Figure 4.5. Summary of cycling clock genes in zebrafish. mRNA levels for each gene are plotted as described in Figure 4.2. Top panel is mRNA collected from zebrafish eye in a 14:10 light:dark cycle. Middle panel is mRNA collected from cultured Z3 cells (see text) in a 12:12 L:D cycle. Bottom panel is mRNA collected from zebrafish eye and whole body in a 14:10 L:D cycle.

protein using the yeast-2-hybrid assay. This was the same approach used to identify the CLOCK partner, BMAL, in the mammalian system (see Chapter 5). Two interacting partners, BMAL1 and BMAL2, were identified using this method. Alignment of the amino acid sequence of each protein revealed that BMAL1 shared highest identity with human BMAL, whereas BMAL2 was more divergent. BMAL1 and BMAL2 share 75% identity with each other, with the most divergence at the *C* terminus. Both proteins, consistent with the mammalian homolog, contain a bHLH domain and PAS domain repeats. Analysis of protein–protein interactions show that BMAL1 interacts with zfCLOCK more efficiently than does BMAL2. BMAL1 and BMAL2 do not interact with each other.

Bmal1 and *Bmal2* mRNA cycle in all tissues tested, with the exception of the testes. Interestingly, peak expression varies both between the *Bmal* genes and across tissues. For example, in the brain, *Bmal1* mRNA peaks at ZT14, and *Bmal2* peaks at ZT10. *Clk* RNA peaks at ZT14–16 in the brain. In the liver, *Bmal1* peaks at ZT9, and *Bmal2* peaks at ZT15, while their *Clk* partner peaks at ZT9–15. These observations could indicate that posttranscriptional modifications may occur and vary across tissues. We have yet to learn whether the *Clk* and *Bmal* proteins cycle and whether they are in phase with their respective RNA cycling or with each other.

Interestingly, as described in Chapters 3 and 5, *Clock*, but not *Bmal1*, cycles in Drosophila, while the reverse is true in mammals (i.e., *Bmal1* cycles), but *Clock* does not. In zebrafish, both are found to cycle. This may represent an intermediate state in evolution or alternatively reflect a specific adaptation in zebrafish.

Period

Zebrafish homologs of each of the three mammalian *per* genes have been cloned. *Per3* mRNA oscillates in embryos in the CNS and in the retina from 40 to 128 hours post-fertilization, with peaks during the early light phase from ZT0 to ZT4. Interestingly, *Per3* mRNA cycling could be detected in unfertilized oocytes and in embryos as early as the 1–4 cell stage. The cycling is independent of LD cycles and persists in constant conditions. These data indicated that embryos synchronize their clocks to the maternal gene products in the oocytes.

zfPer1 and *zfPer2* have not been investigated in vivo, but have been studied in a Z3 cell line established from embryonic tissue. When Z3 cells were subjected to LD cycles, mRNA oscillations of each of the *per* genes could be detected; *zfPer1* peaked at ZT0, *zfPer2* peaked at ZT3, and *zfPer3* peaked between ZT3 and ZT6. The onset of expression for *Per1* and *Per3* could be detected before lights on, but peak *per2* levels were detected only at ZT3. In constant darkness, *Per2* does not cycle and has substantially reduced expression. Oscillations of *Per1* and *Per3*, on the other hand, persisted in DD, with peak expression occurring at CT3. The onset of the expression increase began shortly before the end of the subjective night, similar to the case in LD cycles. Together, these observations indicate that *Per1* and *Per3* are circadian-clock-dependent, and *Per2* is light-dependent.

Cryptochromes

Six CRY genes have been identified in zebrafish. Two pairs of these genes cluster together and show high similarity to human CRY. These are designated *zCry1a*, *zCry1b*, *zCry2a*, and *zCry2b*. A fifth *cry* also shows high conservation with the human CRY, and is named *zCry3*. The last isolated *cry* gene shows fairly low conservation with both human CRY1 and CRY2 and with a *Xenopus* photolyase. This one was designated *zCry4*.

The function of the *zCrys* was determined by using an in vitro luciferase reporter assay. The reporter construct contained a promoter region of the mouse arginine vasopressin gene that carries an E-box consensus sequence, CACGTG, which is the transcriptional activation site recognized by the CLOCK:BMAL heterodimer. Transfecting the cells with CLOCK and BMAL effectively increased the luciferase reporter activity. However, when cells were also transfected with *zCry1a*, *zCry1b*, *zCry2a*, or *zCry2b*, reporter activity was reduced. *zCry3* and *zCry4* had no effect on the luciferase reporter activity. These data indicate that the *zCry1* and *zCry2* pairs share a similar function with the mammalian CRY proteins and inhibit CLOCK:BMAL-mediated transcription. It is possible that *zCry3* and *zCry4* resemble the function of the Drosophila CRY, but this has yet to be determined.

All the zebrafish *Cry* RNAs are expressed throughout the body, brain, and eye, with highest levels of expression in the eye. *zCry* mRNAs all cycle in both eye and brain, but with lower amplitudes in the body. All the *Crys* cycle with different phases: *Cry1a* peaks at ZT0–4, *Cry1b* peaks at ZT9–13, both *Cry2a* and *2b* peak at ZT13–15, *Cry3* peaks at ZT23–1, and *Cry4* peaks at ZT9 (Fig. 4.5).

Others

Several other clock-controlled genes have been identified in zebrafish. One of these is a member of a ligand-activated nuclear receptor family of transcription factors, *rev-erbα*. Because the ligand of this nuclear receptor has not been identified, it is considered an orphan nuclear receptor. *Rev-erbα* mRNA was reported to cycle in zebrafish embryos, with peak expression occurring between ZT20 and ZT0. Although the role of *rev-erbα* is unknown in zebrafish, it was discovered to be an important component of the circadian clock in mammals (see Chapter 5). In this system, *rev-erbα* is positively regulated by CLOCK:BMAL-mediated transcription and feeds back on the clock by binding to a target region in the *Bmal1* promoter to repress its transcription. Thus, *rev-erbα* and *Bmal1* mRNA cycle in opposite phases. As mentioned above, *Bmal1* and *Bmal2* mRNA cycle with different phases across zebrafish tissues. One prediction is that *rev-erbα* would also cycle in the opposite phase, depending on the tissue. It is likely that widespread expression of *rev-erbα* in zebrafish is also an important contributor to peripheral oscillators in this system.

A second example of a circadian controlled gene is the interphotoreceptor retinoid binding protein (IRBP). This protein is thought to be involved in retinoid trafficking between photoreceptors and retinal pigment epithelium. *Irbp* mRNA expression is high during the day and low at night. However, IRBP protein is maintained at constant levels, although turnover of the protein is known to be higher during the day. Thus, oscillation in mRNA expression may compensate for this in order to keep protein levels constant.

ANALYSIS OF CIRCADIAN RHYTHMS IN XENOPUS

Measuring Rhythms in Frogs

Most of what is known about circadian biology in the African clawed frog, *Xenopus laevis*, is based on studies in isolated retinal cultures. The Xenopus eyecup preparation has been advantageous in several respects: (1) with the appropriate culture conditions, output rhythms can be measured from a *single* eyecup (Fig. 4.6), thus eliminating individual differences that can occur within a population; and (2) the retina is known to contain all the features that qualify it as a circadian oscillator. Two of these features are an ability to entrain to

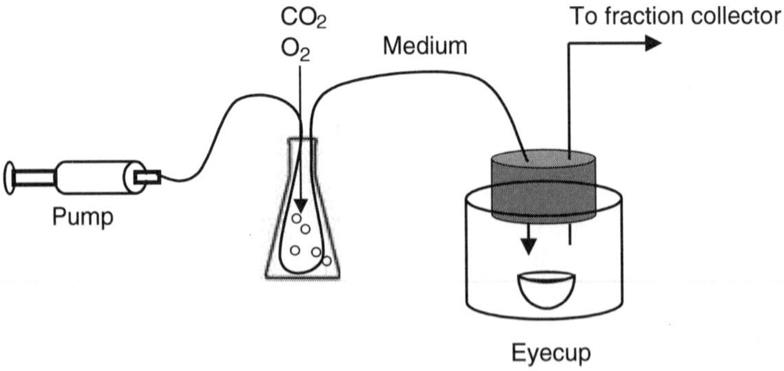

Figure 4.6. Diagram of the Xenopus eyecup preparation. Culture medium is supplemented with 5-hydroxy-L-tryptophan in order to detect melatonin from a single eyecup. The medium is gassed with CO_2 and O_2, and then delivered to a superfusion chamber containing the eyecup. The perfusate is delivered to a fraction collector for quantifying concentrations of melatonin. [Adapted from Anderson and Green (2000): *Microsc Res Tech* **50**: 360–372.]

environmental conditions such as light and a freerunning rhythm in constant conditions. Because of the presence of known clock genes in the retina, this tissue offers a complete circadian system: an input pathway that mediates signals from the environment, a central oscillator, and a regulated output such as melatonin.

A number of circadian outputs have been observed in the retina, including rhythms in the ERG, rhythms in rod disk shedding and cone elongation, and, as mentioned above, rhythms in melatonin synthesis. The rhythm in melatonin release is known to be due to circadian control of enzymes in the synthetic pathway, such as *N*-acetyltransferase (NAT). This enzyme is also strongly inhibited by light and dopamine via D2 receptors. The D2 receptor is a G-protein-coupled receptor that negatively regulates adenylyl cyclase and cAMP. Since NAT activity is induced by cAMP, D2 receptor activation results in a decrease in NAT activity by inhibiting cAMP. Thus, light and darkness control the levels of melatonin and dopamine, and both these environmental and neurochemical factors can control changes in photoreceptor morphology. For example, cone elongation is induced by either darkness or melatonin at the beginning of the night. By the end of the night, melatonin and melatonin metabolites start inducing rod disk shedding, the peak of which is at the light-to-dark transition. Dopamine, on the other hand, is released during the day and induces cone contraction. These processes are known to continue in the absence of a light–dark cycle, indicating an underlying clock-controlled mechanism.

Clock Genes in *Xenopus*

Most of the known clock components have been identified in Xenopus using RT-PCR and subsequent screening of a retinal cDNA library. These components include the positive arm of the feedback loop, CLOCK and BMAL, as well as two *period* genes (*xPer1* and *xPer2*) and three *cry* genes as described below. Precisely how these components function together to produce rhythmic output has not been determined, but the utility of a unique transgenic method called *restriction enzyme-mediated integration* (REMI) is beginning to uncover how each of these components is involved.

Xper was identified using degenerate primers of mouse and Drosophila *per* mRNA. The isolated gene resembles *mPer2*, with high homology throughout the entire protein including the PASB and cytoplasmic localization domains. However, despite the high homology in the *N*-terminal region, no bHLH domain was detected. RT-PCR was performed to determine whether other *per* genes exist, and a partial sequence was identified that resembles the *mPer1* protein. Using this partial sequence as a probe, as well as a piece of the *xPer2* gene, expression patterns were studied both anatomically and temporally.

These studies revealed that *xPer1* and *xPer2* have highest expression levels in the retina and show lower expression throughout the body, including brain, heart, liver, spleen, and testes. Although tissue expression is similar for both mRNAs, temporal expression is quite different. Both *xper* RNAs exhibit robust cycling in LD, but with a 4–8-hour phase difference (Fig. 4.7). *xPer1* continues to oscillate in a circadian manner in DD. *xPer2* does not oscillate in constant conditions, but is expressed at constant peak levels in LL, and at constant trough levels in DD. These observations indicate that as in zebrafish and birds, *xPer2* is regulated by light, and the circadian clock regulates *xPer1*.

Further analysis of the effects of light on *per* indicate that *xper2* mRNA, in particular, shows significant increases 3 hours after the onset of light, and peak expression occurs after 6 hours of lights on. Dopamine (DA), a neurotransmitter that is thought to be a component of the entrainment pathway, had a similar effect on *xper2*, where significant induction was not detected until 3 hours after dopamine treatment. DA induced *xper2* mRNA independent of the time of day. The effect of DA is blocked by the D2 receptor antagonist eticlopride. Interestingly, eticlopride does not block *xper2* induction by light, indicating a separate pathway. Induction of *xper1* by either light or DA, on the other hand, is restricted to the early part of the day. These data support the notion that *xper1* is relatively insensitive to light, whereas *xper2* is affected by phase shifting agents, light, and DA.

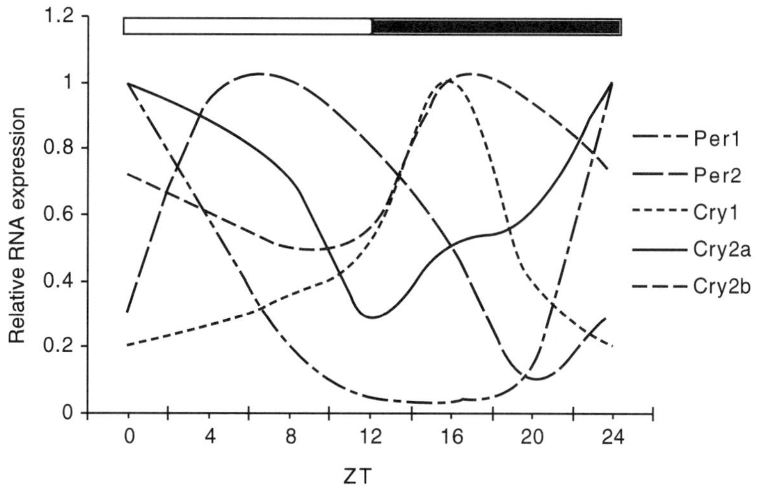

Figure 4.7. Summary of cycling clock genes in Xenopus. mRNA levels for each gene are plotted as described in Figure 4.2. mRNA was collected from Xenopus retina in L:D 12:12. *xClock* mRNA does not cycle and is therefore not plotted here (see text).

Although studies have shown how the *xpers* are regulated (one having a circadian component, and the other by the entrainment pathway), there are no data available showing whether these genes are necessary for circadian behavior. As discussed above, dopamine, itself released in a circadian manner, exerts effects in both melatonin synthesis and photoreceptor morphology. The observation that it also affects a known clock gene suggests an important interaction between the clock and this neurochemical system.

Three *cry* genes have been identified in Xenopus. *xCry1* has 86% identity with *mCry1* and exhibits the highest amplitude in mRNA oscillation with a peak at ZT16. The two other *crys*, *xCry2a* and *xCry2b*, are 90% similar to *mCry2*. It is possible that *xCry2a* and *xCry2b* are duplicate versions of the same gene, since full sequences were not identified, particularly the 5′ end. However, both mRNAs oscillate in both LD and DD, but are 4 hours out of phase with each other. Furthermore, all three *crys* are expressed in the same tissues, including retina, brain, heart, liver, spleen, and testes. *xCry2b* expression is relatively weak, and *xCry1* expression is higher in the brain as compared to the other *cry* genes, indicating a tissue-specific function.

xClock is found in many tissues, including retina, brain, heart, liver, muscle, spleen, and testes. Similar to the mammalian counterpart, *xClk* contains bHLH, PASA, and PASB functional domains, with 94% similarity in these regions. Consistent with its role as a transcription factor, *xClk* contains a glutamine-rich region or a transcriptional activation domain at the *C* terminus. *xClk* mRNA does not cycle in the retina in either LD or DD.

Studies of *xClk* have revealed a similar function for this protein in clock mechanisms. Using the method mentioned above (REMI), Dr. Carla Green and co-workers produced a transgenic frog expressing a dominant negative form of *xClk*. Traditional transgenesis involves injecting plasmid DNA into embryos and looking for surviving adults expressing the transgenic DNA. In frogs, particularly in *Xenopus laevis*, this technique is problematic in that the gestation period is lengthy, and successful insertion into the genome is fairly infrequent. If the transgene is not inserted into the genome, promoter-driven spatial expression does not occur with adequate fidelity, and the embryo often expresses the transgene in a mosaic pattern. REMI, on the other hand, involves introducing plasmid DNA into the male genome. Low concentrations of restriction enzyme are used to integrate the transgene into the frog sperm. Afterward, the sperm nuclei are transplanted into unfertilized eggs, and transgenic embryos are produced.

Dr. Green's group designed a transgene that mimicked the *Clk*^{jrk} mutation described in both Drosophila and mouse. This gene was designed to express all the functional domains of the CLOCK protein, except for that needed for transcriptional activation. The expressed protein therefore competes with the endogenous protein by binding with partners (such as BMAL) required for transcriptional activation. Functional assays in vitro using an mper1-luciferase reporter showed that the mutated *xClk* could effectively block transcription induced by a wild-type *Clk*. When the transgene was expressed in frog retina using the eye-specific promoter, IRBP, melatonin cycling was abolished in a dose-dependent manner. These data were the first evidence that *xClk* is required for circadian regulation of an output. Overall levels of melatonin were not affected, nor were levels of other genes in the melatonin synthetic pathway, although temporal expression of these genes in the presence of the mutant *xClk* transgene was not determined.

Other clock molecules have been identified in Xenopus, one of which is currently being investigated in mammals and insects.

Nocturnin (*noc*) was identified as a cycling transcript in a differential-display PCR screen. *Noc* mRNA expression peaks at night, accounting for its name. Sequence analysis revealed that the *noc* gene encodes a protein containing a leucine repeat domain and a domain homologous to the yeast carbon catabolite repression 4 protein (CCR4), which is a transcription coactivator. *Noc* mRNA expression is restricted to the Xenopus retina. Interestingly, *noc* expression in the mouse is widespread. Not only is it expressed in most tissues examined, but it also cycles in nearly all tissues in a circadian manner. Thus, *noc* is likely an important component that mediates a circadian output. Whether it affects behavior or some other physiological process remains to be determined.

■ CONCLUSIONS

Although the clock machinery in nonmammalian vertebrates is fairly conserved with that in mammals, the organization of their circadian system is more diverse. The model systems outlined above have been useful in identifying novel components of the circadian clock such as *cE4bp4* and *nocturnin*, although a role for the latter in the circadian clock or its output has yet to be determined. Several questions have yet to be addressed in these systems. First, as mentioned above, some of the clock genes cycle in these organisms, while in others they do not. In the previous chapter, we have seen that mRNA cycling may not be necessary for driving rhythmic behavior, but protein cycling is required. Thus, the regulation of the various clock proteins needs to be addressed in these nonmammalian systems.

Furthermore, a key advantage to studying nonmammalian vertebrates is the availability of in vitro preparations that contain a complete circadian system. Many questions in the field remain with respect to every aspect of the circadian clock, including input mechanisms, oscillator mechanisms, and especially outputs. How does the molecular clock signal to the rest of the cell and the entire organism to produce rhythmic behavior? With a variety of measurable outputs available in the in vitro preparations such as the Xenopus eyecup or chick pineal gland, these are ideal systems for addressing these issues.

FURTHER READING

Anderson, FE, Green, CB (2000): Symphony of rhythms in the Xenopus laevis retina. *Micro Res Tech* **50**: 360–372.

Bailey, MJ, Chong, NW, Xiong, J, Cassone, VM (2002): Chickens' Cry2: Molecular analysis of an avian cryptochrome in retinal and pineal photoreceptors. *FEBS Lett* **513**: 169–174.

Cahill, GM (2002): Clock mechanisms in zebrafish. *Cell Tissue Res* **309**: 27–34.

Cahill, GM, Hurd, MW, Batchelor, MM (1998): Circadian rhythmicity in the locomotor activity of larval zebrafish. *Neuroreport* **9**: 3445–3449.

Cermakian, N, Whitmore, D, Foulkes, NS, Sassone-Corsi, P (2000): Asynchronous oscillations of two zebrafish CLOCK partners reveal differential clock control and function. *Proc Natl Acad Sci (USA)* **97**: 4339–4344.

Delaunay, F, Thisse, C, Marchand, O, Laudet, V, Thisse, B (2000): An inherited functional circadian clock in zebrafish embryos. *Science* **289**: 297–300.

Doi, M, Nakajima, Y, Okano, T, Fukada, Y (2001): Light-induced phase-delay of the chicken pineal circadian clock is associated with the induction of *cE4bp4*, a potential transcriptional repressor of *cPer2* gene. *Proc Natl Acad Sci (USA)* **98**: 8089–8094.

Doi, M, Nakajima, Y, Okano, T, Fukada, Y (2002): Light-dependent changes in the chick pineal temperature and the expression of *cHsp90α* gene: A potential contribution of *in vivo* temperature change to the photic-entrainment of the chick pineal circadian clock. *Zool Sci* **19**: 633–641.

Green, CB, Besharse, JC (1996): Identification of a novel vertebrate circadian clock-regulated gene encoding the protein nocturnin. *Proc Natl Acad Sci (USA)* **93**: 14884–14888.

Gwinner, E, Brandstätter, R (2001): Complex bird clocks. *Phil Trans Roy Soc Lond B* **356**: 1801–1810.

Hayasaka, N, LaRue, SI, Green, CB (2002): *In vivo* disruption of Xenopus CLOCK in the retinal photoreceptor cells abolishes circadian melatonin rhythmicity without affecting its production levels. *J Neurosci* **22**: 1600–1607.

Hurd, MW, Debruyne, J, Straume, M, Cahill, GM (1998): Circadian rhythms of locomotor activity in zebrafish. *Physiol Behav* **65**: 465–472.

Kobayashi, Y, Ishikawa, T, Hirayam, J, Daiyasu, H, Kanai, S, Toh, H, Fukuda, I, Tsujimura, T, Terada, N, Kamei, Y, Yuba, S, Iwai, S, Todo, T (2000): Molecular analysis of zebrafish photolyase/cryptochrome family: Two types of cryptochromes present in zebrafish. *Genes Cells* **5**: 725–738.

Li, L, Dowling, JE (2000): Disruption of the olfactoretinal centrifugal pathway may relate to the visual system defect in *night blindness b* mutant zebrafish. *J Neurosci* **20**: 1883–1892.

Li, L, Dowling, JE (2000): Effects of dopamine depletion on visual sensitivity of zebrafish. *J Neurosci* **20**: 1893–1903.

Natesan, A, Geetha, L, Zatz, M (2002): Rhythm and soul in the avian pineal. *Cell Tissue Res* **309**: 35–45.

Okano, T, Yamamoto, K, Okano, K, Hirota, T, Kasahara, T, Sasaki, M, Takanaka, Y, Fukada, Y (2001): Chicken pineal clock genes: Implication of BMAL2 as a bi-directional regulator in circadian clock oscillation. *Genes Cells* **6**: 825–836.

Okano, T, Fukada, Y (2001): Photoreception and circadian clock system of the chicken pineal gland. *Microsc Res Tech* **53**: 72–80.

Pando, MP, Pinchak, AB, Cermakian, N, Sassone-Corsi, P (2001): A cell-based system that recapitulates the dynamic light-dependent regulation of the vertebrate clock. *Proc Natl Acad Sci (USA)* **98**: 10178–10183.

Sanada, K, Hayashi, Y, Harada, Y, Okano, T, Fukada, Y (2000): Role of circadian activation of mitogen-activated protein kinase in chick pineal clock oscillation. *J Neurosci* **20**: 986–991.

Steenhard, BM, Besharse, JC (2000): Phase shifting the retinal circadian clock: *xPer2* mRNA induction by light and dopamine. *J Neurosci* **20**: 8572–8577.

Underwood, H, Steele, CT, Zivkovic, B (2001): Circadian organization and the role of the pineal in birds. *Microsc Res Tech* **53**: 48–62.

Wang, Y, Osterbur, DL, Megaw, PL, Tosini, G, Fukuhara, C, Green, CB, Besharse, JC (2001): Rhythmic expression of *Nocturnin* mRNA in multiple tissues of the mouse. *BMC Dev Biol* **1**: 9.

Whitmore, D, Foulkes, NS, Sassone-Corsi, P (2000): Light acts directly on organs and cells in culture to set the vertebrate circadian clock. *Nature* **404**: 87–91.

Whitmore, D, Foulkes, NS, Strähle, U, Sassone-Corsi, P (1998): Zebrafish clock rhythmic expression reveals independent peripheral circadian oscillators. *Nature Neurosci* **1**: 701–707.

Yamamoto, K, Okano, T, Fukada, Y (2001): Chicken pineal Cry genes: Light dependent up-regulation of cCry1 and cCry2 transcripts. *Neurosci Lett* **313**: 13–16.

Yoshimura, T, Suzuki, Y, Makino, E, Suzuki, T, Kuroiwa, A, Matsuda, Y, Namikawa, T, Ebihara, S (2000): Molecular analysis of avian circadian clock genes. *Mol Brain Res* **78**: 207–215.

Yoshimura, T, Yasuo, S, Suzuki, Y, Makino, E, Yokota, Y, Ebihara, S (2001): Identification of the suprachiasmatic nucleus in birds. *Am J Physiol* **280**: R1185–R1189.

Zhu, H, Green, CB (2001): Three cryptochromes are rhythmically expressed in *Xenopus laevis* retinal photoreceptors. *Mol Vision* **7**: 210–215.

Zhu, H, LaRue, S, Whiteley, A, Steeves, TDL, Takahashi, JS, Green, CB (2000): The Xenopus Clock gene is constitutively expressed in retinal photoreceptors. *Mol Brain Res* **75**: 303–308.

Zhuang, M, Wang, Y, Steenhard, BM, Besharse, JC (2000): Differential regulation of two *Period* genes in the *Xenopus* eye. *Mol Brain Res* **82**: 52–64.

5

GENETIC BASIS FOR CIRCADIAN RHYTHMS IN MAMMALS

J. D. Alvarez

■ INTRODUCTION

In this chapter, studies of the molecular basis of mammalian circadian rhythms are discussed. At the time of this writing, the field is advancing with lightning speed; therefore, by the time readers set their eyes on this book, much more information may be known. Still, a discussion of how mammalian genetics and molecular biology has contributed to a basic understanding of chronobiology is quite useful. Utilizing insect genetics to dissect circadian systems has been enormously successful, and we now know that there is great conservation of the basis for circadian rhythms among all species, including mammals. Mammals are quite a bit different from fruitflies, though, with unique and complex physiologic outputs. A link between a genetic disease in humans and a mutation in a circadian gene has been established underlining the importance of the field.

Using classical genetics to study mammalian circadian systems is not an easy task for a number of reasons. The generation time of mammals is long, especially compared to model systems such as *Drosophila* sp. or *C. elegans*. Another reason is purely logistical—conducting large-scale mutation studies on mammals requires an enormous amount of space and manpower to handle the numbers of necessary animals. Still, the wealth of knowledge that will result from such studies is so vast that a number of groups have initiated large-scale mutagenesis screens in mice. Indeed, it is clear that when applied, forward genetics can be very successful as demonstrated by the elegant identification of the *Clock* gene in mice, the

Molecular Biology of Circadian Rhythms. Edited by Amita Sehgal
ISBN 0-471-41824-2

first cloned mammalian circadian gene. Additionally, sometimes Mother Nature is kind and presents us with a naturally occurring genetic mutant that affects circadian rhythms; the best example is the *tau* mutation in hamsters. Furthermore, although mammalian genetic studies are not as easy as those in lesser organisms, the use of transgenic and gene-targeting technology in mice allows us great control over the genetics in a mammalian system. Such manipulation is already beginning to show us the effects of loss of function mutations of core circadian genes in the murine system. Hopefully, transgenic technology will soon allow us to examine gain of function mutants in these animals as well. In this chapter, we discuss primarily studies that have been performed in rodents, particularly mice, because it is in this model system that most of research on mammalian circadian biology has been performed, particularly with respect to molecular biology. Currently (as of 2003), it appears that the system is conserved in all mammals, but the reader should note that such an assumption is not completely proven yet.

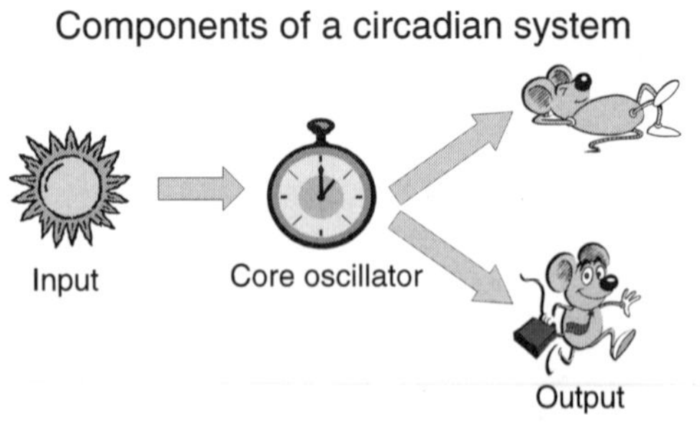

Figure 5.1. Schematic of the three-part model of a circadian system. At the heart of the system is the central oscillator. This is the main timekeeper of the system. The central oscillator is under the influence of input factors. Input is often from environmental factors, such as the light–dark cycle. The central oscillator imposes a rhythm on various aspects of physiology via output pathways. Output can represent both direct neuronal connections and secreted factors.

PROPERTIES OF A MAMMALIAN CIRCADIAN SYSTEM

Components

Historically, before the advent of gene cloning, most research in mammalian circadian biology focused on phenomenology and anatomy. These studies found that circadian systems in mammals consist of three components, just as in nonmammals. These components are the input, the output and the central oscillator (Fig. 5.1). In mammals, a core circadian system is responsible for maintaining the rhythms of the entire organism. However, circadian systems exist in nearly every organ of the body. In fact, evidence is mounting that circadian systems exist in individual cells. How all these systems operate in concert with each other is a very active area of research.

The central oscillator is the timekeeper—it maintains the rhythm. In the mammalian core circadian system, the central oscillator is located in a paired cluster of neurons in the brain called the *suprachiasmatic nuclei* (SCN). Output from the SCN represents the mechanisms by which this core central oscillator regulates the physiology of the whole organism. Examples of such outputs include the sleep–wake cycle and the daily activity of rodents as measured by running on wheels (see text below). The central oscillator is under the influence of the input, which are factors that can reset or "entrain" the central oscillator. The most notable example of an input factor is light, which is detected by the retina of the eye and transduced to the SCN via direct neural connections. An example of entrainment is "jet lag," a condition initially experienced by international travelers when changing time zones, following which the central oscillator becomes reset or entrained to the new light–dark cycle after about a week.

Numerous studies have characterized the effects of many different environmental factors on circadian rhythms, and have led to a defined set of properties that characterize these rhythms.

Circadian Properties Common to All Species

Circadian rhythms display properties that are conserved across species. Mammalian circadian clocks are characterized by the same properties displayed by clocks in all organisms:

1. First and foremost, they cycle with close to a 24-hour period, although there is variability in period length among different species.
2. As mentioned above, mammalian circadian clocks can be "entrained" by environmental cues. Indeed, a single episode of environmental perturbation is enough to shift the phase of the circadian clock. For instance, a brief light pulse given to an animal during the normal "night" time will shift the phase of the circadian rhythm. If the light pulse is given early in the night, the phase will be delayed, and if given later in the night, the phase will be advanced. Moreover, varying the intensity of light to which an animal is exposed affects the freerunning period of the animal. Other environmental factors that phase-shift circadian rhythms include social cues and regulated feeding times.
3. Mammalian circadian clocks are endogenous. In other words, they will persist in the absence of environmental cues, called "freerunning" conditions. This property is demonstrated in mammals by the fact that the circadian rhythm continues even when the organism is housed in constant darkness.
4. Although mammals are able to regulate their body temperature, studies performed with cultured mammalian cells indicates that mammalian rhythms display the requisite property of temperature compensation, in that they are able to maintain a constant period over a wide temperature range.

Physiologic Aspects

Circadian clocks control multiple aspects of physiology. Although we have established rules for circadian rhythms in mammals, we have yet to discuss what activities are actually under circadian control. These output activities allow us to observe the clock in action, and to determine the effects of environmental conditions and genetic mutations on clock function. In truth, it is almost harder to find a physiologic output that does not show a daily variation than one that does. Of note, the word "daily" is used here because in many instances, the observed rhythm has been studied only in the presence of environmental changes, such as light–dark cycles. Therefore, until studies are performed under constant conditions, we should reserve use of the word "circadian." Nevertheless, there are some well-studied examples of circadianly controlled physiologic outputs.

Rest–Activity Cycle

The rest–activity rhythm is the classic example of a mammalian circadian rhythm. Mammals show sustained bouts of activity and inactivity (rest) that follow a circadian rhythm. Normally, these bouts correspond to times when the animal is either asleep or awake. However, the rest–activity rhythm is separate from the sleep–wake cycle. The best example of a rest–activity rhythm is that of wheel running by rodents kept in cages, also called a *locomotor activity rhythm*. Rodents maintained on a light–dark cycle will immediately begin running on a wheel on lights off, and they will stop running on lights on. This activity cycle shows a very precise rhythm of 24 hours, and continues in constant darkness although with the endogenous periodicity of the animal rather than the imposed 24-hour period (Fig. 5.2). As discussed below, this rhythm is useful for the identification of circadian mutants.

Sleep–Wake Cycle

The sleep–wake cycle is probably the most obvious of the circadian rhythms in mammals. As noted above, this cycle is separate from rest–activity rhythms. Under some conditions, mammals will sleep when they are normally active and may be awake when they normally rest. Exciting work is yielding insight into the genetic basis of sleep regulation and is discussed in more detail later in this chapter.

Temperature

Although mammals are warm-blooded, their core body temperature does vary with the time of day. The peak is normally during the animal's active phase, and the trough is during the rest period. Generally, the difference is only a few degrees centigrade, but it is reproducible, is endogenous and can be entrained, making it a true circadian rhythm.

Endocrine Function

The levels of numerous hormones in mammalian serum vary with defined daily

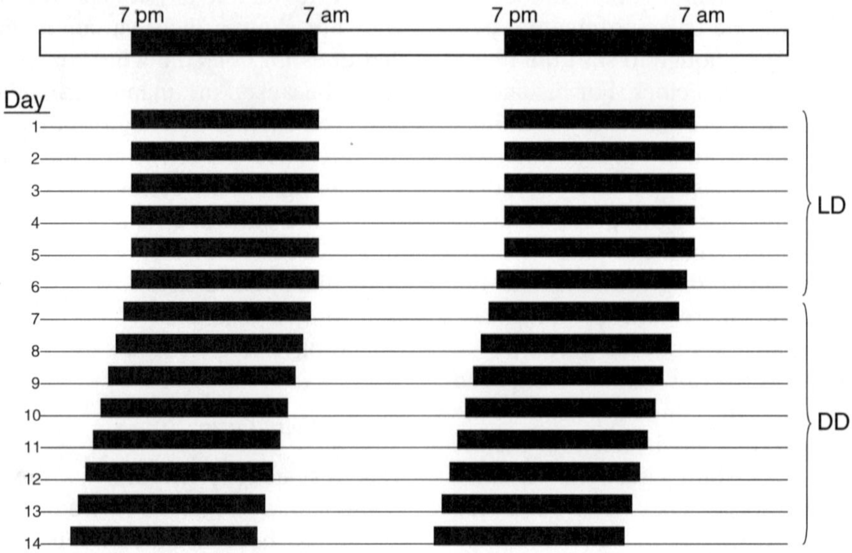

Figure 5.2. Schematic of activity rhythms of a mouse as measured by running on a wheel. The bar at the top shows when lights are on (light bar) and off (dark bar). As indicated, the lights come on at 7 am and turn off at 7 pm. This is a standard 12-hour light–dark (LD) cycle. Activity of the mouse is indicated by the dark bars for each day. Note that the data are double-plotted; that is, day 1 is shown on the first line, followed by day 2. The second line starts with day 2 and ends with day 3. This allows a comparison of days both horizontally and vertically. At the end of day 6, the animal is shifted from an LD cycle to constant darkness (DD). This is called *freerunning conditions*. Note that the periods of activity shift slightly in constant darkness. This is due to the fact that the endogenous periodicity of most mouse strains is slightly shorter than 24 hours.

variations. Some of the best studied are melatonin and glucocorticoids. These are discussed further below. Also, aldosterone and renin, responsible for salt and water retention, are circadianly regulated which may account for observed circadian functions of the kidney.

Reproduction

The levels of most reproductive hormones, such as testosterone, are circadianly regulated in mammals. However, most rhythms governing reproduction are primarily infradian (>24 hours). Still, the circadian system does play a major role in regulating these infradian rhythms. One such rhythm is the estrus cycle of a rat, which is 4 days long. Each day brings defined changes in hormone levels and in the response of the female reproductive tract. The changes of each day occur at the same time of that day. For example, ovulation occurs at the same time of day each estrous cycle, so it is clear that circadian rhythms influence ovulation. Additionally, many mammals are reproductively capable only at specific times of the year. An awareness of the time of year requires measurement of day length, which again is a function related to circadian regulation.

■ LIGHT: THE PRIMARY INPUT FACTOR CONTROLLING MAMMALIAN CLOCKS

Light Detection by the Retina

Input in mammals consists primarily of light. The eye is responsible for light detection; enucleation of various mammals disrupts circadian responses to light. The retina is the part of the eye that detects light. Photic signals from the retina are conveyed to the SCN directly via a neural pathway called the retinohypothalamic tract (RHT). Additionally, an indirect neural pathway courses through the intergeniculate leaflet (IGL) (Fig. 5.3). Still, the big questions are what retinal cell, and what molecule in that cell, act to detect light and to convey this information to the clock? In other words, what is the circadian photo-

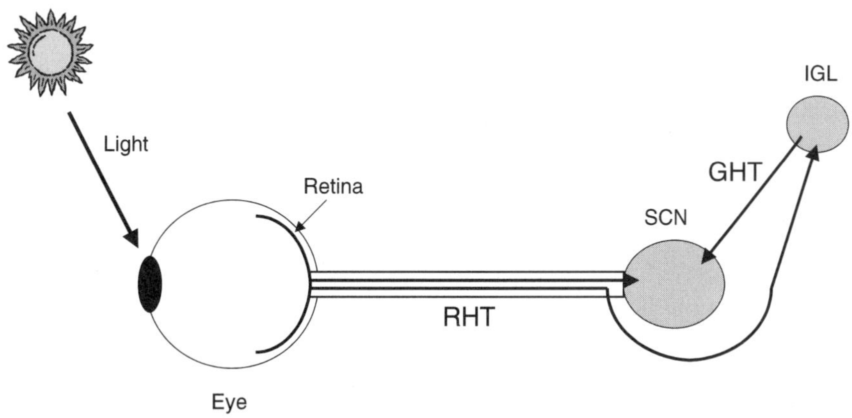

Figure 5.3. Schematic of photic input pathways to the clock. Light from the environment is detected by cells of the retina. This photic signal is transduced to the SCN through two main pathways. The first is direct, which connects the retina to the SCN via the retinohypothalamic tract (RHT). The second pathway is indirect and first connects to the intergeniculate leaflet (IGL), which in turn connects to the SCN via the geniculohypothalamic tract (GHT).

receptor? With respect to vision, there are two types of photoreceptor cells in the retina: rods and cones. Rods are the primary photoreceptor and contain the photopigment rhodopsin. Rods are responsible for vision in dim light. Cones contain the photopigment opsin. Cones are responsible for color vision and for vision in bright light. Frustratingly, neither rods nor cones are necessary for circadian photoreception.

Rods and Cones

Genetic mutations are useful in studies directed at identifying the circadian photoreceptor. These mutations have allowed us to exclude certain candidates but unfortunately have not fingered a culprit yet.

Rods

The naturally occurring mouse mutant, retinal degeneration (*rd*), carries an autosomal recessive mutation that results in the complete loss of rods by 10 weeks of age. These mice also lose a majority of their cones, and are blind. Despite these losses, *rd* mice are normally responsive to light with respect to circadian responses. Another natural mutation, retinal degeneration slow (*rds*), also leads to massive rod and cone loss, but, like *rd*, does not appear to affect circadian responses to light. Additional information comes from studies using mice that express the *rdta* transgene. This transgene contains the diphtheria toxin gene under the control of the human rhodopsin promoter. These *rdta* transgenic mice have a massive early loss of rods coupled with a slow degeneration of cones. Despite these losses, *rdta* mice are still responsive to light. Taken together, these studies indicate that rods are not solely responsible for circadian photoreception.

Cones

There are no naturally occurring mutant mice that completely lack cones. However, mice that lose their cones were produced using transgenic methodology. These mice express the *cl* transgene, which contains the diphtheria toxin gene under the control of the human opsin promoter. As is the case with the rodless mutant mice, *cl* mice do not show deficits in circadian photoreception. Finally, by breeding *cl* mice with either *rd* or *rdta* mice, mouse lines that completely lack both rods and cones were produced. However, these double mutant lines are normal in circadian photoreception, further underlining the conclusion that neither rods nor cones are necessary for circadian photoreception.

Novel Photopigments

Because neither rods nor cones are the answer to the circadian photoreceptor quest, it is likely that other retinal cells contain novel photopigments that are the true circadian photoreceptor. A number of interesting candidates have been described, but, to date, none have been specifically linked to circadian photoreception in mammals.

Cryptochrome

One interesting development was the discovery of mammalian cryptochrome (*cry*) genes. These genes code for putative flavin binding proteins, which, in other systems, are responsive to blue light and to UVA light. This type of photopigment was originally identified in plants, but homologs in animals exist. In insects, there is a single cryptochrome gene that appears to represent the true circadian photoreceptor. In mammals, however, the story is not as clear because the *cry* genes are apparently part of the central clock. A more in-depth discussion of the *cry* genes is found later in this chapter.

Melanopsin

Melanopsin is a novel opsin-related photopigment that is found in a subset of retinal ganglion cells (RGCs). RGCs act primarily as relay cells that convey information from rods and cones to various parts of the brain involved in vision. Some of these RGCs connect with the SCN. It has been found that the subset of RGCs that connect with the SCN is the same subset of RGCs that contain melanopsin. Furthermore, these melanopsin-containing RGCs are responsive to light. In fact, these cells are not involved in image formation, but rather are sensitive to the level of light, called *luminance*. Tying these lines of evidence together leaves us with a small group of melanopsin containing RGCs that are responsive to luminance and also synapse with neurons in the SCN. However, mice lacking melanopsin are still responsive to light.

Of course, there may not be a single circadian photoreceptor in mammals. It is possible that this system is so important that redundancy is in place. In other words, cryptochromes and melanopsin, and possibly others, may all be involved. Indeed, mice that lack cryptochromes and rods and cones are partially impaired in their responses to light. Moreover, mice that lack rods, cones, and melanopsin completely lack photic responses. Therefore, redundancy does appear to function in mammalian circadian photoreception.

Circadian Response to Light

Light entrains the clock, and a light pulse given to an animal during the night resets the clock by either delaying or advancing the phase of the circadian rhythm (Fig. 5.4). Characterization of the signal transduction pathways responsible for light responsiveness is actively being pursued. Like many aspects of circadian biology, however, much clarification is still necessary.

Recall that photic signals from the retina are conveyed along the RHT to the SCN. The primary neurotransmitter of this pathway is glutamate. On reaching the SCN, the glutamatergic signal activates a glutamate specific receptor called the *N*-methyl-D-aspartate (NMDA) receptor. Activation of this G-protein-coupled receptor leads to an increase in intracellular calcium, which in turn activates mitogen-activated protein kinase (MAPK). Activation of this kinase leads to phosphorylation of a transcription factor called *cyclic AMP response element binding* (CREB) *protein*. The phosphorylated form of CREB is the active form. Strikingly, CREB activity is upregulated by a light pulse given during the subjective night, but not during the subjective day, indicating that CREB is part of the signal transduction pathway that mediates phase resetting. Binding sites for CREB are found in the promoters of some circadian genes that are core components of the circadian clock. These genes, called *Period-1* (*Per1*) and *Period-2* (*Per2*), are normally induced by light pulses given in the middle of the night and are discussed in further detail below (see also Fig. 5.5).

Adding complexity to the circadian response to light is the fact that both CREB activity and MAPK activity are normally circadianly regulated in the SCN. The use of antibodies specific for the phosphorylated form of CREB demonstrated that the level of active (phosphorylated) CREB is rhythmic in the SCN. The active form of CREB is at its highest level in the middle of the night. Correspondingly, the activity of CREB in the SCN is also circadianly regulated, as demonstrated using a transgenic reporter construct that is responsive to activated CREB. Expression of the reporter

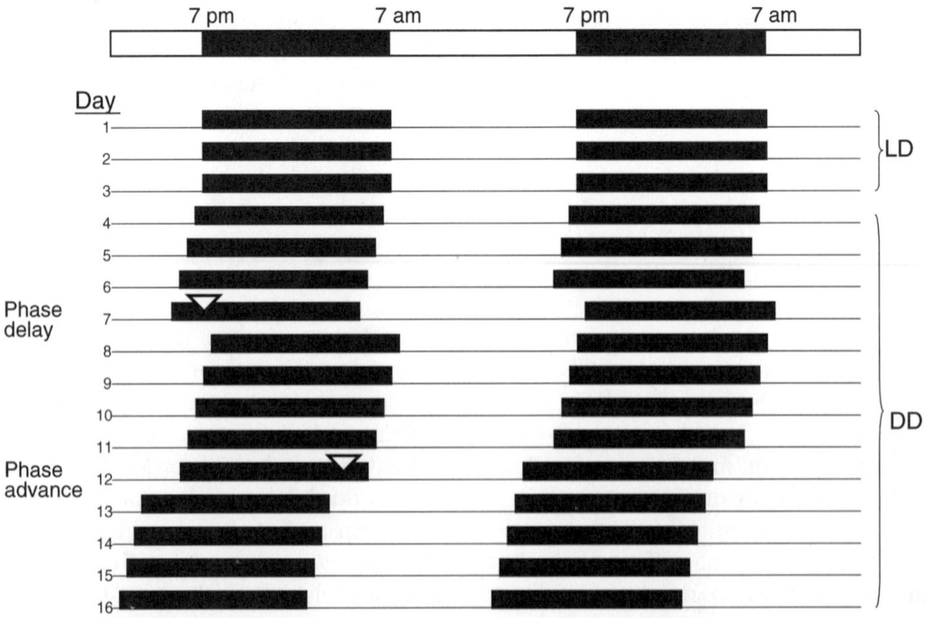

Figure 5.4. Schematic of activity rhythms of a mouse showing effects of a light pulse to generate either a phase delay or a phase advance. The mouse is first entrained to a LD cycle and then allowed to freerun in constant darkness. When a light pulse (indicated by the upside-down triangle) is administered at the beginning of the night, a phase delay occurs as indicated by the delay in the active period during the next night. The overall period stays the same, however. Similarly, a light pulse administered late in the subjective night results in a phase advance as indicated by the earlier onset of activity. Again, note that the overall period stays the same. Often at the onset of a phase change, there is a transient period of near arrhythmicity, which is not illustrated here. Presumably, this occurs while the clock is being "reset."

construct was highest in the middle of the day and lowest in the middle of the night. The lag time between the peak of active CREB levels (middle of the night) and the peak of reporter gene expression (middle of the day) is due to the fact that it takes about 12 hours to see the full effect of transcriptional activity. MAPK activation is also endogenously rhythmic in the SCN. Like CREB, MAPK can be phosphorylated, and this phosphorylation event activates the enzyme. Using antibodies specific for phosphorylated MAPK, it was found that the level of active MAPK peaks at the end of the subjective day. Note that this peak is actually out of phase with the peak of phosphorylated CREB, suggesting that additional signaling cascades may be at work.

Light responses of the clock are even more complex than indicated above because there may be more than one pathway at work. When a light pulse is given to an animal at the beginning of the night, a phase delay occurs. The pathway outlined above is concerned primarily with such delays. In contrast, a light pulse delivered toward the end of the night results in a phase advance. There is some evidence that the pathways for phase delays and phase advances are different. In phase advances, activation of the NMDA receptor leads to an increase in nitric oxide. Nitric oxide is a relatively recently described

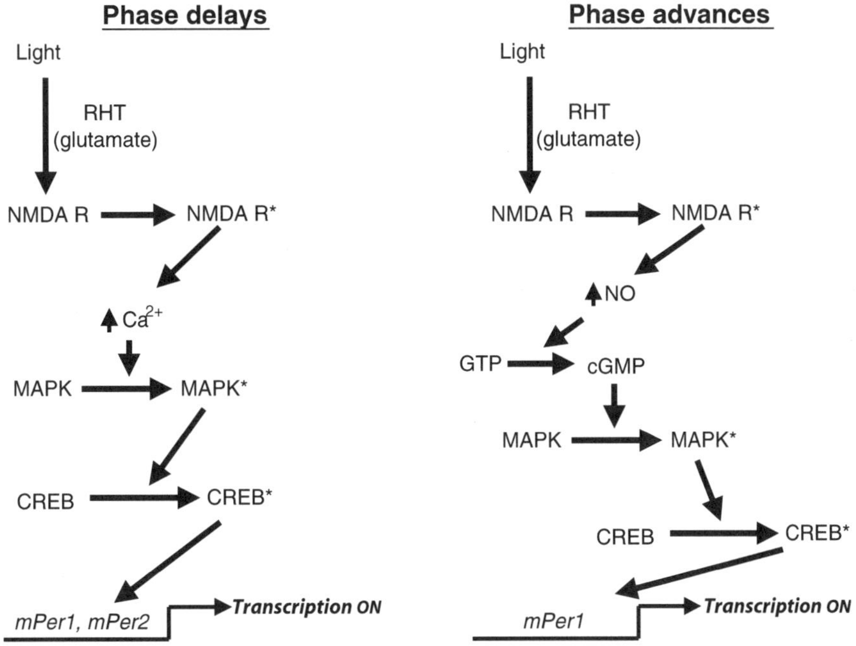

Figure 5.5. Molecular pathway for either phase advances or phase delays in response to a light pulse during the subjective night. During a phase delay, the RHT releases glutamate, which is bound by the NMDA receptor. The activated receptor (indicated by the asterisk) leads to an increase in calcium. Through a series of proteins not shown here, the calcium leads to phosphorylation and activation of MAP kinase (asterisk). Active MAPK leads to activation of the transcription factor CREB by phosphorylation (asterisk). Active CREB then induces transcription of the circadian clock genes, *Period-1* and *Period-2*. A light pulse given at the latter part of the subjective night also results in activation of the NMDA receptor. In contrast, though, at this time of night, the active NMDA receptor leads to an increase in nitric oxide (NO). NO activates an enzyme, guanylyl cyclase, which converts GTP into cyclic GMP. The rise in cGMP levels also results in activated MAPK and eventually activated CREB. However, transcription of only *Period-1* is induced. The reason underlying the dichotomy of responses is not clear, but presumably there are additional factors necessary for transcriptional activation that are available in the early night, but not in the late night.

second messenger that activates guanylyl cyclase, an enzyme responsible for the production of cyclic GMP (cGMP). High levels of cGMP activate cGMP-dependent kinase, which in turn activates MAPK. As pointed out above, activation of MAPK leads to activation of CREB. Note that both the phase advance pathway and the phase delay pathway, while different, utilize activated MAPK (Fig. 5.5). How this similar outcome leads to two different effects depending on the time of night is still an area of extensive study.

Involvement of Neuropeptide PACAP in Photic Input

Although glutamate is the primary neurotransmitter of the RHT, the neuropeptide pituitary adenylate cyclase activating polypeptide (PACAP) is also important for circadian clock function, especially with regard to input pathways. PACAP is formed in the same retinal ganglion cells that produce melanopsin, as discussed earlier. The nerve terminals of these cells synapse with SCN neurons and express PACAP in a circadian manner. Levels are highest at night and lowest during the day. There is both in vitro and in vivo evidence that PACAP functions in the clock. PACAP induces phase shifts in the SCN in much the same way as in response to light. This includes induction of the *Per1* and *Per2* genes. PACAP may actually act as a modulatory molecule because it modulates glutamate effects on the clock. PACAP appears to dampen the glutamate derived phase advance seen late at night and enhance the glutamate-driven phase delay seen in the early night. Still, precisely how PACAP figures into the input pathway is not yet clear.

There are three receptors for PACAP. All are G-protein-coupled receptors. Two of these, VPAC1 and VPAC2, bind both PACAP and another neuropeptide, vasoactive intestinal peptide (VIP), equally well. The third receptor is PAC1, which essentially is specific for PACAP. Mice lacking PAC1 were created using gene targeting technology. Experimentation with these knockout mice revealed that PAC1 is not absolutely essential for entrainment of circadian rhythms by light. When confronted with a light pulse in the early night, PAC1-deficient mice responded with phase delays greater than that observed with wild-type mice. The induction of *Per1* and *Per2* by light was not affected in PAC1-deficient mice, which may account for the fact that these mice still respond to light. The only notable deficiency displayed by the knockout mice was an inability to generate phase advances in response to a light pulse given late at night. In contrast, such a light pulse caused PAC1-deficient mice to respond with phase delays. The relatively mild phenotype of PAC1-deficient mice suggested that one or both of the other PACAP receptors, VPAC1 or VPAC2, is the key receptor for light entrainment or that redundancy among these receptors exists.

Functions of VIP and VPAC Receptors

VIP (see above), which is related to PACAP, is also implicated in photic input. Unlike PACAP, which is produced in retinal ganglion cells, VIP is produced by neurons in the SCN that receive input from the retina. These neurons synapse with other regions of the hypothalamus. VIP is implicated in the phase shifting response of the SCN to light both in vitro and in vivo. As mentioned above, VIP shares receptors with PACAP, namely, VPAC1 and VPAC2. Of these, only VPAC2 is expressed in the SCN. Indeed, in the nervous system, the SCN is the area of highest VPAC2 expression. Moreover, expression in the SCN is circadian, as it is highest during the day and lowest at night. The importance of VPAC2 in photic input was established by the creation of transgenic mice that overexpress the human homolog of VPAC2. These transgenic mice reset their activity rhythms more quickly in response to an advance in the light–dark cycle. The creation of a knockout mouse lacking VPAC2 validated the function of this receptor in photic input. Mice lacking VPAC2 fail to induce *mPer1* and *mPer2* in response to light; therefore, it is clearly necessary for the response to photic signals. VPAC2 is involved in more than just photic input, though. Evidence indicates that it is necessary for maintaining core circadian activity (see text below).

Other Retina–SCN Pathways

Light is not the only input onto the central oscillator in the mammalian SCN. As mentioned above, other pathways indirectly connect the retina to the SCN. Because these pathways have relay stations, they do not solely transmit information from the retina. These pathways integrate information from other parts of the brain and thus also transmit nonphotic input to the SCN. One pathway from the retina to the SCN first synapses in the IGL. The IGL innervates the SCN via the geniculohypothalamic tract (GHT); the main neurotransmitter is neuropeptide Y. The IGL also functions in mediating nonphotic input to the SCN. Induced physical activity shifts the phase of an animal's circadian rhythm, and the IGL is necessary for this activity-induced shift. Another indirect pathway from the retina to the SCN courses through the raphe nuclei. Serotonin is the primary neurotransmitter of this pathway. The serotonergic input may have a modulatory effect on input from the RHT and may also function in nonphotic phase shifting. Finally, hormones have input onto the SCN; the most widely studied is melatonin. Melatonin is produced cyclically by the pineal gland under the control of the SCN, and is regarded primarily as an output molecule (see next section). However, melatonin can feed back and influence the SCN, so it is mentioned here. Still, very little is known about the mechanisms underlying any nonphotic input; therefore, they are not discussed further.

THE INTRACELLULAR MAMMALIAN CENTRAL OSCILLATOR

Housing of the Mammalian Central Oscillator in the SCN

As mentioned above, the SCN is home to the mammalian central clock. This structure is located just above the optic chiasm in the anterior hypothalamus. Several lines of evidence established the SCN as the central pacemaker of mammals. Neuroanatomical studies mapped the abovementioned connections from the retina to the SCN, and other research demonstrated strong circadian rhythms of glucose utilization and neuronal firing rate in the SCN. Additionally, ablation of the SCN in rats resulted in a loss of circadian rhythms as measured by adrenal corticosterone secretion, drinking behavior, and locomotor activity. Moreover, transplanting SCN cells from one animal into an animal with an ablated SCN reversed the loss of circadian rhythms. In fact, the period of the rhythms of the transplanted animal always followed those of the donor.

Intracellular Rhythms of the Core Circadian Oscillator

A circadian rhythm of neuronal firing has been demonstrated in the SCN using brain slices. However, many elegant studies have established that the basis for rhythm generation is a cumulation of intracellular rhythms from individual SCN neurons. Reppert's group showed that dissociated rat SCN neurons plated on microelectrode arrays display sustained rhythms in their electrical activity. These findings were extended using a mutant hamster line called *tau*, which has a shortened circadian period of wheel-running and a mutant mouse line, *Clock*, which has a long period in the heterozygous state and becomes arrhythmic when homozygous. Dissociated SCN neurons from these mutants display electrical activity that correlates well with the altered behavioral rhythms. The *tau* mutant cells had shortened periods of electrical activity. Heterozygous *Clock* mutant SCN cells had long periods, and homozygous cells were arrhythmic. Results with these mutant animals indicated that the circadian clock is cell-autonomous.

Notably, the circadian rhythms of individual cells are more variable than the behavioral rhythms. It is hypothesized that the periodicity of the overt rhythm is the average of the periods displayed by individual neurons. This is supported by elegant studies utilizing chimeric mice in which the SCN are composed of a mixture of cells from two different mouse strains of different period lengths. The period of the rest–activity rhythm depends on the relative ratio of the two cell types. How the individual cells communicate with each other to generate a unified rhythm is unknown at this time, and is one of many fascinating, unanswered questions in the field.

Multifaceted Synchronization of SCN Neurons

How the individual neurons of the SCN are synchronized is unclear. The primary neurotransmitter of these neurons is γ-aminobutyric acid (GABA). GABA normally acts to inhibit neuronal activity. Within the SCN, the exact function of GABA is unclear. One report suggested that GABA excites SCN neuron firing during the day, and inhibits such firing at night. Unfortunately, this finding has been difficult to replicate, and it is currently believed that GABA only acts in an inhibitory manner in the SCN. Still, work with dissociated SCN neurons in culture suggests that GABA is important for intercellular communication. GABA synchronizes the phase of firing rate rhythms in SCN cells in culture. Also, GABA treatment can phase-shift the electrical activity of these cells. The direction of the phase shifting, either advancing or delaying, is dependent on the circadian time.

Because the SCN is composed of so many closely apposed cells, it is possible that direct cell–cell contacts may contribute to intercellular communication. The SCN expresses a specialized neural cell adhesion molecule (NCAM) that contains a high level of polysialic acid (PSA-NCAM). NCAMs are normally important for cellular contact and intracellular signaling in neurons. The presence of polysialic acid moieties on NCAM is found primarily in the developing nervous system and in plastic areas of the brain. Therefore, PSA-NCAM is postulated to allow changes in tissue architecture. The importance of PSA-NCAM in the SCN was demonstrated using mice deficient for this molecule. Under freerunning conditions, these mice show a significantly shorter period of activity rhythm that degenerates into arrhythmicity. Additionally, enzymatic removal of PSA from the SCN of adult animals similarly shortens the freerunning period.

GABA and PSA-NCAM are just two examples of molecules necessary for synchronization of clock neurons. These indicate that both secreted factors and direct intercellular contact are involved in establishment of the unified core clock. A number of other factors, both secreted and nonsecreted, are suggested as being important for this clock. The evidence for the majority of these is incomplete; therefore, they are not discussed further.

A SCN Cell Line

Utilizing rat fetal precursors of the SCN, a SCN cell line was established in 1999 by immortalization with the adenovirus E1A gene. This cell line demonstrates many properties found in the intact SCN, including circadian rhythmicity of glucose utilization and neurotrophin expression. Moreover, transplantation of this cell line into the brains of rats with SCN lesions led to the reestablishment of circadian locomotor activity. In contrast, a fibroblast cell line, which can be induced to display circadian expression of clock genes (see section below), is unable to reestablish such rhythms. Also, the SCN cell line is able to drive circadian rhythms of glucose uptake and circadian gene expression in a cocultured fibroblast cell line. Therefore, much like the intact SCN, this cell line can regulate rhythms of other cells via secreted factors. Hopefully, this SCN cell line will be beneficial in understanding the intra/intercellular nature of the circadian clock.

■ THE CENTRAL OSCILLATOR IS COMPOSED OF A FEEDBACK LOOP

It is clear that phenomenological research in chronobiology has been quite successful in leading to an understanding of both the properties of, and the anatomical basis for, mammalian circadian rhythms. However, since the early 1990s molecular biology and genetics have provided insight into the molecular mechanisms underlying the central clock in mammals. The molecular basis of the mammalian circadian clock is a cyclic transcription/translation feedback loop that is based on the one originally described in drosophila (Fig. 5.6). Two genes, *period* (*per*) and *cryptochrome* (*cry*), are cyclically transcribed over the course of 24 hours; the peak transcript levels occur in the middle of the day and the trough occurs in the middle of the night. Translation is coupled to transcription; therefore, as the level of transcript rises during the day, more and more protein is produced. As the levels of these proteins rise in the cell, they heterodimerize and translocate to the

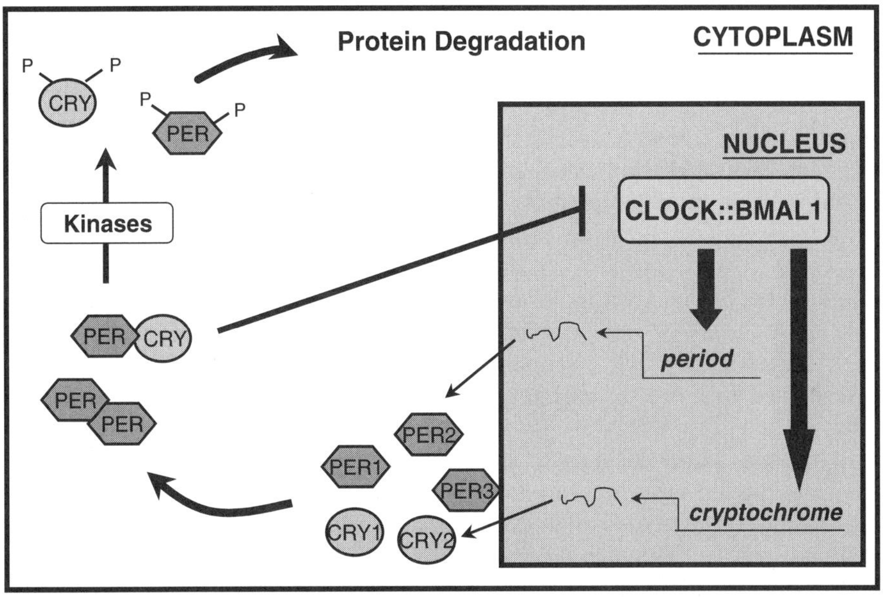

Figure 5.6. Schematic of the feedback loop. The feedback loop consists of four main components: PERIOD (PER), CRYPTOCHROME (CRY), CLOCK, and BMAL1. CLOCK and BMAL1 are positively acting transcription factors. They heterodimerize and activate transcription of the *Per* and *Cry* genes. There are actually three isoforms of PER and two isoforms of CRY as indicated in the figure. The PERs form both homodimers and heterodimers and also heterodimerize with the CRYs. On dimerization, PER and CRY translocate to the nucleus and inhibit the activity of the CLOCK:BMAL1 heterodimer. Evidence suggests that CRY is the dominant repressor. Suppression of CLOCK:BMAL1 activity causes transcription of *Per* and *Cry* to be shut off. The feedback loop is not permanently disrupted, however, because the PER and CRY proteins are targeted for degradation. The targeting signal is probably phosphorylation. The PERs are phosphorylated by casein kinase Iε. The CRYs are also phosphorylated, although currently the kinase responsible for this is unknown. On degradation of PER and CRY, transcriptional repression is relieved and the cycle begins anew. The end result of the feedback loop is daily oscillation of both mRNA and protein levels of the core components. The exception is CLOCK, which is expressed at the same level throughout the day.

nucleus. In the nucleus, the PER:CRY heterodimer blocks the activity of two transcription factors, CLOCK and BMAL1. In the absence of PER and CRY, CLOCK and BMAL1 form a heterodimer that activates transcription of the *per* and the *cry* genes. In other words, when levels of PER and CRY are high, transcription of *per* and *cry* is repressed. Transcription is not permanently halted; however, because the PER and CRY proteins are eventually targeted for degradation, which decreases the protein levels in the cell. When the levels fall enough, the transcriptional repression is relieved and the cycle begins anew. Notably, in mammals, there are actually three homologs of *per*: *Per1*, *Per2*, and *Per3*; and two homologs of *cry*: *Cry1* and *Cry2*. The reasons for having multiple homologs are not completely clear at this time, and are discussed further below. An important point is that the central clock machinery is apparently at work in many tissues, not just the brain; that is, cyclic expression of *Per*, *Cry*, and *Bmal1* is found in organs such as liver and kidney. *Clock* expression is constant in both the SCN and peripheral organs. The function of such peripheral clocks is discussed in more detail later in this chapter.

As noted above, both PER and CRY proteins are targeted for degradation, which allows the molecular circadian cycle to be continually reset. The signal to the cell's degradation machinery is hypothesized to involve protein phosphorylation. Research has shown that casein kinase Iε (CKIε) is responsible for phosphorylation of PER. CKIε and its role in the circadian clock are discussed in further detail below. Regarding the degradation of CRY, however, the story is not as clear as with PER. Interestingly, in Drosophila, TIMELESS (TIM), not CRY, partners with PER to affect transcriptional repression, and glycogen synthase kinase (GSK) phosphorylates TIM (Chapter 3). As yet, it has not been determined if GSK has a similar role in mammals.

■ THE *Clock* GENE

Clock was the first cloned mammalian circadian gene.

A Forward Genetic Screen

The murine *Clock* (for circadian locomotor output cycles kaput) gene was the first mammalian circadian gene to be cloned. Its isolation is a beautiful example of the power of forward genetics in a mammalian system. The strategy to identify this gene was based on that of Konopka and Benzer, who identified the first circadian gene, *period*, in Drosophila. The basic idea is to induce genetic mutations in the germ cells of an organism by exposing it to a chemical mutagen. The exposed organisms are then bred to nonexposed organisms. If a mutation occurs in a gene important for circadian rhythms, the offspring that inherit that mutation will show alterations in circadian behaviors. Because the offspring will only be heterozygous for any induced genetic mutations, this type of screen identifies dominant mutations. Takahashi and his colleagues undertook such an ambitious screen, and published their results in 1994. In their original screen, C57BL/6J male mice were exposed to the mutagen, *N*-ethyl-*N*-nitrosurea (ENU). These mice were bred to nonmutated female mice, and the progeny were screened for alterations in circadian behavior, specifically wheel running. Laboratory mice normally are most active at night and show sustained wheel-running activity with an average period of slightly less than 24 hours. Approximately 300 progeny animals were screened, and the mean period of wheel-running activity was 23.7 hours. However, the period of one animal slowly lengthened over a month and stabilized at 24.8 hours. Breeding of this animal showed that the long period phenotype was inherited as an autosomal, semidominant mutation. Homozygous *Clock* animals have a mean period of 27.3 hours when first moved to constant darkness, but become arrhythmic over time (Fig. 5.7). These data indicated that *Clock* is a core component of the mammalian circadian system. The successful isolation of *Clock* by positional cloning proved that such a technique could be used to find behavioral genes in mammals.

A PAS-Containing Protein Revealed in *Clock* Cloning

The discovery that *Clock* represented a single gene allowed its location to be mapped by linkage analysis. Genetic crosses

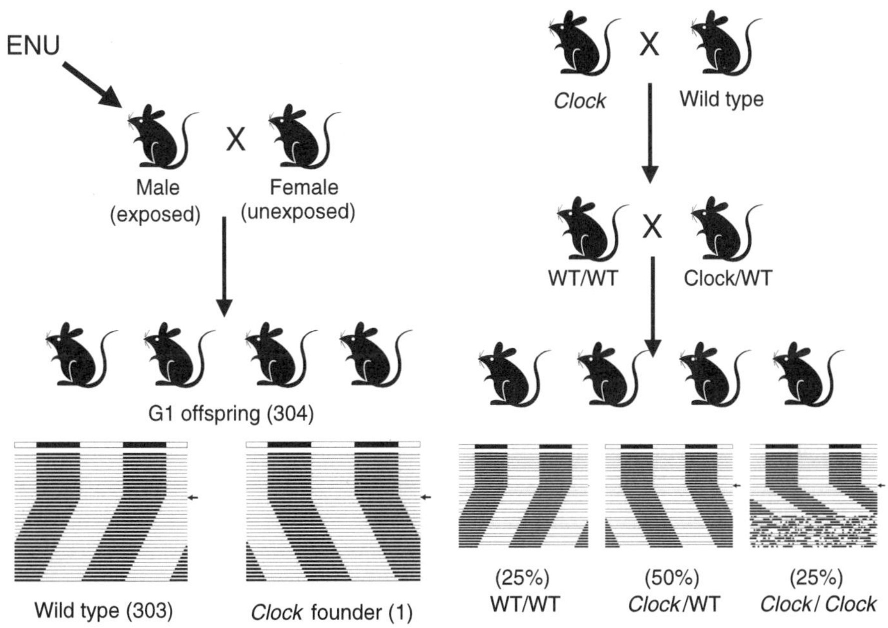

Figure 5.7. Mutation strategy used to generate *Clock* mutant mice. Male mice were exposed to the mutagen, ENU. These mice were then bred to unexposed female mice, and the offspring (G1) were screened for disorders of circadian locomotor activity rhythms. The vast majority of mice had a normal period length of 23.7 hours, but one mouse had a period of approximately 25 hours. Schematics of the locomotor activity records are shown. The arrow indicates the point where the mice are moved from a light–dark cycle to constant darkness. The long-period mouse was bred to wild-type females and the offspring were screened for locomotor periods. Approximately half of these offspring had wild-type periods, and half had the long-period phenotype, indicating that the mutation was a single-gene autosomal dominant. The long-period phenotype animals were interbred to generate homozygous animals. The offspring fell into three groups phenotypically with respect to locomotor activity period: wild type (25%), parental (50%), and an even longer period of approximately 28 hours (25%). These represented homozygous wild type, heterozygotes, and homozygous *Clock* mutation, respectively. Such data indicates that the *Clock* gene is a single, semidominant gene. Notably, the homozygous *Clock* mutant animals degenerate into arrhythmicity after ~2 weeks in constant darkness.

mapped the *Clock* locus in relation to simple sequence length polymorphisms (SSLPs) and assigned the locus to chromosome 5. The actual cloning of *Clock* was a bit more difficult and was reported in a pair of landmark papers in 1997. Two complementary approaches were used: standard positional cloning and transgenic rescue. With regard to positional cloning, first, high-resolution genetic mapping using SSLPs narrowed the region of interest on chromosome 5 to approximately 400 kb (thousand base pairs). This area was further subdivided by constructing yeast artificial chromosomes (YACs) and bacterial artificial chromosomes (BACs). Shotgun sequencing of these constructs was used to identify candidate genes. Strikingly, one of these candidates had sequence similarity to the PAS domain. This domain is found in many proteins, including certain basic helix–loop–helix (bHLH) transcription factors. More importantly, the PAS domain had already been found in the Drosophila circadian protein PERIOD. The link to a known circadian protein was quite compelling, and certainly suggested that this gene represented the *Clock* gene. Additionally, northern blot analysis of hypothalamus RNA showed that *Clock*/*Clock* homozygous mice expressed a lower level of this candidate gene transcript than did wild-type mice, further strengthening the hypothesis that this candidate was the actual *Clock* gene.

As a complementary approach to positional cloning, BAC clones covering the 400-kb area were isolated and used to make transgenic mice. Lines of mice that were homozygous for the *Clock* mutation, but also contained individual BAC transgenes, were generated. One of these mouse lines had a wild-type phenotype with respect to locomotor activity rhythms. In other words, this BAC transgene was able to rescue the *Clock* phenotype. Therefore, the actual *Clock* gene had to be contained within this BAC clone. This BAC clone contained only three candidate genes. One of these corresponded to the bHLH-PAS protein that was identified by positional cloning. Therefore, it was clear that this gene was able to rescue the *Clock* mutant phenotype, and proved that this candidate gene indeed was the *Clock* gene.

Sequencing of the entire candidate *Clock* gene showed it to be quite large. It spans 100 kb of genomic DNA and contains 24 exons. The open reading frame encodes a protein of 855 amino acids with a glutamine-rich carboxyl terminus and a bHLH domain common to many transcription factors. Sequencing of this gene in *Clock* mutant mice showed an A-to-T transversion at the splice donor site of intron 19. This mutation results in a splicing change that removes exon 19 and therefore leads to a loss of 51 amino acids in the glutamine-rich region. These groundbreaking experiments proved that this PAS-containing protein was the product of the *Clock* gene, and fully established the use of positional cloning to isolate mammalian clock genes. Subsequent work placed CLOCK into the transcriptional translational feedback loop that makes up the central oscillator.

■ BMAL1

BMAL1 binds to *Clock* and regulates circadian transcription.

A Dominant Negative Clock

Given the fact that the CLOCK protein had domains common to many transcription factors, such as the bHLH domain and the glutamine-rich region, it was hypothesized that CLOCK functioned in the circadian clock by controlling transcription. Indeed, cyclic expression of *mPer1*, *mPer2*, and *mPer3* is blunted in *Clock* mutant mice, thus providing further evidence that CLOCK controls transcription of circadian genes. Furthermore, since CLOCK contains a PAS domain for protein dimerization, the obvious experiment was to search for the CLOCK partner. The first partner isolated was BMAL1 (brain–muscle ARNT-like protein 1), also known as MOP3 (member of PAS superfamily 3). Strikingly, this protein is also a bHLH transcription factor that contains a PAS domain.

Members of the family of bHLH transcription factors bind as dimers to the DNA sequence, CACGTG, also known as an *E box*. Interestingly, at the time of CLOCK and BMAL1 isolation, E boxes were beginning to be found in the promoter regions of many circadian genes. Moreover, extensive analysis of the Drosophila *period* gene, which was the best characterized clock gene, revealed that these E boxes were necessary for proper transcription. The hypothesis was that PAS-bHLH proteins bound to the E boxes of circadian gene promoters and regulated transcription of these genes. This hypothesis was confirmed by the demonstration that a CLOCK:BMAL1 heterodimer bound to an E box found in the promoter of the Drosophila *period* gene. Moreover, the CLOCK:BMAL1 heterodimer activates transcription of a reporter gene fused to the promoter sequence of the *mPer1* gene, which also contains E boxes. The establishment of transcriptional activity for CLOCK nicely explained the observed phenotype of *Clock* mutant mice. Recall that the *Clock* mutation results in a deletion of 51 amino acids from the glutamine-rich carboxyl terminus. Such regions often act as transcription activation domains. Therefore, it appears that the mutant CLOCK protein is able to heterodimerize with BMAL1 and bind to the E box, but this heterodimer cannot activate transcription. In other words, the protein product of the *Clock* mutation acts as a dominant negative (Fig. 5.8).

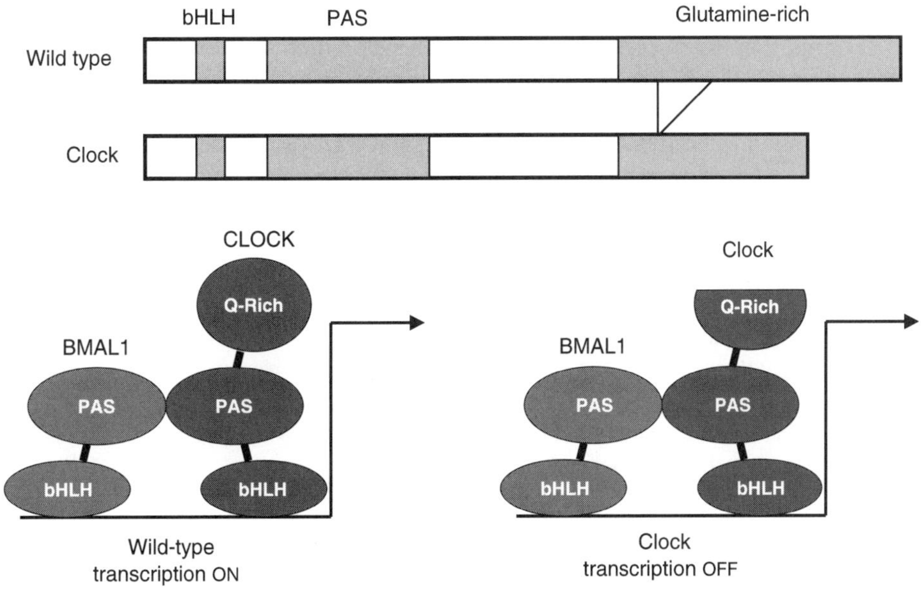

Figure 5.8. The *Clock* mutation results in a dominant negative transcription factor. The *Clock* mutation leads to alternative splicing, resulting in a protein that lacks exon 19. This exon is part of the glutamine-rich region that is important for transcriptional activation. CLOCK normally partners with BMAL1 by direct interaction of the PAS domains. The CLOCK:BMAL1 heterodimer is transcriptionally active. The CLOCK mutant protein can also heterodimerize with BMAL1. However, this complex is not transcriptionally active because of the truncated glutamine-rich region.

Function of BMAL1 in Circadian Rhythms in Mice

The finding that BMAL1 binds to CLOCK and activates transcription of the drosophila *period* gene strongly suggested that this protein was a core component of the mammalian circadian clock. Furthermore, *Bmal1* expression is cyclically expressed in the SCN, as would be expected of a core circadian component. Additionally, the highest expression of *Bmal1* occurs during the latter portion of the night. This is the opposite phase to *mPer1* and *mPer2*, consistent with the idea that BMAL1, along with its partner, CLOCK, activates the transcription of these two genes. Additional data indicate that BMAL1 itself functions in a loop where its protein product regulates the levels of its mRNA. PER2 positively regulates *Bmal1*, resulting in the interlocking of the BMAL1 loop with the PER/CRY loop. Finally, BMAL1 also plays a role in the light input control over circadian rhythms. For example, the actual protein level of BMAL1 is reduced in response to a light pulse delivered in the middle of the night. In contrast, the *Bmal1* transcript in the SCN shows no change in level on exposure to such a light pulse. Such a result indicates that regulation of BMAL1 by light is posttranslational.

To fully characterize the role of BMAL1 in mammalian circadian rhythms, a genetic knockout in mice was created. In light–dark conditions, these mice run on wheels with regular periodicity. However, when moved to constant darkness, these knockout mice display an immediate and complete loss of circadian rhythms (Fig. 5.9). Such a result proves that BMAL1 is a central clock component. Moreover, the knockout animals have lost rhythmic expression of *mPer1* and *mPer2* in the SCN. In fact, overall expression levels of these genes are very low. These findings proved that BMAL1 is necessary for *mPer1* and *mPer2* transcription. Interestingly, rhythmic expression of other genes is also abolished. As we will discuss later in this chapter, peripheral organs also show rhythmic expression of many genes, the function of which is not at all clear. In the peripheral organs of BMAL1 knockout animals there is low-level, nonrhythmic expression of many of these genes.

The BMAL1 knockout animals had additional, although more subtle, phenotypes. For example, while the animals show rhythmic behavior in light–dark conditions, the phase of activity onset is variable. Also, a good portion of the activity occurs during the lights-on time. These phase and activity disturbances indicate that BMAL1 knockout mice have difficulty in responding to light, which suggests that BMAL1 does play a role in entrainment, as had been suggested earlier. Moreover, these mice do not respond to a light pulse in the middle of the night, either, further supporting that hypothesis.

Examination of the BMAL1 knockout animals reveals an additional phenotype. These animals show an overall lower amount of activity than do their wild-type littermates. Such a result suggests that loss of BMAL1 disrupts not only the core clock but also the clock's communication with the organism's physiology as a whole. In other words, BMAL1 may have a role in circadian output pathways, such as behavioral or metabolic outputs. Clearly, BMAL1 is one of the most important circadian proteins, but its function is still far from being completely understood.

■ THREE MAMMALIAN HOMOLOGS OF PERIOD

So far, our discussions have focused on CLOCK and BMAL1, which together activate transcription of circadian genes. However, recall that, in accordance with models derived from insects, there are negative regulators of the circadian clock; the primary ones are PERIOD and

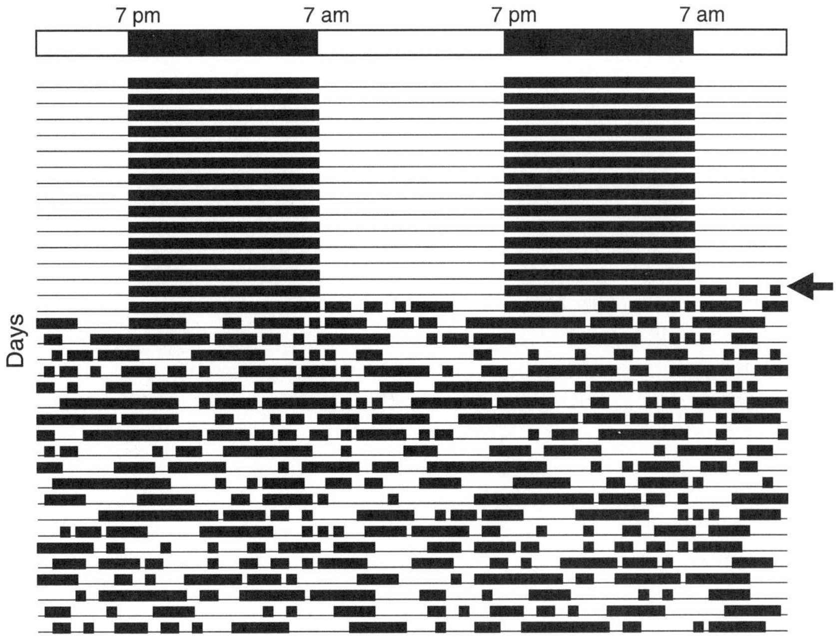

Figure 5.9. Schematic activity record of *Mop3* knockout mice. The knockout mice entrain to a light–dark cycle without difficulty and show the standard 24-hour period. However, on placement in constant darkness, the mice immediately become arrhythmic. Note that this result contrasts to the *Clock* mutant animals, which display long periods in constant darkness for approximately 14 days before becoming arrhythmic. The immediate arrhythmicity observed in the *Mop3* knockout animals indicates that this gene is indispensable for maintaining circadian rhythms in mice.

TIMELESS. Now we turn our attention to the negative regulators found in the mammalian circadian clock.

Three PER Homologs

The Drosophila *period* gene was originally identified in a mutation screen by Konopka and Benzer in 1971 and was eventually cloned in 1984. However, despite numerous attempts to clone a mammalian homolog, none was found. This lack of success led many to believe that the entire basis for the mammalian circadian clock was unrelated to that found in insects. After a decade of waiting, though, a mammalian *period* homolog was finally isolated. In fact, the field was rewarded with three homologs. These are named *Per1*, *Per2*, and *Per3* for no particular reason, except for the order of isolation. The presence of multiple homologs does indicate, not surprisingly, that mammalian chronobiology is a bit more complex than that found in insects.

Two groups originally reported the isolation of the first mammalian *period* gene, *Per1*. One group fortuitously found *Per1* when analyzing cDNAs derived from human chromosome 17. One of these cDNAs had a PAS domain. Complete cloning of this gene, which was originally called *RIGUI*, demonstrated that it was highly homologous to the Drosophila *period* gene. The murine homolog, called *m-rigui*, was also isolated. The expression of this gene was shown to oscillate in both the retina and the SCN of mice. These oscillations not only continued in constant darkness, but they were also phase-shifted by a pulse of light. These lines of evidence strongly suggested that *RIGUI* represented the functional homolog of Drosophila *period*.

The other group that first reported the sequence of *Per1* utilized a PCR-based screen to amplify products from human genomic DNA that had regions homologous to the PAS domain of Drosophila *period*. One of the amplified genomic products was used to isolate a corresponding cDNA. This cDNA had high homology to Drosophila *period*, and was called *hPER*. Using *hPER* as a probe, the mouse homolog *mPer* (identical to *m-rigui*) was also cloned. Like the first group, this second group also showed that *mPer* was expressed in the SCN, and that the levels of the transcript varied throughout the day. It was quite clear that the search for a mammalian counterpart for drosophila *period* was over.

Possible Functional Differences

The other two PER homologs were found in a much simpler manner than that used for PER1. Reppert's group knew that more and more sequence data were being deposited into a public database daily. By continuing to run computer searches of the database for homologs to Drosophila PER and hPER1, two additional genes were identified. These were called *Per2* and *Per3*, and their identification suggested that in mammals, families—rather than individual members—represented the circadian proteins.

Reasons for having multiple homologs were not at all clear at the time of discovery, and while by no means complete, some key differences are emerging. The transcript levels of all three homologs cycle in the SCN, albeit with staggered phases. Still, levels of all three are highest during the day and fall to a nadir during the night. One primary difference among the genes involves the response to light. The expression of both *mPer1* and *mPer2* is induced on exposure to light. However, *mPer1* induction is much stronger and more rapid than that for *mPer2*. These differences may have functional consequences. Mice lacking the *mPer1* gene do not phase-advance their locomotor rhythms in response to a light pulse given early in the night. In contrast, mice lacking the *mPer2* gene do not phase-delay their locomotor rhythms in response to a light pulse given in the latter part of the night. In contrast to *mPer1* and *mPer2*, expression of *mPer3* is immune to light induction, suggesting that it has no role in the input to the circadian clock.

Another key difference between the proteins is the expression pattern in various tissues. Strikingly, there is differential expression of the *Per* genes within the SCN itself. Silver and colleagues have shown that, in hamsters, one region of the SCN expresses *Per1*, *Per2*, and *Per3* in a rhythmic manner. Another region displays constant *Per3* expression and light-induced expression of *Per1* and *Per2*. These results suggest functional differences between the PER homologs. Additionally, there are differential expression patterns among nonneuronal tissues as well. As mentioned earlier, peripheral tissues rhythmically express the core circadian genes. The *mPer* genes are not exceptions. However, each of these genes is expressed in specific subsets of peripheral tissues and at different levels in these tissues. The reasons for this differential expression are unknown, but suggest that the *Per* homologs may have tissue-specific functions. Clearly, deciphering the individual functions of the different homologs will give great insight to the function of the clock in physiology.

Knockout Mice

As we indicated, one advantage that mammals have over insects is the ability to manipulate the mammalian genome to generate animals that lack certain genes. To date, the PER genes have all been targeted for deletion, and such knockout animals prove that the PER proteins are core components of the central clock. Moreover, evidence from knockout mice now suggests that mPER1 acts posttranscriptionally, and that mPER2 is a transcription regulator. Also, mPER3 is hypothesized to be an output molecule.

PER2

The first gene that was deleted was *mPer2*, and it was actually the first circadian gene deleted in such a manner. In this original knockout, the PAS domain of *mPer2* was targeted for deletion, creating an allele called *mPer2*^{brdm1}. These mice display a shorter period than do wild-type mice when measured for wheel-running activity. This shortened period gradually degrades into arrhythmicity when the mice are housed in constant darkness, much like the *Clock* mutant mice. These findings definitely established PER2 as a component of the central clock. Interestingly, this *mPer2* knockout also showed diminished RNA levels of the *mPer1* and *mPer2* genes in the SCN, indicating that it may control transcription of these genes in vivo either directly or indirectly. Surprisingly, such a result does not fit with the hypothesis that the PER proteins act as transcriptional repressors as discussed above. In fact, the opposite hypothesis is suggested; that is, PER2 has a positive effect on transcription of circadian genes. This result was repeated by another group that produced a *mPer2* knockout animal. The phenotype and the effect on transcription is the same. Therefore, while it is clear that mPER2 is a core clock component, its role with respect to transcription is not clear.

PER1

The most recent of the *Period* genes to be genetically deleted is *mPer1*. Deletion of *mPer1* was carried out by three different groups, and surprisingly, two interesting but slightly different phenotypes are found. All groups found that, much like the *mPer2* knockout animals, the circadian rhythm of *mPer1* knockout mice was shorter (approximately one hour) than wild type when measured by wheel running. Additionally, all groups found that, unlike the *mPer2* mutant animals, the *mPer1* mutant mice maintained their rhythms in constant darkness, at least for some time. Extremely interesting, though, is the fact that the *mPer1* mutant animals are quite variable in the length of their periods. Specifically, it appears that mPER1 is important for maintaining the precision of the circadian period, and not necessarily the rhythm itself. Reppert's group found that their *mPer1* knockout mice (*mPer1*^{ldc}) completely lose rhythmicity after about 2 weeks in constant darkness; however, the other two groups did not observe any loss of rhythm. Currently, it is unclear as to why there is a difference in the phenotypes of the three lines of knockout mice; how-ever, the targeting strategy for the *mPer1*^{ldc} mouse included disrupting the start codon. This mouse presumably makes no transcript. In contrast, the other two groups disrupted primarily the PAS domain of the *mPer1* gene. Indeed, in both of these mouse lines a small partial transcript is detected. It is possible that these transcripts code for partial proteins that have some residual mPER1 function.

Despite their differences, all three groups found that there is no apparent effect on circadian gene transcription in the SCN of *mPer1* knockout animals; namely, the *mPer2*, *Bmal1* and *mCry1* (see text below) genes are still rhythmically expressed at high levels. However, despite the lack of effect on transcription, the levels of these proteins are strikingly reduced in the SCN. Thus, the primary function of mPER1 may be to mediate protein stability rather than transcriptional control. This phenotype is in striking contrast to that found with the *mPer2* knockout mice, which have strong effects on circadian RNA levels. Drawing comparisons between the *mPer1* and *mPer2* knockout animals nicely demonstrates the power of gene targeting technology to allow a better understanding of specific functions of two proteins.

PER3

Finally, a *mPer3* knockout mouse has also been produced. Surprisingly, mutation of this gene has very little effect on circadian rhythms. The only observed effect is a moderate shortening (0.5 hour) of the circadian wheel-running period when the mice are held in constant darkness. It is possible that mPER3 has a more specialized role in the mammalian circadian clock, or even that it is completely dispensable for clock function. The short-period phenotype of PER3 knockout mice is similar to that of mice lacking the output gene, albumin D-element binding protein (*Dbp*) (see discussion below). Therefore, some hypothesize that PER3 acts as an output molecule possibly by regulating transcription. Complicating interpretation of this study, though, is the fact that a partial transcript is made in these mice. Therefore, similar to two of the *mPer1* knockout mouse lines, the *mPer3* knockout mouse may make a partial protein. If such a protein is at least partially active, it may hide the true null phenotype.

Multiple Knockouts

Another complication of these knockout studies is the possibility that the PER proteins may be functionally redundant. That is, if one is missing, another can take its place, and substitute for its function. The only way to address this issue is to generate "double knockout" animals, that is, animals that lack two genes (or more). These types of studies have been accomplished. Double-knockout animals of *mPer1* and *mPer2* show a much stronger phenotype than do either single-knockout types alone. These mice display an immediate loss of wheel-running rhythmicity when placed in constant darkness. This type of profound phenotype is similar to the BMAL1 mutant animals described above. In contrast, *mPer1*/*mPer3* and *mPer2*/*mPer3* double-knockout animals do not show any phenotypic differences from the single knockout of either *mPer1* or *mPer2*. This result suggests that *mPer3* is not really a core component of the circadian clock, although with the caveat that the *mPer3* knockout animal makes a partial transcript.

Dimerization

Just like CLOCK and BMAL1, the PER proteins have PAS domains for protein dimerization. Recall that in Drosophila, PER utilizes its PAS domain to heterodimerize with TIM, enter the nucleus and repress transcription by the insect homologs of CLOCK and BMAL1. If mammals were as easy as insects, the PER proteins would interact with mammalian TIM homologs and repress transcription by CLOCK and BMAL1. However, the knockout of the only known mammalian TIM homolog is embryonically lethal, making it difficult to address its function in the clock. Also, although the hypothesis exists that PER proteins function in transcriptional repression, current data suggest that only mPER2 regulates transcription.

Since all the PER homologs have PAS domains, it is logical to postulate that they can interact with themselves and with each other. Indeed, this is the case; multiple groups have shown that the PER homologs can homodimerize and heterodimerize in all combinations in both tissue culture and the SCN. Still, such studies leave us a long way from determining the function of these interactions. Indeed, the PER proteins actually accomplish their functions by heterodimerizing with the CRYs, which were mentioned above. Cryptochromes are discussed in greater detail in the next section, but for now, recall that there are two cryptochrome homologs, CRY1 and CRY2. It appears that heterodimerization with the CRYs enhances protein stability of the PER proteins. Additionally, such heterodimerization results in nuclear entry of the PER proteins, although there is also evidence of CRY-independent nuclear entry. In short, the homo- and heterodimerization of the various PER protein isoforms is important for their subcellular localization and for their function. However, precisely defining the specific functions of specific dimer combinations in the circadian clock is a work in progress.

■ CRYPTOCHROMES IN THE MAMMALIAN CORE CLOCK

Two Mammalian Homologs of Cryptochrome

As mentioned above in the section regarding light input to the circadian clock, a novel class of molecules, called *cryptochromes* (CRYs), has been recently described in mammals. CRYs are homologs of blue-light photoreceptors originally found in plants and are related to the family of proteins called *UV-dependent photolyases*, which function to repair DNA. Photolyases bind the cofactors methenyltetrahydrofolate (MTHF), which "harvests" light, and flavin-adenine dinucleotide (FAD), which is necessary for the catalytic function of photolyases (Fig. 5.14). CRYs also bind MTHF and FAD; however, they do not have photolyase activity. In plants, CRYs act as blue-light photoreceptors that regulate flowering and circadian photoentrainment of gene expression. In flies, there is one CRY homolog that functions as the actual circadian photoreceptor molecule. Continuing with the theme, though, the situation is much less clear in mammals. There are two mammalian homologs of CRY: CRY1 and CRY2. In the mouse, the two *Cry* genes are clearly important for circadian rhythms. Gene expression of both *mCry1* and *mCry2* cycles in the SCN with message levels highest at the end of the day. This is delayed from the peak of the *Per* genes' expression. Protein expression of CRY1 and CRY2 also cycles in the SCN, with the peak following approximately 3 hours after the mRNA peak. The cyclic expression of the *Cry* genes is compelling; however, proof of the CRYs' function in the clock came from knockout mice.

Mammalian CRYs Not Necessary for Circadian Photoreception

Obviously, the discovery that there were mammalian homologs of the plant and insect circadian photoreceptor set off an intense search to prove that these homologs had similar functions in mammals. Data from knockout animals indicate that this may not be the case. The first of the two genes that were deleted was *mCry2*. In light-dark conditions, the *mCry2* knockout mouse displayed a normal 24-hour period of wheel running. However, this period lengthened by about an hour when the mice were kept in constant darkness. The *mCry1* knockout animals also showed a 24-hour period of wheel running in light–dark conditions. However, in contrast to the *mCry2* knockout animals, the *mCry1* knockout animals had wheel-running periods that were shortened by about 2 hours in constant darkness. Taken together, these results indicate that the two mCRY proteins have opposite effects on the timing of the clock. Also, the fact that these animals could be driven by light–dark conditions indicates that neither of these is absolutely essential for light-driven rhythmicity.

Of course, it was possible that in a single-knockout animal, the nondeleted homolog takes over for the disrupted gene. To address this possibility, double-knockout animals were generated. Strikingly, animals lacking both mCRY1 and mCRY2 have an immediate loss of rhythms in constant conditions. This finding proves that the CRY proteins are necessary to generate core circadian rhythms. In addition, the mCRYs cannot be the sole circadian photoreceptors because these double-knockout mice do entrain to light. Furthermore, induction of *mPer1* and *mPer2* expression in the SCN in response to a nocturnal light pulse still occurs in double-knockout animals. As mentioned above, however, *rd* mice that lack both *Cry* genes show deficits in entrainment not seen in either the *rd* mice themselves or in the *mCry1/mCry2* double knockouts, leaving open the possibility that the mCRYs contribute at least partially to circadian photoreception.

Mammalian CRYs in the Core Circadian Clock

The fact that *mCry1/mCry2* double-knockout animals are arrhythmic indicates that they are part of the core circadian clock mechanism. Interestingly, in the mCRY1/mCRY2 double-knockout animals, *mPer1* and *mPer2* are no longer cyclically transcribed; the level of expression is constitutively high. This finding suggested a transcriptional repressor role for CRY1 and CRY2. Furthermore, transcription of *mCry1* and *mCry2* is attenuated in *Clock* mutant mice, suggesting that CLOCK acts as an activator of *mCry1* and *mCry2* transcription. A role for CRY in the circadian transcriptional feedback loop was further strengthened by the finding that both mCRY1 and mCRY2 block transcription of *mPer1* by CLOCK:BMAL1 heterodimers. This finding applies to both mouse and human cryptochromes. Moreover, data suggest that this repression occurs through direct interaction of the mCRYs with BMAL1. Also, both mCRY1 and mCRY2 can physically interact with all three mPER protein isoforms, particularly with mPER2. Taken together, in mammals CRYs have apparently replaced TIMELESS as the PER partner. In other words, CRYs and PERs heterodimerize and block the activity of CLOCK and BMAL1. In this model, the CRYs are the dominant repressors, and the PERs are necessary for nuclear localization.

■ THE *Tau* GENE

Tau is an example of a naturally occurring circadian mutation.

Short Circadian Periods

The isolation of *Clock* is a great example of identifying circadian components by forward genetics. The story of *tau* is an example of how nature occasionally conducts mutagenesis experiments for us. The *tau* mutation was originally isolated as a spontaneously arising mutation in a male Syrian hamster. This male had a very short period of 22 hours in constant darkness as measured by wheel running. To further characterize the circadian phenotype, this male was bred with normal female hamsters, and the wheel-running periods of the offspring were determined (Fig. 5.10). These offspring fell into two distinct groups without regard to the animals' sex. One group had a normal period of 24 hours, and the other had a period of 22 hours. Such a pattern was consistent with the *tau* mutation representing a single, semidominant, autosomal mutation. To produce homozygous animals, the offspring with the shorter period were interbred. Again consistent with *tau* representing a single autosomal

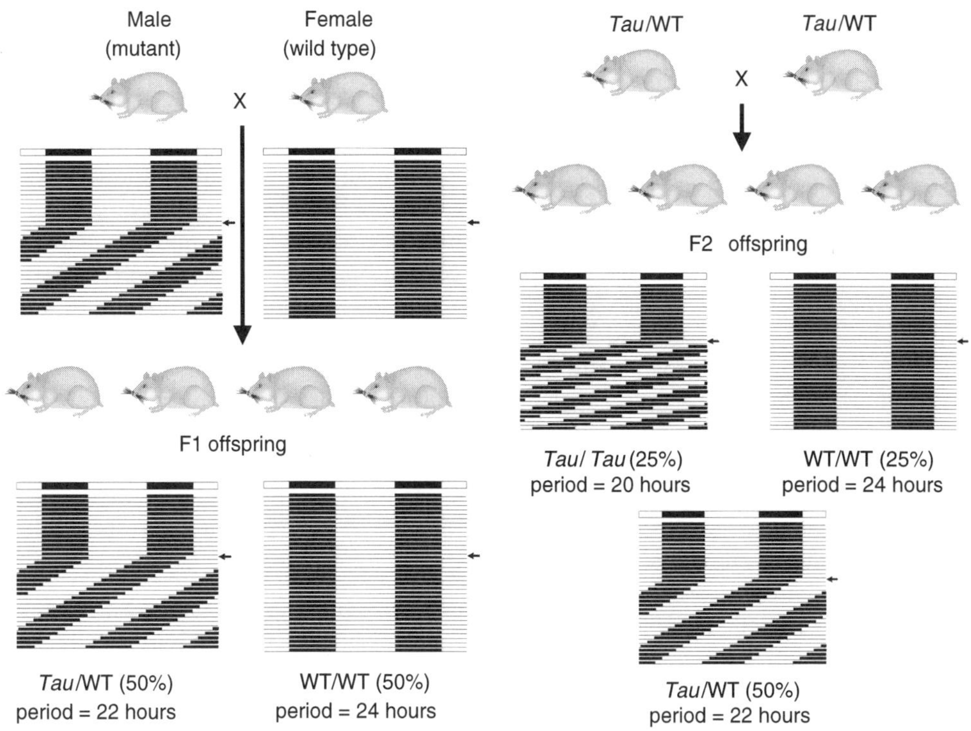

Figure 5.10. Isolation of the *tau* mutant hamster line. One male with a locomotor period of 22 hours was isolated as a spontaneously generated mutant in a group of Syrian golden hamsters. Note that the normal period for hamsters is 24 hours, so that the period stays the same even in constant darkness. Mating of this mutant hamster male, called *tau*, to wild-type females yielded a 50:50 ratio of short and wild-type period phenotypes, irregardless of sex. Such a result indicates that the *tau* mutation represents a single, dominant, autosomal gene. Mating of two heterozygote animals generated three phenotypes with respect to locomotor periods: 20, 22, and 24 hours. These were in a ratio of 1:2:1 indicative of classic Mendelian inheritance of a single semidominant gene.

gene, the phenotype of the offspring fell into three groups with respect to period length: 24, 22, and 20 hours. Presumably, the group with the shortest period represented homozygous animals. The reporting of the *tau* mutation in 1988 was extremely useful to the study of mammalian circadian rhythms. The use of hamsters homozygous for the *tau* mutation further established the SCN as the seat of circadian rhythms in mammals, as SCN transplants between wild-type and *tau* mutant hamsters showed that periodicity of wheel-running activity was conferred exclusively by the SCN. Despite its importance, the *tau* gene was not cloned until 2000, because of the lack of genetic markers in hamsters. The development of an elegant cloning strategy solved that problem.

Cloning Difficulties

Cloning of the *tau* gene was difficult because of the lack of a genetic map in hamsters. In other words, classical methods of positional cloning such as those used to isolate *Clock*, would not be useful because not enough genetic markers were known in the hamster genome. This difficulty was circumvented by using a technique called *genetically directed representational difference analysis* (GDRDA), which allows the identification of molecular markers linked to a trait without a genetic map. First, the *tau* mutant hamster strain was crossed to a wild-type hamster strain of a significantly different genetic background. The F1 generation was interbred to produce F2 animals. These crosses produced animals that were genetically heterogeneous. Phenotype analysis showed that, as expected, the F2 progeny fell into three distinct groups in a mendelian ratio with respect to the *tau* phenotype. DNA was extracted from homozygous wild type and from homozygous *tau* F2 animals, and was compared by using differential hybridization and the PCR to detect polymorphisms. Two polymorphisms that segregated with the *tau* mutation were isolated. Sequencing of hamster-derived lambda phage clones containing these polymorphisms revealed a microsatellite repeat that was also polymorphic. This microsatellite was conserved in mice, and mapped to chromosome 15. Analysis of the mouse genetic map around this microsatellite revealed a 15-centimorgan region of chromosomal synteny between mouse and hamster. According to previously established maps, this region was also syntenic with human chromosomes 8, 12, and 22. Surprisingly, the human casein kinase Iε (CKIε) gene is found on chromosome 22. This was an extremely exciting finding because mutations in the *doubletime* (*dbt*) gene, the Drosophila homolog of CKIε, alter circadian rhythms. Therefore, CKIε was a strong candidate for the *tau* gene.

To confirm that CKIε was the product of the *tau* locus, the hamster CKIε was cloned. The mouse homolog of CKIε had not been mapped yet; therefore, primers designed from the human CKIε sequence were used to isolate the hamster CKIε gene in both wild-type and *tau* mutants by the RT-PCR using RNA extracted from the SCN of these animals. Comparison of the sequences revealed that *tau* hamsters have a C-to-T transition that causes an arginine to cysteine substitution at residue 178. This residue is highly conserved among casein kinase isoforms from many species, suggesting that this transition represented the actual *tau* mutation. Although there was no apparent difference in amount of CKIε mRNA or protein in the SCN of *tau* and wild-type hamsters, the *tau* CKIε isoform had diminished kinase activity. Taken together, these data strongly indicated that CKIε was the product of the *tau* locus and that the point mutation was responsible for the observed phenotype.

Regulation of Cyclic Activity

The *tau* mutant phenotype was explained on the basis of work done in drosophila. In Drosophila, it appears that DBT phosphorylates PER, and this phosphorylation leads to PER degradation. Notably, wild-type CKIε phosphorylates both mPER1 and mPER2 in vitro. *Tau* mutant CKIε binds to both mPER1 and mPER2, despite the presence of the C-to-T transition. This is similar to what was shown for the Drosophila DBT mutant in its interaction with PER. However, as mentioned above, the mutant CKIε was much less efficient at phosphorylation. Finally, in *tau* mutant hamsters the level of *Per1* RNA in the SCN is reduced, and the period of RNA cycling is shortened. This finding is consistent with the shortened circadian period demonstrated by *tau* mutant animals.

The means by which CKIε affects the clock is still actively being studied. Some groups have used tissue culture models to show that CKIε phosphorylates both mPER1 and mPER2, and suggest that this phosphorylation event leads to protein degradation. However, there is also evidence that the phosphorylation state of the PERs affects their localization within a cell: nuclear or cytoplasmic. Of course, targeted degradation and subcellular localization are not mutually exclusive methods of regulating protein activity.

CKIε also phosphorylates mCRY1, mCRY2, and BMAL1. It appears that CKIε, the PERs, and the CRYs all form a large complex and that phosphorylation by CKIε regulates the nucleocytoplasmic localization of the dimer. Presumably, this will affect the repressor function of the PER:CRY complex. In contrast, the effect of phosphorylating BMAL1 is believed to increase BMAL1-mediated transcription. Indeed, the CLOCK:BMAL1 heterodimer is apparently always bound to DNA, but the phosphorylation state of BMAL1 is circadianly regulated. Therefore, cyclic phosphorylation of the transcription factors may regulate their cyclic activity.

OTHER POSSIBLE COMPONENTS OF THE CLOCK

So far, we have discussed the main components of the core circadian clock. These are the transcriptional activators CLOCK and BMAL1, and the transcriptional repressors, PER and CRY. As we have pointed out, in mammals, the complexity of the core clock is reasonably high because multiple homologs of PER and CRY exist. In this section, we will expand on that complexity by discussing numerous additional molecules for which there is evidence regarding a function in the mammalian clock.

REV-ERBα

The four proteins, PER, CRY, CLOCK, and BMAL1 constitute a feedback loop with the two former proteins acting as negative components and the two latter proteins acting as positive components. However, there are additional levels of control: (1) BMAL1 inhibits its own transcription and (2) PER2 actually activates *Bmal1* transcription. Work by Ueli Schibler's laboratory identified REV-ERBα as a link between the positive and negative limbs of the feedback loop by establishing its role in cyclic *Bmal1* expression and by demonstrating its regulation by PER proteins.

REV-ERBα is a member of the ligand activated nuclear receptor superfamily of transcription factors. Because its activating ligand is unknown, REV-ERBα is called an "orphan" nuclear receptor. REV-ERBα acts as a transcriptional repressor and is implicated in the regulation of adipogenesis and of metabolism. However, Schibler's group found that REV-ERBα is a transcriptional repressor of *Bmal1*. Additionally, *Rev-erbα* expression is cyclic in the SCN, and *Rev-erbα* knockout mice lack cycling of *Bmal1* in the SCN. Surprisingly, though, the circadian phenotype of these animals is subtle. Knockout mice have shortened activity rhythm period lengths, and a greater diversity in these lengths, suggesting that REV-ERB*α* is necessary for maintaining precision of the clock. Such a phenotype is in contrast to *Bmal1* knockout mice, which lack circadian activity rhythms altogether. Furthermore, data suggest that the PER proteins are negative regulators of *Rev-erbα*, and that *Rev-erbα* is activated by the CLOCK:BMAL1 heterodimer. Thus, the negative effect of BMAL1 on its own transcription would be through upregulation of its own repressor (REV-ERB*α*).

DEC1 and DEC2

Two bHLH transcription factors that do not contain PAS domains were implicated in circadian processes. These proteins, called DEC1 and DEC2 (so named because they were identified in differentiated human embryo chondrocytes) are related to the

HAIRY and ENHANCER OF SPLIT proteins of Drosophila that function in neural development by repressing transcription. Expression of both *DEC* genes cycles in the SCN and in peripheral tissues. The phase of the *DEC1* and *DEC2* RNA oscillation in the SCN lags slightly behind *mPer1*. Consistent with this fact, DEC1 and DEC2 can strongly repress CLOCK:BMAL1 activation of *mPer1* in tissue culture. Interestingly, *DEC1* expression is activated by a light pulse much like expression of *mPer1*, suggesting that the DECs may connect input pathways to the core clock. Still, there is no direct evidence yet implicating either DEC protein in the core circadian clock or in peripheral oscillators.

CKIδ

Evidence is accumulating that CKIε is not the only kinase important for circadian functions. All the core clock proteins previously described are phosphorylated in a circadian-dependent manner. Casein kinase Iδ (CKIδ) is highly homologous to CKIε and is also highly expressed in the SCN, although not rhythmically. It phosphorylates all three mPER isoforms in vitro, and is coimmunoprecipitated with CLOCK, BMAL1, mPER1, mPER2, mCRY1, and mCRY2 in vivo. In terms of circadian functions, CKIδ is probably not completely redundant with CKIε, or presumably there would not be as much of an effect on rhythms in *tau* mutant animals. Therefore, the functional significance of phosphorylation by CKIδ is still being investigated.

VPAC2

As indicated above in the section on input molecules, the PACAP receptor, VPAC2, may have a role in maintaining the core clock. In mice expressing the human VPAC2 receptor transgene, the activity rhythm was shortened to 8 hours in constant darkness. This finding indicates that the core oscillator is compromised in these transgenic mice. Additionally, VPAC2 knockout mice show a unique circadian phenotype as assessed by locomotor behavior. These mice did have a circadian cycle of locomotor activity in constant darkness, although the onset and offset of wheel running were not well defined and the overall amplitude of the rhythm was lower than in the wild type. Strikingly, however, the phase of the rhythm in the knockout animals was completely opposite that of wild-type animals in freerunning conditions. The mechanism underlying the reverse phase of activity is not clear, but the authors hypothesize that pacemakers in the brain other than the SCN may be driving these rhythms. Although there is little evidence for such a postulation, it is clear that the oscillator in the SCN is compromised. Molecularly, cyclic expression of the core genes, *mPer1*, *mPer2*, and *mCry1* is abolished in the knockout mice. Moreover, expression of these genes is at baseline (trough) levels indicating that CLOCK:BMAL1 activity is compromised. Exactly how loss of VPAC2 results in a compromised core oscillator is not clear, but Hastings and colleagues propose that intercellular communication within the SCN is important for maintaining the core rhythm. Recall that VPAC2 binds VIP in addition to PACAP. VIP is released from SCN neurons. Therefore, VIP binding to VPAC2 may regulate SCN neurons in a paracrine manner to maintain the core oscillator.

Microarray Analysis

Microarray experiments show that many genes cycle in the SCN. The development of microarray technology allows the simultaneous assessment of transcript expression of literally tens of thousands of genes. It is little wonder, then, that this technique is being applied to discover new genes that may be important for clock function. At the time of this writing, two laboratories have reported microarray analysis of SCN tissue collected from mice at various times of day. Both of these groups found that literally hundreds of genes cycled in their expression in the SCN. The types of genes could be grouped by functionality and included groups important for transcription, metabolism, protein synthesis, protein processing, and protein secretion. Additionally, a number of genes of unknown function were found to cycle in the SCN. Both of these laboratories also reported microarray analysis of gene expression in mouse livers collected at different times of day. Again, literally hundreds of genes showed cyclic expression in this tissue, but surprisingly, the number of genes that cycled in both the SCN and the liver was less than 30 in both cases. Such a finding indicated that cyclic expression is a tissue-specific property.

Although the wealth of data obtained from microarray analysis is extensive, to date it has not proved to be helpful for identifying new core clock genes. One of the major problems has been a lack of reproducibility of the data; that is, while both laboratories identified hundreds of genes that cycle in the SCN, there is surprisingly little overlap between the two reported results. Additionally, although one would expect that the limited overlapping subset of cycling genes in the SCN and the liver might reveal new core clock genes, even here there is little similarity between the two laboratories' results. In truth, the only genes that are reported by both laboratories to cycle in both the SCN and the liver are the core clock genes already mentioned. Therefore, while microarray technology surely will give us a greater understanding of circadian biology, currently we await additional experiments to fully realize these benefits.

QTL Analysis

Within the mouse species, different strains are variable in the quantitative aspects of circadian biology. For example, C57BL/6 mice have a period of 24 hours, but Balb/c mice have a period of slightly less than 23 hours, and CS mice have a period longer than 24 hours. A number of laboratories have reasoned that the allelic differences at certain genes underlie this strain variability. Determining the genes would therefore provide additional insight into how genes affect a complex behavior, such as circadian locomotor activity. Such a task is easier said than done, though. The genetic variability among strains is so high that multiple variants might affect the behavior being studies. Quantitative trait locus (QTL) analysis is a genetic technique that allows study of such polygenic effects.

QTL is really a specialized form of genetic mapping. Two disparate strains of mice that differ in a circadian measurement (e.g., period length) are mated. It is important that a large number of genetic markers exist that differentiate these two strains. With mice, this is not a problem given the large number of SSLPs that have been identified. The F1 offspring are intercrossed to generate F2 offspring. These F2 progeny are assessed for the trait being tested. Given the heterogeneous genetic makeup of these F2 progeny, there is much more variability in this trait. The genomes of these mice are scanned for the genetic markers. Utilizing computer programs, one can determine the likelihood that a trait segregates with a specific marker. Such markers are called QTLs, and represent areas of the genome that have a high probability of containing genes affecting the trait (Fig. 5.11). Candidate genes from these areas can then be tested to determine if they regulate the trait under study.

QTL analysis has been used to isolate QTLs that affect wheel-running behavior in

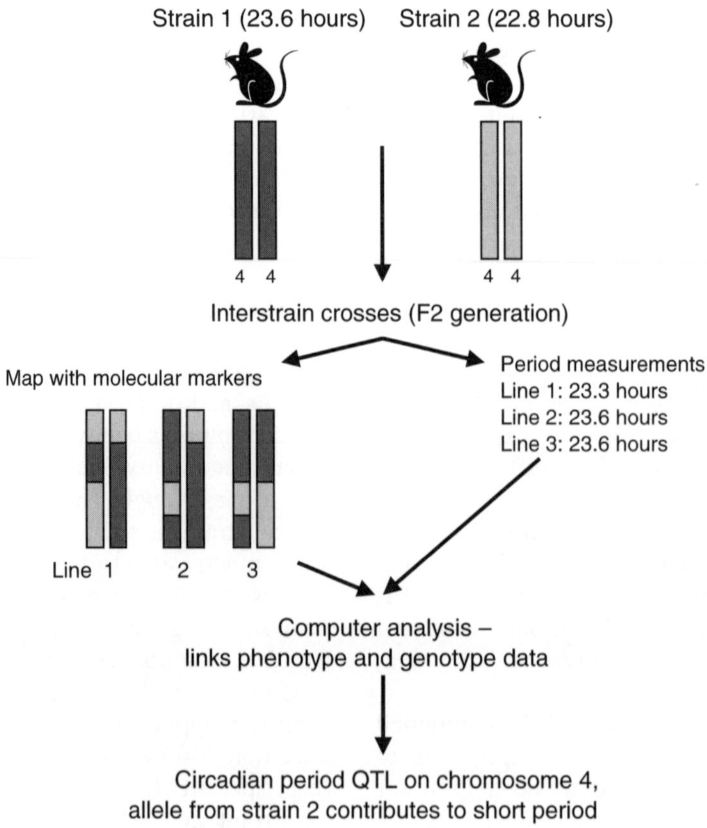

Figure 5.11. Schematic representation of QTL analysis for short-period phenotype in mice. Two mouse strains that differ in circadian period are interbred to produce an F2 generation. The genotypes of these animals are determined by mapping with molecular markers. The circadian period of each of these is determined and correlated with the genotypic data. Only chromosome 4 is shown here for both strains, as it is the one with the relevant QTL in this case. In this simplified example, an F2 animal that is homozygous for a strain 2 allele at the top of the chromosome has a shorter period than do F2 animals that are either heterozygous or lack the allele. Analysis of numerous F2 animals determines whether this correlation is significant. If so, then a QTL is established at the tip of chromosome 4. [Schematic is based on Salathia, N, Edwards, K, Miller AJ (2002): QTL for timing: A natural diversity of clock genes. *Trends in Genetics* **18**: 115–118].

mice. Few of these map to the known circadian genes. One hypothesis is that the core circadian genes are so essential to circadian rhythms that little allelic variability can be tolerated. In any case, identification of these QTLs gives researchers new directions to identify additional genes involved in regulating circadian rhythms. Hopefully, the genes associated with some of these QTLs will be cloned soon.

Other Agents

Even though we now have a good understanding of the molecular basis of the circadian clock, it is clear from both microarray and QTL studies that additional components may have subtle functions affecting the clock. Additionally, the advent of gene targeting technology has revealed new functions of many genes. Many knockout mice have disturbances in circadian rhythms, although a number of these have underlying neural defects. Some compounds for which there is experimental evidence supporting a circadian clock function are briefly mentioned below.

Serotonin

Serotonin is a neurotransmitter implicated in many behavioral processes in mammals. As mentioned above, serotonergic neurons innervate the SCN in rodents via the indirect photic input system. There are multiple receptors for serotonin, making study of the function of this molecule in circadian biology somewhat complicated. Nevertheless, many studies using various serotonin receptor agonists have shown suppression of light-induced responses at both the behavioral and the molecular level in hamsters. Moreover, depletion of serotonin enhances light-induced phase shifts in hamsters. These results suggest the hypothesis that serotonin is inhibitory for photic entrainment. This story is complicated by the fact that similar results are not found in mice. Therefore, whether serotonin has a general function in mammalian circadian rhythms, particularly with respect to input, is not clear.

GRP

Gastrin releasing peptide (GRP) is a neurotransmitter found in some neurons in the SCN, which respond to a light pulse by increasing expression of c-Fos, a protein marker of cellular activity. Also, GRP administration in vivo elicits a phase shift similar to a light pulse given at night. Such an effect is not seen in mice lacking a functional GRP receptor. GRP may function directly in the light-response pathway because GRP administration increases *mPer1* and *mPer2* expression in the SCN much as a light pulse does.

Fyn Kinase

Fyn kinase is a member of the Src family of nonreceptor tyrosine kinases that is highly expressed in the brain. Mice lacking fyn kinase show significantly longer periods of wheel-running behavior under freerunning conditions, although they can entrain to light. However, these animals have additional behavioral deficits, and the anatomic morphology of the SCN in these mice is abnormal. Therefore, effects of the fyn kinase deletion may be secondary to neuronal defects.

CaMKII

CREB is a known component necessary for the induction of *mPer1* and *mPer2* expression in response to light. As outlined above, CREB is phosphorylated in response to an increase in intracellular calcium. The calcium/calmodulin-dependent kinase II (CaMKII) is one kinase that activates CREB. CaMKII itself is phosphorylated in the SCN in response to a light pulse. Intracerebral injection of a CaMKII inhibitor into hamsters inhibited light-induced phase delays of wheel-running rhythms. Also, this inhibitor attenuated the induction of hamster *Per1* and *Per2* by a light pulse. Therefore, CaMKII may have some function in the circadian responses to light.

TIM Homologs

In mammals the role of Drosophila TIMELESS is apparently played by mCRY1 and mCRY2. In mammals, there does not appear to be a direct homolog of TIM. However, mammalian homologs of a TIM-related Drosophila gene, called *timeout*, do exist. In mice, expression of this mammalian TIM (mTIM) is found in the SCN, albeit at moderate levels. Whether *mTim* expression cycles in the SCN is unclear. Reports exist supporting both cyclic and noncyclic expression. It is not completely clear why disparate results exist, but *mTim* may produce alternative transcripts which result in two forms of mTIM. Only one of these may cycle.

Although there are some data suggesting a function for mTIM in the circadian clock, such a hypothesis has been difficult to study because the knockout of mTIM is embryonically lethal. Embryos homozygous for a disrupted *mTim* gene display cellular disorganization and necrosis as early as embryonic day 5.5, which is around the time of implantation into the uterus. Obviously, the function of mTIM is essential for early development of an embryo. Indeed, expression of *mTim* is found in many tissues in the developing embryo and the adult. More recent work suggests that mTIM may function in regulating development of tissues that undergo branching morphogenesis of tubules and ducts, such as the kidney.

In sum, the circadian function of mTIM is still open to question. The answer may lie in developing tissue-specific knockouts of this gene. Interestingly, in flies, glycogen synthase kinase (GSK) phosphorylates TIM, and regulates its levels and its subcellular localization. However, no such function for GSK has been found in mice.

A FAMILY OF bHLH PAS DOMAIN PROTEINS

As we pointed out above, BMAL1 and CLOCK are both PAS-domain-containing bHLH transcription factors. It turns out that a family of these proteins exists. Not all of these are involved in circadian rhythms, but they do seem to have functions in environmental sensing.

NPAS2

NPAS2, a new PAS-domain-containing protein, functions in the forebrain. Steve McKnight's group searched the GenBank database for previously unidentified proteins that contained PAS domains. They identified two which were selectively expressed in the mammalian brain (forebrain) and spinal cord. These were called neuronal PAS domain protein (NPAS) 1 and NPAS2. These have also been termed *member of PAS* superfamily (MOP) 5 and MOP4, respectively (Table 5.1). Expression of both of these genes appeared to correlate with development of the brain. *NPAS1* expression was first detected at the commencement of brain organogenesis, and *NPAS2* expression began shortly after birth. In the adult, NPAS2 is expressed primarily in the mammalian forebrain, an area involved in memory formation. A high degree of similarity between NPAS2 and CLOCK was noted and NPAS2 was found to activate transcription by heterodimerizing with BMAL1, much like CLOCK. This observation raised the possibility that NPAS2 is involved in circadian rhythms. Furthermore, in tissue culture cells, expression of NPAS2 and BMAL1 leads to induction of *mPer1*, *mPer2*, and *mCry1*, but repression of *Bmal1* expression. However, *NPAS2* is not expressed in the SCN, which would suggest that it does not function in the central oscillator. McKnight's group went on to show that normally, *mPer1*, *mPer2*, *mCry1*, and *Bmal1* all cycle in the mammalian forebrain. Notably, *Bmal1* cycles out of phase to the former three. A knockout mouse of NPAS2 was created and originally reported to have deficits in memory formation. However, further examination of the knockout mouse revealed that cyclic expression of *mPer2* in the forebrain is disrupted. In areas of

TABLE 5.1. Known PAS Proteins

Name	Class of Protein	Function	PAS Protein Binding Partners	Other Names
AHR	Transcription factor	Response to planar aromatic hydrocarbons (PAHs)	ARNT	
HIF1α	Transcription factor	Hypoxic response	ARNT, ARNT2	MOP1
HIF2α	Transcription factor	Hypoxic response	ARNT, ARNT2	EPAS1, HLF, HRF, MOP2
HIF3α	Transcription factor	Hypoxic response	ARNT, ARNT2	MOP7
ARNT	Transcription factor	Hypoxic response, PAH response	AHR, HIF1α, HIF2α, HIF3α	HIF1β
ARNT2	Transcription factor	Hypoxic response, neuronal development?	HIF1α, HIF2α, HIF3α, SIM1	
AHRR	Transcription factor	Inhibit AHR signal transduction	ARNT	
BMAL1	Transcription factor	Circadian clock	CLOCK, NPAS2	MOP3, ARNT3
CLOCK	Transcription factor	Circadian clock	BMAL1, MOP9	
PER1	Transcription factor	Circadian clock	PER1, PER2, PER3	
PER2	Transcription factor	Circadian clock	PER1, PER2, PER3	
PER3	Transcription factor	Circadian clock	PER1, PER2, PER3	
NPAS1	Transcription factor	Unknown		MOP5
NPAS2	Transcription factor	Peripheral clocks, memory formation	BMAL1	MOP4
NPAS3	Transcription factor	Unknown		MOP6
SIM1	Transcription factor	Neuronal development	ARNT2	
SIM2	Transcription factor	Neuronal development		
PASKIN	Serine/threonine kinase	Regulation of glycolysis?	Unknown	
MOP9	Transcription factor	Regulation of blood clotting?	CLOCK	CLIF, BMAL2
SRC1	Nuclear receptor coactivator	Potentiate transcriptional response to steroid hormones		
TIF2	Nuclear receptor coactivator	Potentiate transcriptional response to steroid hormones		GRIP1
RAC3	Nuclear receptor coactivator	Potentiate transcriptional response to steroid hormones		ACTR, pCIP, AIB1, TRAM1

the brain where NPAS2 is not expressed, *mPer2* levels continued to cycle. These data indicate that NPAS2 is involved in timekeeping, but presumably in non-SCN rather than central tissues. More evidence to support this hypothesis is reported below. Of note, little additional information has been reported regarding the function of NPAS1 since its identification in 1997.

Possible Function of MOP9 in Peripheral Tissues

Another bHLH-PAS protein was cloned by a number of different groups in 2000. This has been called *BMAL2*, *MOP9*, and *cycle-like factor* (CLIF). MOP9 is highly homologous to BMAL1. It is expressed in multiple tissues including the SCN. *Mop9* expression does not cycle either in the SCN or peripheral tissues, much like *Clock*. It is clear, however, that MOP9 functions as a transcription factor. Multiple groups showed that MOP9 interacts with CLOCK to activate expression of reporter constructs in tissue culture cells. Moreover, a MOP9/CLOCK heterodimer activates expression of plasminogen activator inhibitor 1 (PAI1). PAI1 is a circadianly expressed gene that regulates clotting activity in blood vessels. Therefore, it is possible that MOP9 is more involved with peripheral oscillators (see text below) than with the central clock, possibly with respect to hemostasis.

Other PAS-Domain-Containing Proteins

As of today (2003), there is a whole family of PAS-domain-containing bHLH transcription factors. Not all of these are involved in circadian rhythms; however, they all appear to have functions in environmental sensing. These include the hypoxia inducible factors (HIFs), which activate genes in response to low oxygen tension, and endothelial PAS domain protein 1 (EPAS1), which is found preferentially in vascular endothelial cells. Another group of bHLH-PAS proteins act as coactivators for nuclear hormone receptors. While none of the aforementioned proteins are yet proven to be directly involved in circadian biology, it may turn out that they interact with the central clock to provide a link to peripheral organ physiology.

LARGELY UNKNOWN OUTPUT PATHWAYS

Output Control by SCN

The preceding discussion demonstrates that much progress has been made towards determining the molecular bases underlying both the input to the clock and the core oscillator. However, the final part of the circadian system, the output pathway, is still somewhat of a mystery. How the SCN controls the physiology of an organism is largely unknown. Complicating discussion of this subject is the fact that outputs are measured by numerous, very disparate functions. For example, these can be hormonal (Chapter 10), such as daily variations of testosterone; or behavioral, such as the daily locomotor activities of rodents. Moreover, such outputs may be directly controlled by the SCN or may be regulated indirectly through SCN-controlled intermediaries. Such a dichotomy is borne out through both anatomical and molecular studies. Anatomically, distinct neural connections from the SCN to other areas of the brain exist. These SCN-connected areas then connect to different parts of the body and regulate physiology. While such neuroanatomical studies are quite useful, they do not contain complete answers, because surprisingly, the SCN can also control rhythms via secreted factors. Molecularly, output is defined primarily as substances that are rhythmically expressed by the

SCN. Genes regulated in this fashion are referred to as "clock-controlled genes" (CCGs), and studies indicate that some are transcriptionally regulated by the core pacemaker. In this section, we discuss specific candidates for regulating distinct output pathways, and attempt to place them in a greater context with respect to the SCN.

SCN Connection to Distinct Brain Areas

Just as there are a number of different input pathways to the SCN, there are also a number of different neuronal output pathways. Most of these pathways connect to other areas of the brain, particularly within the hypothalamus. It is the connection to these areas that causes rhythmic release of hormones that act as output molecules. Retrograde tracing and immunohistochemistry have allowed the construction of an anatomic map of the SCN output pathways. The main output pathway connects the SCN to the subparaventricular zone (SPZ) below the hypothalamic paraventricular nucleus (PVN). A few fibers also connect to the dorsal medial nucleus of the hypothalamus (DMH). Additional efferent connections run to the lateral geniculate nucleus (LGN) and the raphe nuclei. These connections may act as a modulatory influence on the input signals that originate in the retina. We discuss the importance of some of these connections in the next section. Of course, there are additional efferents from the SCN, the function of which is still unknown. For example, connections to the paraventricular nucleus of the thalamus (PVT) exist. The PVT regulates activity; therefore, the SCN may regulate locomotor activity through this connection. However, such a link has not been conclusively determined.

Hormone Release Regulated by SCN via Distinct Neuroanatomic Pathways

Many hormones show distinct circadian or diurnal rhythms. Output pathways demonstrating SCN control have been described for some of these. A more detailed discussion of hormonal rhythms is found in Chapter 10; however, we briefly mention some of these here to illustrate examples of SCN output pathways (Fig. 5.12).

Melatonin

As mentioned above, melatonin can act as an input factor on the SCN. But it also acts as an output factor, and primarily translates the length of the photoperiod to the rest of the organism. Melatonin is cyclically produced by the pineal gland with levels highest at night. The pathway that controls melatonin production is multisynaptic. The SCN synapses with the PVN, which in turn connects to the intermediolateral column of the spinal cord. This structure connects with the superior cervical ganglion, which then sends connections to the pineal gland. Melatonin apparently has different effects depending on the species of mammal studied. In many species, seasonal variations in photoperiod, and thereby melatonin production, regulate reproductive activity. Humans, however, are not photoperiodic, and the function of melatonin is largely unknown, although it does have some sleep promoting effects. More information on melatonin can be found in Chapter 10.

Corticosterone

Corticosterone levels are circadianly regulated in mammals. Corticosterone is the immediate precursor of aldosterone in mammals and therefore regulates sodium retention and blood pressure. It is released by the adrenal glands under the control of adrenocorticotropic hormone (ACTH),

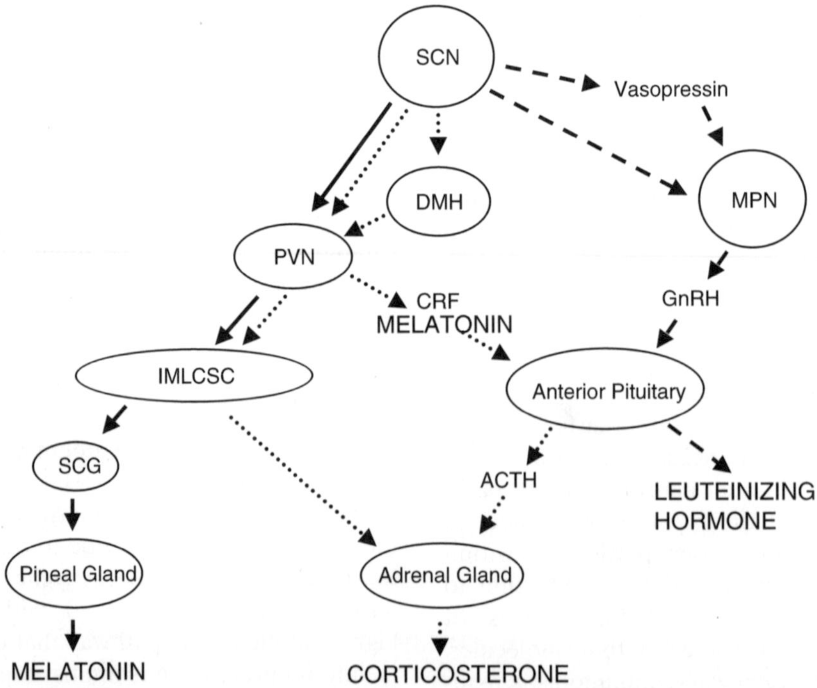

Figure 5.12. Output pathways of the SCN that regulate hormonal rhythms. The three most widely described hormonal outputs of the circadian system are melatonin, corticosterone, and leuteinizing hormone production. The melatonin pathway is shown connected by solid arrows. The SCN connects to the paraventricular nucleus (PVN), which in turn connects to the interomediolateral column of the spinal cord (IMLCSC). The IMLCSC connects to the pineal gland via the suparcervical ganglion (SCG). The corticosterone pathway is shown by dotted arrows. There are multiple pathways controlling the corticosterone rhythm: (1) similar to the melatonin pathway, the SCN connects to the PVN, which in turn connects to the IMLCSC—however, the IMLCSC connects directly to the adrenal glands; (2) the PVN also regulates corticosterone release by producing corticotroph releasing factor (CRF), which stimulates the anterior pituitary to release adrenocorticotrophic hormone (ACTH, which stimulates the adrenals); and (3) the SCN also connects to the CRF, producing neurons of the PVN indirectly via the dorsal medial nucleus of the hypothalamus. The pathway regulating leuteinizing hormone release is shown by dashed arrows. LH is released by the anterior pituitary in response to gonadotrophin releasing hormone (GnRH) from the medial preoptic nucleus (MPN) of the hypothalamus. The SCN regulates the MPN both directly and indirectly via rhythmic vasopressin production.

which itself is released from the anterior pituitary gland. In turn, ACTH release is under the control of neurons of the PVN that release corticotroph releasing factor (CRF). There are several routes of SCN control over the corticosterone rhythm: (1) there are direct synapses from the SCN to the CRF producing neurons of the PVN; (2) the SCN synapses with the DMH, which in turn synapses with the PVN; and (3) there is a pathway that connects the PVN directly to the adrenals via the interomediolateral column of the spinal cord. Together these three pathways work to establish the corticosterone rhythm.

Gonadotropins

Luteinizing hormone (LH) is released by the anterior pituitary in a circadian fashion. This hormone has different effects in males and females, but primarily regulates reproductive functions. The medial preoptic nucleus (MPN) of the hypothalamus releases gonadotropin releasing hormones (GnRH) in response to vasopressin from the SCN. There are also direct connections from the SCN to estrogen-containing neurons in the MPN. Therefore, the SCN has both direct and indirect ways of regulating the gonadotropin rhythm.

Output Control by SCN Using Secreted Factors

As mentioned above, SCN transplants from one animal can restore rhythms in an animal that has an ablation of the SCN. In 1996, Silver and colleagues extended these findings to show that transplanted SCN do not require synaptic connections to affect rhythm control. SCN transplants from *tau* mutant hamsters were encapsulated in a semipermeable membrane prior to transplant into wild-type animals with ablated SCN. These encapsulated transplants did not form synaptic connections. However, recovery of circadian rhythms, as judged by locomotor activity, was still accomplished. The recovered rhythms matched that of the donor. These experiments proved that a secreted factor from the SCN could at least regulate circadian locomotor activity. Of course, these results did not rule out the necessity of synaptic connections for the control of other circadian rhythms. For example, certain rhythms are never restored by any transplants. These include reproductive responses to daylength and certain endocrine rhythms. Therefore, synaptic connections are necessary to maintain some rhythms. However, this classic study proved that a secreted factor from the SCN was capable of controlling locomotor output.

Regulation of Locomotor Activity by TGFα

Building on the abovementioned study, Weitz' group sought to identify factors that regulated wheel-running behavior. To accomplish this goal, they first isolated numerous peptides that are normally secreted by the SCN in hamsters. These peptides were tested for effects on locomotor activity rhythms by infusing them into the third ventricle of hamsters and observing their wheel-running rhythms. One peptide, transforming growth factor α (TGFα), completely and reversibly blocked wheel running by the hamsters, suggesting that it acts as an inhibitory factor for locomotor activity. TGFα is a peptide signaling molecule best known for its function in cellular growth control and differentiation. The receptor for TGFα is the epidermal growth factor receptor (EGFR). Strikingly, epidermal growth factor (EGF) also blocked locomotor activity in hamsters, further suggesting that TGFα acts through the EGFR. Additional work showed that *tgfα* is rhythmically expressed in the SCN, and that EGFR is expressed in the SPZ. As mentioned above, this area of the brain is innervated by the SCN. Lesions of the SPZ demonstrate its importance for locomotor rhythms. It was hypothesized that rhythmic release of TGFα from the SCN acts on EGFRs in the SPZ to regulate locomotor activity. However, mice with a mutation in the EGFR called *waved-2*, which reduces receptor activity by 90%, were completely rhythmic in constant darkness. In LD conditions, *waved-2* mice were rhythmic, but were abnormally active during the light time. The phenotype of these EGFR mutant mice indicates that this receptor may actually mediate masking effects of light rather than the persistence of circadian rhythms under freerunning conditions.

Regulation of Locomotor Behavior by PK2

Another molecule that was implicated in locomotor output in rodents is the secreted protein, prokineticin 2 (PK2). PK2 was originally identified as a protein that causes contraction of gastrointestinal smooth muscle. General examination of the expression pattern of *PK2* revealed that it is expressed in the hypothalamus. Further study showed circadian expression in the SCN, peaking during the first half of the subjective day. Genomic DNA analysis revealed E boxes upstream of the *PK2* gene. These E boxes are necessary for transcription. Moreover, CLOCK and BMAL1 can activate transcription of *PK2*, and this activation is blocked by the PERs and the CRYs. The oscillations of *PK2* expression are disrupted in *Clock* mutant mice and *mCry1*/*mCry2*-deficient mice, indicating that these core circadian proteins regulate *PK2* expression in vivo. Of course, as mentioned above, microarray experiments show that expression of many genes cycles in the SCN. Presumably, not all of these necessarily have output functions. However, the receptor of PK2 (PK2R) is expressed in numerous areas where the SCN has its connections, although this does not include the SPZ. The most compelling evidence that PK2 is truly an output protein came from studies involving the intracerebroventricular injection of PK2 in rats. Administration of PK2 in this way suppressed the wheel-running behavior of rats during the subjective night. This effect is much like that seen for TGFα (Fig. 5.13).

Arginine Vasopressin

The classic example of a CCG is *arginine vasopressin* (AVP), also known as *antidi-*

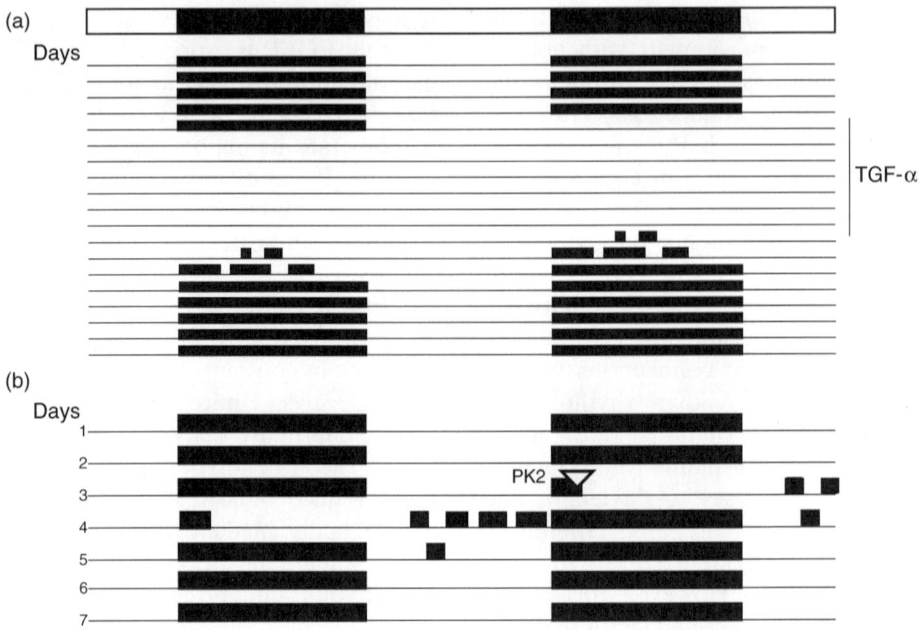

Figure 5.13. TGFα and prokineticin 2 both suppress locomotor activity. (a) Hamsters were entrained to a 14–10 light:dark cycle for two weeks and then moved to constant darkness. At the indicated time, TGFα was constantly infused into the third ventricle through an intraventricular cannula. Locomotor activity almost immediately stopped, and very little activity occurred while infusion was being carried out. When the infusion was stopped, the animals were active again. (b) Similarly, rats subjected to recombinant human PK2 injection into the ventricular system almost immediately stop running on wheels. Interestingly, there is abnormal activity observed the day after the PK2 injection.

uretic hormone (ADH). AVP is a neuropeptide that is synthesized in the PVN and the supraoptic nuclei (SON) of the hypothalamus and then released from the posterior pituitary. AVP causes an organism to retain water and sodium; therefore, it regulates osmotic pressure and blood pressure. AVP is also found in SCN neurons where both its synthesis and release are circadianly regulated. Synthesis of AVP is controlled at the transcriptional level in the SCN. Indeed, the *avp* gene contains E boxes, which are transcriptionally regulated by the CLOCK:BMAL1 heterodimer. Moreover, in *Clock* mutant mice, the level of *avp* is markedly diminished and nonrhythmic. In contrast, the generation of *avp* in the SON is unchanged in these mutant mice, indicating that different regulatory elements are at work in different areas of the brain. It could be argued that rhythmic expression of *avp* in the SCN implicates it in the generation of the core oscillator. However, in a rat model that lacks *avp*, locomotor and drinking rhythms still exist, although there are slight differences in the amplitude and entrainability of the rhythms. As such, AVP is considered a CCG.

D-Element Binding Protein

D-Element binding protein (DBP) is another classic example of a CCG. Albumin D-element binding protein (DBP) is a member of the PAR-domain-containing basic leucine zipper family of transcription factors. Members of this family contain an amino-terminal activation domain that is rich in proline and acidic residues, called the *PAR domain*. Transcription of *dbp* oscillates strongly in the SCN, which originally suggested that it may be a component of the core oscillator. Also, DBP can bind to the *mPer1* promoter and cooperate with CLOCK and BMAL1 to activate transcription. However, knockout mice lacking the *dbp* gene still show robust circadian gene expression. In addition, although these mice have defects in circadian locomotor activity (a 30-minute shorter period than wild type and overall less activity), they are still rhythmic in freerunning conditions. Such findings indicate that *dbp* is more likely an output gene. Indeed, the *dbp* promoter contains E boxes, and transcription of *dbp* is dependent on CLOCK. Further supporting an output function for DBP is the fact that expression of this gene cycles not only in the SCN but also in numerous peripheral tissues. Moreover, disruption of DBP causes a lack of circadian expression of other genes in the liver, many of which function in metabolism of steroid hormones. Therefore, DBP may serve to amplify circadian signals in nonneuronal tissues by regulating circadian expression of other genes in these tissues. This type of function introduces the next topic: peripheral oscillators.

■ CLOCKS EVERYWHERE: PERIPHERAL OSCILLATORS

Peripheral Oscillators: Organs with Their Own Clocks

As briefly mentioned numerous times above, one of the most intriguing aspects of circadian outputs is the presence of tissue-specific clocks in many organs other than the brain. Evidence that peripheral oscillators exist comes from the fact that all the known circadian genes that cycle in the SCN—*Bmal1*, *Per1*, *Per2*, *Per3*, *Cry1*, and *Cry2*—are also rhythmically expressed in nearly every organ examined. Additionally, as observed in the SCN, microarray experiments identified hundreds of genes that cycle in peripheral tissues. The purpose of these peripheral oscillators presumably is to impart local control of circadian functions in a tissue-specific manner.

Peripheral oscillators do not normally function in a vacuum, however. The core circadian pacemaker in the SCN is hypothesized to maintain control over these peripheral clocks. Several lines of evidence support this hypothesis: (1) peripheral oscillation of clock gene expression is disrupted in SCN-lesioned animals; (2) the phase of rhythmic expression of the clock genes is delayed with respect to the phase of expression in the SCN, suggesting that the SCN controls their expression; and (3) peripheral oscillation of clock gene expression in explanted tissues dampens after about a week in culture. In contrast, sections of SCN will continue to cycle indefinitely. These lines of evidence have yielded the theory that the SCN houses the "master oscillator" which is in control of numerous "slave oscillators" found in peripheral tissues. The mechanisms by which these two types of oscillators interact represent another great unknown area of circadian biology. Notably, however, there are instances in which the slave oscillators can operate independently of the SCN. Such a phenomenon illustrates that the peripheral oscillators can override the SCN control when physiologically advantageous (Fig. 5.14).

Peripheral Oscillators: Genes that Cycle in the Periphery

Just as is seen in the SCN, literally hundreds of genes cycle in peripheral organs of the rodent. This is borne out by microarray studies. Strikingly, again, the subset of genes that cycles in one tissue is quite different

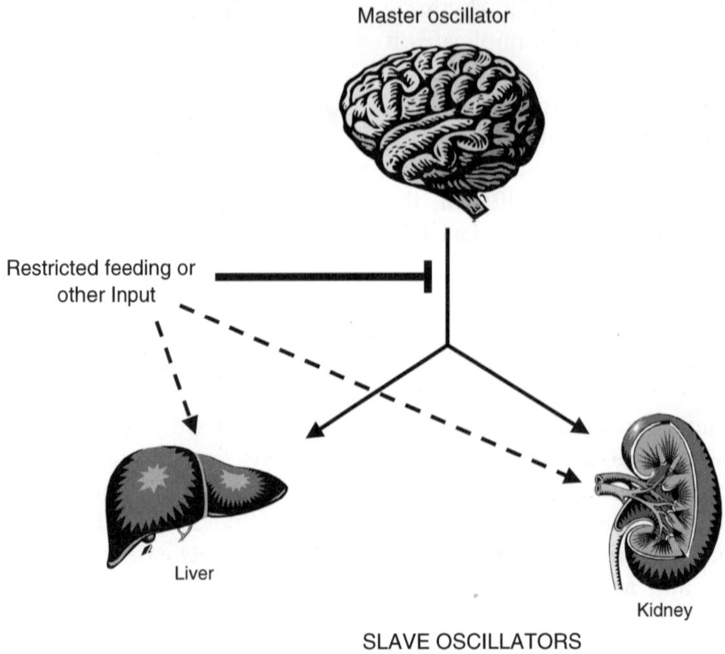

Figure 5.14. Master–slave oscillator relationship. Under normal conditions, the core clock in the SCN elicits control over all other peripheral oscillators, represented by the liver and the kidney. Control may be by indirect neuronal connections, secreted factors, or a combination of both. These peripheral oscillators have a phase that lags the core clock by approximately 4 hours. Under conditions of stress, such as restricted feeding, the slave oscillators can be liberated from the master oscillator. The phase of oscillations in peripheral organs will change, but the phase of oscillations in the SCN remains unchanged. The mechanisms underlying both the uncoupling event and the imposition of a new phase are unknown.

from that in another tissue, further underlining the possibility of tissue specific functions for the peripheral oscillators. Some of the best studied genes that show oscillatory expression in peripheral tissues are discussed below. Notably, although many of these also cycle in the SCN, they are considered output genes.

PAR Proteins

The classic example of a PAR protein is DBP. As discussed above, expression of this gene cycles in the SCN, but also cycles robustly in most organs of the body. Expression of two other PAR domain genes, hepatic leukemia factor (*hlf*), and thyrotroph embryonic factor (*tef*), also cycles robustly in the liver (and in the SCN), suggesting that PAR proteins function generally in circadian output. However, the exact function of PAR proteins in peripheral oscillators is not completely clear. Although expression of *dbp* is regulated by CLOCK, such regulation has not been shown for other PAR genes.

E4BP4

E4BP4 is a basic leucine zipper transcription factor originally isolated as an E4 adenovirus E4 promoter binding protein. E4BP4 lacks a PAR domain, but its DNA binding domain is homologous to those in DBP, TEF, and HLF. E4BP4 is the mammalian homolog of the Drosophila gene *vrille*, which is necessary for circadian rhythms. Examination of *e4bp4* expression revealed that it is circadianly expressed in both the SCN and that of peripheral tissues, much like the PAR genes. However, E4BP4 cycles in a phase opposite DBP. Moreover, E4BP4 acts as a transcriptional repressor and competes for the same DNA binding site as the PAR proteins. According to those data, it is hypothesized that E4BP4 and the PAR proteins act in opposite directions to "fine-tune" circadian gene expression in the periphery. There are some data from studies of chickens that E4BP4 may also function in entrainment by light; however, such experiments have not yet been performed in mammals.

REV-ERBα

The orphan nuclear receptor REV-ERBα was discussed above as a possible component of the core oscillator in the SCN. However, like many core clock genes, *rev-erbα* is circadianly expressed in peripheral tissues. Such peripheral circadian expression has been best studied in the liver. Indeed, REV-ERBα was originally implicated in lipoprotein metabolism and adipogenesis. Possibly REV-ERBα regulates circadian aspects of metabolism in the liver. It is clear from knockout mice that core clock gene expression is disrupted in the liver of these knockout mice. However, no information is yet available regarding any disruption to liver function in these mice.

RORβ

The retinoid-related orphan receptor β (RORβ) is another orphan nuclear receptor. RORβ is expressed primarily in the central nervous system, although it is also found in the retina and in the pineal gland. Interestingly, although expressed in the SCN, *rorβ* expression does not cycle there. However, there is robust oscillation of *rorβ* in both the retina and the pineal gland, suggesting that it has a function in peripheral oscillators in these tissues. Mice lacking *rorβ* were generated. These mice have a neuronal phenotype that may result from developmental defects. Additionally, RORβ-deficient mice show retinal degeneration; however, they are still able to entrain to light dark cycles, indicating that a functional input pathway exists. In freerunning conditions, the locmotor activity period of these knockout mice increased by approximately 25 minutes. Such a phenotype may result from a defect intrinsic to the core oscillator or to defective feedback from the pineal gland. However, measurements of pineal function, namely, melatonin production, have not been performed. Such measurements are necessary to fully elucidate the function of RORβ in peripheral oscillators. Notably, the RORβ-related proteins, RORα and RORγ, may also regulate circadian gene expression in the liver because they form complexes with promoter elements in a cyclic manner. As such, the ROR family of proteins may be general regulators of chronobiology in the periphery.

Microarray Experiments

Microarray experiments have been performed on peripheral tissues collected over a 24-hour period just as has been performed on the SCN. Many tissues have been profiled, and it is a safe assumption that more profiling is occurring as this is written. While most of these experiments need to be repeated and individual gene cycling confirmed, a few conclusions can be drawn: (1) in each tissue, literally hundreds of genes show circadian expression in individual tissues; (2) there is little overlap between the subsets of genes that cycle in the different tissues; (3) many genes that are involved in specific cellular processes, such as glycolysis, all show cyclic expression in the same tissue; and (4) the number and diversity of genes that show cyclic expression suggest that the core clock transcription factors are not sufficient for directly driving rhythmic expression in peripheral tissues. As more and more microarray experiments are performed, a greater understanding of how cyclic expression is established will emerge.

Oscillators Revealed in Every Cell Through Tissue Culture

Recall that the SCN is composed of multiple oscillators in individual cells. Strikingly, this is the case for individual cells in tissue culture. Schibler's group in 1998 demonstrated that a serum shock delivered to both a Rat-1 fibroblast cell line and a rat hepatoma (liver) cell line induced cyclic oscillations of the rat *Per1* and *Per2* genes. Moreover, cyclic expression of the transcription factors *rev-erbα*, *tef*, and *dbp* was also induced by serum shock. The periods of these rhythms were 22.5 hours and oscillatory expression could be maintained for at least three cycles in vitro. Strikingly, the temporal sequence of the individual rhythms mimicked that found in the liver. That is, normally in the liver the peak of *rev-erbα* expression precedes that of *Per1* and *Per2*. The same result was found in the cell lines following the serum shock. Taken together, these results indicated that the clockworks were intrinsic to many cells in the body. Additional studies have shown that numerous mammalian cell lines, including murine fibroblasts and vascular endothelial cells, display rhythmic clock gene expression on serum shock. The establishment of circadian gene expression in tissue culture cells was an extremely important finding for it opened up a new, easily manipulated model system to study peripheral oscillators.

Retinoid Acid Receptor (RAR)

Sometimes in science, a laboratory searching for an answer to one question fortuitously discovers an answer to another. FitzGerald and colleagues made just such a discovery that lent great insight into peripheral oscillator mechanisms. To examine mechanisms that control vascular smooth muscle contraction, a search for protein binding partners of the retinoid X receptor (RXR) was initiated. This screen yielded NPAS2. Additional experiments showed that RXR also interacted with CLOCK. This interaction was specific. Only CLOCK and NPAS2 interacted with RXR. BMAL1 did not. Also, interaction was found only with RXR and the highly related retinoic acid receptor (RAR). Binding to other steroid receptors such as peroxisome proliferator-activated receptor (PPAR) and thyroid hormone receptor (TR) did not occur. The interaction with the clock proteins was greatly enhanced by retinoic acid (RA) binding to either the RXR or the RAR, suggesting that RA might affect a function of the interacting protein complex. A functional importance for this interaction was provided by the finding that RAR and RXR both blocked transcriptional activation by the NPAS2:BMAL1 and CLOCK:BMAL1 heterodimers in tissue culture cells. Expression of all the core clock genes cycles in the vascular smooth-muscle cells. Most importantly, this cyclic gene expression is phase-shifted on injection of retinoic acid into mice. These experiments established a direct mechanism by which a steroid hormone could affect circadian gene transcription by the core clock components.

Given the abovementioned circadian functions of the REV-ERB and ROR families of nuclear receptors, the possibility is emerging that the nuclear receptor protein family has a general function in peripheral oscillators, much like the PAR family of proteins. Additional data supporting this hypothesis come from studies regarding the glucocorticoid receptor, discussed in the next section.

Glucocorticoids

Glucocorticoids can phase-shift peripheral circadian gene expression. The finding that a serum shock induces circadian gene expression in fibroblasts suggested that a bloodborne factor could signal circadian time to organs in the body. Schibler's group reasoned that glucocorticoids were good candidates for such signaling factors because they are expressed in circadian cycles. Consistent with this hypothesis, dexamethasone induces circadian gene expression in fibroblasts grown in culture. Moreover, dexamethasone injection into mice shifts the phase of circadian RNA expression in liver, kidney and heart. No shift was seen in the SCN, which does not express the glucocorticoid receptor. To prove their hypothesis, Schibler's group used a mouse line that has an inactive glucocorticoid receptor only in the liver. Dexamethasone injection into these mice phase shifted circadian gene expression in heart and kidney, but not in liver. These studies were quite beneficial because they link a specific cellular pathway to changes in peripheral circadian gene expression. The importance of this pathway under physiological conditions has yet to be determined however.

Restricted Feeding

Restricted feeding can uncouple peripheral oscillators from SCN control. Restricted feeding had long been known to entrain locomotor activity in rodents. Normally rodents feed at night, when they are active. However, if food availability is restricted to the daytime, these animals will become active a few hours prior to food availability, such that they are now active during the light phase. Two groups examined the effects of restricted feeding on the molecular oscillations seen in peripheral tissues and in the SCN. Both groups found similar results. Altered feeding regimens shifted the phase of cyclic expression of both core clock genes and other genes in peripheral tissues. Of note, the phase shift occurred faster in the liver than in other organs such as the lung and the kidney. It is unclear why this differential response occurs. It is possibly due to differences in sensitivity to an entraining factor. Notably, the liver is intimately related to feeding; therefore it makes intuitive sense that such an organ would be more sensitive to feeding regimens.

Possibly the most striking finding from these experiments, however, was that restricted feeding did not induce phase shifts of gene oscillations in the SCN. This finding established that peripheral oscillators could be uncoupled from the SCN. Therefore, while the SCN normally does act as a master pacemaker over the peripheral, slave oscillators, these oscillators can be "liberated" under the right conditions. Such a dichotomy may allow an organism local control over specific organs and thus to better respond to abrupt shifts in environmental conditions.

Because glucocorticoid release is intimately tied to feeding, it was possible that restricted feeding-induced phase shifting was acting through glucocorticoids. However, several lines of evidence indicate that this is not the case: (1) a line of mice that have a mutated glucocorticoid receptor specifically expressed in the liver can be entrained by restricted feeding as measured by cyclic gene expression in the liver and (2) repeated injections of corticosterone into mice fed ad libitum at times when this hormone peaks in feeding-restricted animals do not shift oscillations in the liver. The function of glucocorticoids may actually be to inhibit food-induced phase shifting of peripheral oscillators when animals are fed in the daytime. Consistent with this hypothesis, adrenalectomized mice respond with more rapid phase shifting of circadian gene expression in the liver in response to food restriction than did nonmanipulated animals. Of course, numerous hormonal and signaling pathway changes occur in response to feeding, so there may not be a single signal that is responsible for food-induced shifting of peripheral oscillators. Alternatively, changes in cellular metabolic state in response to feeding may be involved, as is discussed in the next section.

The Cellular Redox State

Recall from above that NPAS2 is a homolog of CLOCK that can heterodimerize with BMAL1 and activate transcription in peripheral tissues. Using a neuroblastoma cell line, McKnight's group identified a number of target genes activated by the NPAS2:BMAL1 heterodimer. One such gene is lactate dehydrogenase A (*LdhA*), which catalyzes the reduction of pyruvate to lactate. While not obviously interesting, McKnight's group realized that LDHA utilizes reduced nicotinamide adenine dinucleotide (NADH) as a cofactor, generating the oxidized form, NAD^+. They reasoned that these cofactors might have influence over the activity of the NPAS2:BMAL1 heterodimer. They found that DNA binding activity was enhanced by NADH (reduced form) and inhibited by NAD (oxidized form). Because the absolute cellular concentration of cofactor is constant, it was hypothesized that binding is a function of the ratio of the reduced to the oxidized form. Consistent with this hypothesis, maximal binding was found when the ratio of NADH to NAD was at least 75%. Interestingly, the binding of CLOCK:BMAL1 heterodimers was also enhanced by NADH. However, NADH had no effect on DNA-binding of BMAL1: BMAL1 homodimers. Therefore, it appears that only NPAS2 and CLOCK are responsive to NADH. Indeed, using deletion mutations, the responsive site in NPAS2, and presumably CLOCK, was localized to the bHLH domain; the PAS domain was dispensable for responsiveness.

The oxidation state of NADH may affect transcriptional activity in two ways that are not mutually exclusive: (1) it enhances binding of heterodimers to DNA and therefore presumably enhances transcriptional activity, and (2) NADH apparently promotes the formation of NPAS2:BMAL1 heterodimers at the expense of BMAL1:BMAL1 homodimers. This sets up an elegant system such that when the reduced form is high, transcriptional activation occurs. However, when the oxidized form is high, BMAL1:BMAL1 homodimers, which are transcriptionally inactive, bind, and transcriptional repression is the result.

The finding that redox state can affect DNA binding by the core circadian transcription factors raises intriguing questions regarding food-induced entrainment. The presence of food causes cells to activate metabolic pathways, which results in higher concentrations of reduced cofactors. For example, feeding activates glycolysis and thus generates NADH. In contrast, food restriction results in utilization of reduced cofactors. Although food-induced entrainment may be influenced by a number of different mechanisms, redox state is a novel possibility.

■ CLOCK COMPONENT MUTATION LEADING TO SPECIFIC DISORDERS

Most of the preceding sections focused on describing the current state of knowledge regarding the molecular basis for circadian rhythms. Obviously, much has been learned since, but still more remains to be understood. One aspect of circadian biology that has not been adequately addressed is how all the molecular data fit into the larger picture of organismal physiology and behavior. Everyone seems to agree that circadian rhythms are important. Indeed, it is hard to believe that Mother Nature would put in place a conserved timekeeping system in organisms from cyanobacteria to mammals, unless such a system was vital. It would stand to reason, therefore, that disruption of this system, à la murine gene knockouts, would have profound effects on the animal. Surprisingly, this does not seem to be the case. Aside from running on wheels in the middle of the day, so far the animals with artificially disrupted circadian rhythms appear to be just fine. Therefore, we are left with the question, is a circadian clock necessary? However, remember that utilizing mammals under strict laboratory conditions does not necessarily reproduce what would happen in the wild. To fully understand the importance of an intact circadian clock, field studies must be performed. Such heroic studies are actually being undertaken using SCN-lesioned ground squirrels released into the wild. These studies suggest that the SCN-lesioned animals are more likely to fall prey to predators than are their non-lesioned counterparts possibly because of excessive nocturnal activity. However, while we await results from more field studies, familial genetics and careful laboratory analyses are at least proving to us that disruption of specific clock components can have profound effects on an organism.

Sleep Disorders

One of the most exciting findings is the demonstration that humans with a mutated *Per2* gene display a specific sleep disorder. Patients with familial advanced sleep phase syndrome (FASPS) are characterized by a normal sleep pattern, but it is advanced by approximately 4 hours. That is, these patients fall asleep at around 7:30 pm every evening and spontaneously awake at around 4:30 am every morning. Patients with this disorder also have advanced melatonin and temperature rhythms. FASPS is inherited as a simple autosomal dominant disorder, indicating that it is a single-gene mutation. Ptácek and colleagues originally mapped the FASPS gene to the telomeric portion of the long arm of chromosome 2 in a family afflicted with this disorder. Interestingly, the human *Per2* gene resides in the same area as the putative FASPS gene. Because it is a circadian gene, *Per2* was an excellent candidate for the FASPS gene. Sequencing of the *Per2* gene from affected individuals showed a transition from A to G that changed a serine to a glycine. This substitution occurs in the binding site for the human casein kinase Iε protein, which, you may recall, is the homolog of the hamster TAU protein. Moreover, this mutation leads to hypophosphorylation of PER2. Taken together, these findings strongly suggest that the PER2 mutation is the cause of FASPS in this afflicted family, and provide the first link between circadian gene function and human disease.

PER2 as a Possible Tumor Suppressor

Another extremely exciting finding is that PER2 may have a role in oncogenesis, specifically as a tumor suppressor. Lee and colleagues observed that PER2 knockout mice were more prone to develop tumors with age than were wild-type mice. These mice specifically developed salivary gland hyperplasia, genital teratomas, and lymphomas. Moreover, these knockout mice are more sensitive to γ irradiation than are wild-type mice as assessed by tumor formation, primarily lymphoma. Also, thymocytes from the knockout mice are resistant to apoptosis induced by γ irradiation. Normally, thymocytes will die by apoptosis in response to irradiation. If they do not, they may accumulate genetic defects that lead to cancer formation. Mechanistically, Lee and colleagues found that the tumor suppressor, p53 is not induced by irradiation in PER2-deficient thymocytes. Such a finding suggested that PER2 functions in the transcriptional response to irradiation. Consistent with this hypothesis, all the core circadian clock genes are induced by irradiation, but such induction is abrogated in PER2 knockout animals. Additionally, the oncogene, *c-myc*, is normally expressed in a cyclic fashion in peripheral tissues. In PER2 knockout animals, though, *c-myc* expression is still cyclic, but is greatly upregulated. This finding led to the proposal that PER2-deficient mice are more susceptible to cancer formation because of an inability to properly regulate genes involved in cell growth control.

■ CONCLUSION

This chapter has attempted to introduce the reader to the rapidly expanding world of molecular circadian biology in mammals. Notably, mammalian clockworks are very similar to those of insects. However, in many instances it appears that the issue is more complicated. Also, this chapter is very rodent-centric. As pointed out in the introduction, it is a fair assumption that there are differences in aspects of circadian biology between mice and humans. These differences are illustrated by the mere fact that one is nocturnal and the other diurnal. However, the rodent model has been extremely useful in our understanding of mammalian circadian biology and no doubt will continue to be so.

FURTHER READING

Entries are listed in order of relevance or priority, not alphabetically.

Photoreception and Input

Ginty, DD, Kornhauser, JM, Thompson, MA, Bading, H, Mayo, KE, Takahashi, JS, Greenberg, ME (1993): Regulation of CREB phosphorylation in the suprachiasmatic nucleus by light and a circadian clock. *Science* **260**: 238–241.

Obrietan, K, Impey, S, Storm, DR (1998): Light and circadian rhythmicity regulate MAP kinase activation in the suprachiasmatic nuclei. *Nat Neurosci* **8**: 693–700.

Shigeyoshi, Y, Taguchi, K, Yamamoto, S, Takekida, S, Yan, L, Tei, H, Moriya, T, Shibata, S, Loros, JJ, Dunlap, JC, Okamura, H (1997): Light-induced resetting of a mammalian circadian clock is associated with rapid induction of the mPer1 transcript. *Cell* **91**: 1043–1053.

Albrecht, U, Sun, ZS, Eichele, G, Lee, CC (1997): A differential response of two putative mammalian circadian regulators, mper1 and mper2, to light. *Cell* **91**: 1055–1064.

Shearman, LP, Zylka, MJ, Weaver, DR, Kolakowski, LF, Jr., Reppert, SM (1997): Two period homologs: Circadian expression and photic regulation in the suprachiasmatic nuclei. *Neuron* **19**: 1261–1269.

Central Oscillator

Moore, RY, Eichler, VB (1972): Loss of a circadian adrenal corticosterone rhythm following suprachiasmatic lesions in the rat. *Brain Res* **42**: 201–206.

Welsh, DK, Logothetis, DE, Meister, M, Reppert, SM (1995): Individual neurons dissociated from rat suprachiasmatic nucleus express independently phased circadian firing rhythms. *Neuron* **14**: 697–706.

Low-Zeddies, SS, Takahashi, JS (2001): Chimera analysis of the Clock mutation in mice shows that complex cellular integration determines circadian behavior. *Cell* **105**: 25–42.

Core Components

Vitaterna, MH, King, DP, Chang, AM, Kornhauser, JM, Lowrey, PL, McDonald, JD, Dove, WF, Pinto, LH, Turek, FW, Takahashi, JS (1994): Mutagenesis and mapping of a mouse gene, Clock, essential for circadian behavior. *Science* **264**: 719–725.

King, DP, Zhao, Y, Sangoram, AM, Wilsbacher, LD, Tanaka, M, Antoch, MP, Steeves, TD, Vitaterna, MH, Kornhauser, JM, Lowrey, PL, Turek, FW, Takahashi, JS (1997): Positional cloning of the mouse circadian Clock gene. *Cell* **89**: 641–653.

Antoch, MP, Song, EJ, Chang, AM, Vitaterna, MH, Zhao, Y, Wilsbacher, LD, Sangoram, AM, King, DP, Pinto, LH, Takahashi, JS (1997): Functional identification of the mouse circadian Clock gene by transgenic BAC rescue. *Cell* **89**: 655–667.

Gekakis, N, Staknis, D, Nguyen, HB, Davis, FC, Wilsbacher, LD, King, DP, Takahashi, JS, Weitz, CJ (1998): Role of the CLOCK protein in the mammalian circadian mechanism. *Science* **280**: 1564–1569.

Thresher, RJ, Vitaterna, MH, Miyamoto, Y, Kazantsev, A, Hsu, DS, Petit, C, Selby, CP, Dawut, L, Smithies, O, Takahashi, JS, Sancar, A (1998): Role of mouse cryptochrome blue-light photoreceptor in circadian photoresponses. *Science* **282**: 1490–1494.

Ralph, MR, Menaker, M (1988): A mutation of the circadian system in golden hamsters. *Science* **241**: 1225–1227.

Lowrey, PL, Shimomura, K, Antoch, MP, Yamazaki, S, Zemenides, PD, Ralph, MR, Menaker, M, Takahashi, JS (2000): Positional syntenic cloning and functional characterization of the mammalian circadian mutation tau. *Science* **288**: 483–492.

Additional Components

Panda, S, Antoch, MP, Miller, BH, Su, AI, Schook, AB, Straume, M, Schultz, PG, Kay, SA, Takahashi, JS, Hogenesch, JB (2002): Coordinated transcription of key pathways in the mouse by the circadian clock. *Cell* **109**: 307-320.

Ueda, HR, Chen, W, Adachi, A, Wakamatsu, H, Hayashi, S, Takasugi, T, Nagano, M, Nakahama, K, Suzuki, Y, Sugano, S. Iino, M, Shigeyoshi, Y, Hashimoto, S (2002): A transcription factor response element for gene expression during circadian night. *Nature* **418**: 534–539.

Output Pathways

Silver, R, LeSauter, J, Tresco, PA, Lehman, MN (1996): A diffusible coupling signal from the transplanted suprachiasmatic nucleus controlling circadian locomotor rhythms. *Nature* **382**: 810–813.

Peripheral Oscillators

Akhtar, RA, Reddy, AB, Maywood, ES, Clayton, JD, King, VM, Smith, AG, Gant, TW, Hastings, MH, Kyriacou, CP (2002): Circadian cycling of the mouse liver transcriptome, as revealed by cDNA microarray, is driven by the suprachiasmatic nucleus. *Curr Biol* **12**: 540–550.

Storch, KF, Lipan, O, Leykin, I, Viswanathan, N, Davis, FC, Wong, WH, Weitz, CJ (2002): Extensive and divergent circadian gene expression in liver and heart. *Nature* **417**: 78–83.

Balsalobre, A, Damiola, F, Schibler, U (1998): A serum shock induces circadian gene expression in mammalian tissue culture cells. *Cell* **93**: 929–937.

McNamara, P, Seo, SP, Rudic, RD, Sehgal, A, Chakravarti, D, FitzGerald, GA (2001): Regulation of CLOCK and MOP4 by nuclear hormone receptors in the vasculature: A humoral mechanism to reset a peripheral clock. *Cell* **105**: 877–889.

Balsalobre, A, Brown, SA, Marcacci, L, Tronche, F, Kellendonk, C, Reichardt, HM, Schutz, G, Schibler, U (2000): Resetting of circadian time in peripheral tissues by glucocorticoid signaling. *Science* **289**: 2344–2347.

Damiola, F, Le Minh, N, Preitner, N, Kornmann, B, Fleury-Olela, F, Schibler, U (2000): Restricted feeding uncouples circadian oscillators in peripheral tissues from the central pacemaker in the suprachiasmatic nucleus. *Genes Dev* **14**: 2950–2961.

Yamazaki, S, Numano, R, Abe, M, Hida, A, Takahashi, R, Ueda, M, Block, GD, Sakaki, Y, Menaker, M, Tei, H (2000): Resetting central and peripheral circadian oscillators in transgenic rats. *Science* **288**: 682–685.

Rutter, J, Reick, M, Wu, LC, and McKnight, SL (2001): Regulation of Clock and NPAS2 DNA binding by the redox state of NAD cofactors. *Science* **293**: 510–514.

Disorders

DeCoursey, PJ, Walker, JK, Smith, SA (2000): A circadian pacemaker in free-living chipmunks: essential for survival? *J Comp Physiol A—Sens Neural Behav Physiol* **186**: 169–180.

Toh, KL, Jones, CR, He, Y, Eide, EJ, Hinz, WA, Virshup, DM, Ptacek, LJ, Fu, Y-H (2001): An h*Per2* phosphorylation site mutation in Familial Advanced Sleep Phase Syndrome. *Science* **291**: 1040–1043.

Fu, L, Pelicano, H, Liu, J, Huang, P, Lee, C (2002): The circadian gene *Period2* plays an important role in tumor suppression and DNA damage response in vivo. *Cell* **111**: 41–50.

Part III

MOLECULAR CONTROL OF CIRCADIAN RHYTHMS: FROM CYANOBACTERIA TO PLANTS

6

CIRCADIAN RHYTHMS IN CYANOBACTERIA

Nirinjini Naidoo

INTRODUCTION

Cyanobacteria: The Simplest Organisms Known to Possess Circadian Clocks

The circadian clock was initially assumed to exist only in eukaryotes. Prokaryotic cells, either multicellular or unicellular, were considered too simple to express circadian behavior. Despite the fact that no evidence existed to support such an idea, it persisted and became accepted dogma. However, since the early 1990s, circadian rhythms have been observed and reported in several strains of cyanobacteria, making them the simplest organisms known to possess a circadian clock. Most of the studies on cyanobacterial rhythms have been carried out on a unicellular strain called *Synechococcus elongatus*.

Cyanobacteria as Photoautotrophs and Diazotrophs

Cyanobacteria are aquatic, prokaryotic organisms capable of photosynthesis. These organisms possess chlorophyll a and they split water and reduce CO_2 to produce oxygen and carbohydrate, respectively:

$$6CO_2 + 12H_2O \xrightarrow[\text{chlorophyll}]{\text{visible light}} (C_6H_{12}O_6) + 6H_2O + 6O_2$$

They also possess two other pigments, blue phycocyanin and red phycoerythrin. Cyanobacteria, often referred to as *blue-green algae* because of the presence of phycocyanin, are in fact bacteria. They comprise one of the largest and most important groups of bacteria on earth.

Molecular Biology of Circadian Rhythms. Edited by Amita Sehgal
ISBN 0-471-41824-2

They are among the oldest known fossils at around 3.5 billion years and have played an important role in the evolutionary and ecological history of the Earth (Fig. 6.1). The oxygen atmosphere was generated by cyanobacteria during the Archaean and Proterozoic eras. In addition, cyanobacteria residing in plant cells gave rise to the modern chloroplast. Some cyanobacteria are unicellular; whereas others are filamentous. Many of both types grow in colonies (Fig. 6.2).

In addition to producing oxygen, many species of cyanobacteria also fix nitrogen. These cyanobacteria are photoautotrophs and diazotrophs. Diazotrophs are organisms that can convert atmospheric nitrogen (N_2) into biologically necessary compounds

Figure 6.1. *Stromatolites*—3.5-billion-year-old cyanobacterial fossils. Stromatolite layers were produced as calcium carbonate precipitated over the growing mat of bacterial filaments. The minerals, along with grains of sediment precipitating from the water, were trapped within the sticky layer of mucilage that surrounds the bacterial colonies. The colonies continued to grow upward through the sediment to form new layers. [Pictures courtesy of Isao Inouye (University of Tsukuba), Mark Schneegurt (Wichita State University), and Cyanosite (www.cyanosite.bio.purdue.edu).]

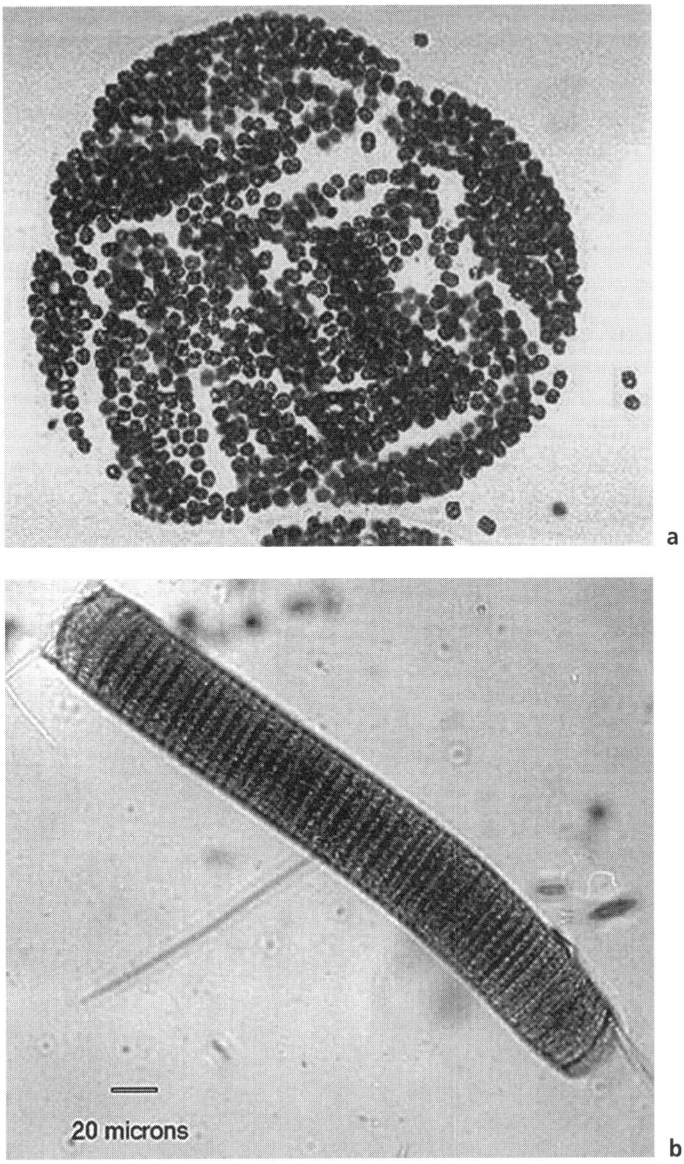

Figure 6.2. (a) Colony of *Microcystis* sp. Cyanobacteria are quite small and usually unicellular, although they often grow in colonies large enough to see. [Picture courtesy of Isao Inouye (University of Tsukuba), Mark Schneegurt (Wichita State University), and Cyanosite (www.cyanosite.bio.purdue.edu).] (b) *Oscillatoria* filament. Many cyanobacterial species are filamentous. [Pictures courtesy of Roger Burks (University of California at Riverside), Mark Schneegurt (Wichita State University), and Cyanosite (www.cyanosite.bio.purdue.edu).]

through a process called *nitrogen fixation*. At the heart of the nitrogen fixation process is the enzyme nitrogenase that catalyzes the following reaction:

$$N_2 + 8H^+ + 8e^- \rightarrow 2NH_3 + H_2$$

Nitrogen Fixation Versus Photosynthesis

Nitrogen fixation and photosynthesis need to be separated from each other. The nitrogen fixing enzyme nitrogenase is extremely sensitive to oxygen and is inactive in the presence of even small amounts of oxygen, creating a design problem for these photosynthetic diazotrophs. In addition, nitrogenase is a rather slow enzyme, reducing only three molecules of nitrogen gas per second. Because of its weak activity, nitrogen fixing cells need to maintain large amounts of the enzyme. As much as 5% of the protein in these cells may be nitrogenase.

Some filamentous nitrogen fixing cyanobacteria have overcome the dilemma of nitrogenase sensitivity to oxygen by developing specialized cells called *heterocysts* for fixing nitrogen, thereby spatially separating nitrogen fixation from photosynthesis (Fig. 6.3). Vegetative cells participate in photosynthesis, produce oxygen, and do not fix nitrogen. Nitrogen fixation takes place in the heterocysts that switch off photosynthesis when the nitrogenase enzyme is expressed. Heterocysts supply fixed nitrogen compounds to the vegetative cells and survive on sugar supplied by them.

Spatial segregation of photosynthesis and nitrogen fixation is not possible or practical in unicellular or nonheterocystous species. An alternative strategy, temporal separation, has been utilized by these cyanobacteria.

Temporal Separation

Unicellular and nonheterocystous cyanobacteria temporally separate photosynthesis and nitrogen fixation. In 1985,

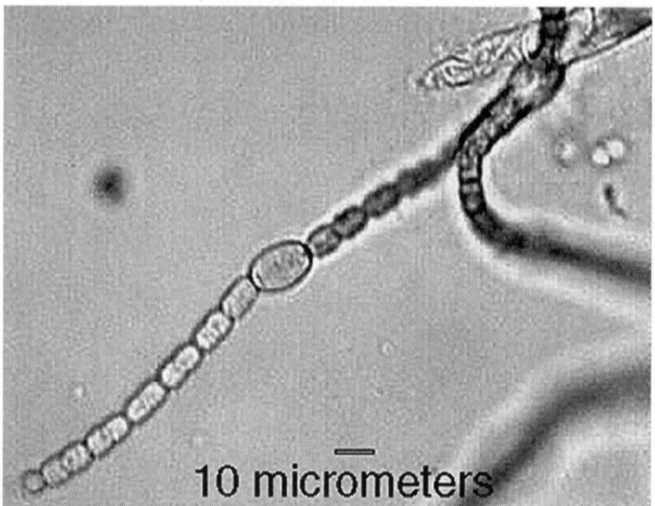

Figure 6.3. Filaments of *Anabaena* and *Nostoc* showing heterocysts. Heterocysts are specialized cells in which nitrogen fixation takes place, spatially separating it from photosynthesis. [Pictures courtesy of Roger Burks (University of California at Riverside), Mark Schneegurt (Wichita State University), and Cyanosite (www.cyanosite.bio.purdue.edu).]

nitrogenase activity in the nonheterocystous, filamentous cyanobacterium *Oscillatoria* sp. was shown to display daily rhythmicity, with a peak at night. These rhythms persisted even after the organism was transferred to continuous light (LL). Subsequently, in 1986, Mitsu and co-workers described the daily oscillations of nitrogenase activity in the unicellular marine cyanobacterium, *Synechococcus*. They demonstrated that nitrogen fixation and photosynthesis were temporally separated. Photosynthesis took place during the day or light period, while nitrogen fixation occurred in the dark. They further showed that carbohydrate synthesis, respiratory oxygen uptake and cell cycle exhibited daily rhythms in constant light (LL).

Robust oscillations of nitrogenase activity in light–dark (LD) conditions were also reported for a freshwater species of *Synechococcus* RF1, isolated from rice fields. Rhythms of photosynthesis and nitrogen fixation were demonstrated to continue in constant light and were entrained by prior light–dark cycles. These scientists recognized that these daily rhythms were reminiscent of eukaryotic circadian rhythms, further demonstrated temperature compensation of the circadian period in *Synechococcus*. Temperature-compensated rhythms of nitrogen fixation and amino acid uptake have also been reported. In the marine cyanobacterium, *Synechococcus* WH7803, rhythms of cell division that are temperature-compensated over the 16–22°C range have been reported. In another marine cyanobacterium, *Synechococcus* sp. Miami BG43511, rhythms of nitrogenase activity, photosynthesis, and cell division maintain a 24-hour period over the 24–30°C temperature range.

In cyanobacteria circadian rhythms, we see the three criteria for circadian clock function satisfied: (1) oscillation of rhythms under constant conditions, (2) phase resetting, and (3) temperature compensation. This indicates that cyanobacteria, although lacking obvious cellular organization, possess a circadian clock.

Circadian Control of Carbohydrate Storage

Carbohydrate storage in *Cyanothece* occurs in a circadian fashion. *Cyanothece* sp., a marine, unicellular diazotrophic organism, studied by Sherman and co-workers, temporally separates photosynthesis and nitrogen fixation. When this cyanobacterium is grown under conditions of alternating light and dark (LD), photosynthetic oxygen evolution is limited to the light phase and the fixation of nitrogen occurs during discrete periods in the dark phase. Nitrogen fixation is an energy dependent process, and as photosynthesis cannot directly supply the energy and reducing power required in the dark, this organism relies on stored carbohydrate granules to provide the requisite energy. The carbohydrates, which are stored as granules between the photosynthetic membranes, accumulate toward the end of the light phase, with the number of granules and amount of carbohydrate peaking prior to the onset of nitrogen fixation. The number of granules and level of carbohydrates is greatly reduced at the end of the dark phase. When *Cyanothece* is grown under conditions of constant light, nitrogen fixation activity peaks during the subjective night and, despite the continuous light, photosynthetic activity is reduced during this phase. In addition, the cells still appear to use stored carbohydrates to fuel nitrogen fixation even though light energy is available. Carbohydrate conversion to energy requires oxygen in a process called *respiration*. It has been suggested that the use of carbohydrates, as a fuel source during nitrogen fixation, lowers the oxygen concentration within the cell, thereby protecting the nitrogenase enzyme from inactivation (see Fig. 6.4 for a model of circadian pattern).

Many species of cyanobacteria do not grow without light. *Cyanothece* can how-

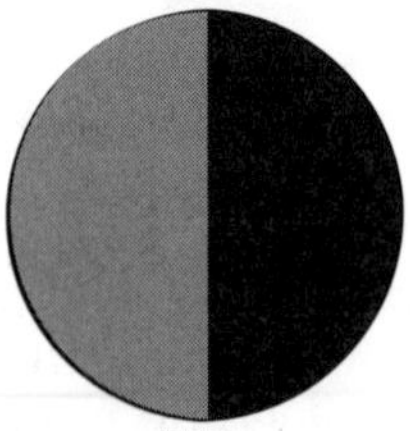

Figure 6.4. Model of circadian pattern in *Cyanothece*, showing the temporal separation of photosynthesis and nitrogen fixation.

ever be grown under conditions of continuous darkness (DD) when provided with glycerol, a source of carbon. Photosynthetic activity is reduced during this period. Nitrogen fixation occurs within the subjective dark phase. Carbohydrate granules are accumulated during the subjective light phase and dissipated during the period of nitrogen fixation. The molecular mechanisms of *Cyanothece* circadian oscillations are relatively unknown and a molecular clock has yet to be found.

■ MOLECULAR BASIS OF THE CLOCK IN CYANOBACTERIA

Synechococcus elongatus

Synechococcus elongatus is a model system for molecular genetic dissection of the cyanobacterial circadian clock. The research groups of Kondo, Johnson, and Golden have developed cyanobacteria as a prokaryotic model for molecular analysis of circadian rhythms. For their study they chose *Synechococcus elongatus* (previously known as *Synechococcus* sp. strain PCC7942), a nondiazotrophic freshwater strain for which genetic transformation is well established. Although this strain does not fix nitrogen, it has many advantages for genetic analyses. It is transformable by linear or circular DNA, which recombines at homologous sites, can receive DNA by conjugation from *Escherichia coli*, can express reporter genes, and has a small genome. Studies of rhythms in cyanobacteria have relied on the use of a luciferase reporter that provides a rapid, quantitative assay for circadian function.

Bioluminescence Assays for Circadian Regulation

Bioluminescence produced by a luciferase reporter is a convenient parameter to monitor as an assay for circadian function. Bacterial luciferase catalyzes the oxidation of reduced flavin mononucleotide (FMN) and a long-chain aldehyde by molecular oxygen to yield FMN, the corresponding acid, water, and light. The aldehyde is not essential to the reaction, but it has an activating effect on the luminescence:

$$\begin{aligned}&FMNH_2 + n\text{-decanal} + O_2 \rightarrow \\ &\quad FMN + n\text{-decanoic acid} + H_2O + \text{light}\end{aligned}$$

Bacterial colonies expressing the luciferase gene glow. Researchers exploited this property of luciferase, together with its short half-life and real-time reporting, to monitor circadian regulation of gene expression. A promoterless segment of the *Vibrio harveyi* luciferase structural gene, *luxAB*, was introduced downstream of the promoter of the *psbA1* gene (Fig. 6.5a). *psbAI* is a photosynthetic gene that encodes D1, a major protein component of the photosystem II reaction center. The *psbA1* mRNA is expressed at high levels and cycles rhythmically in constant light following entrainment to a LD cycle. The psbAI promoter–luciferase reporter fusion construct was integrated at a specific target site in the chromosome of *Synechococcus elongatus* to generate the reporter strain AMC149. The reporter strain exhibited robust circadian bioluminescence rhythms, under both light–dark and constant-light conditions, when *n*-decanal, the aldehyde

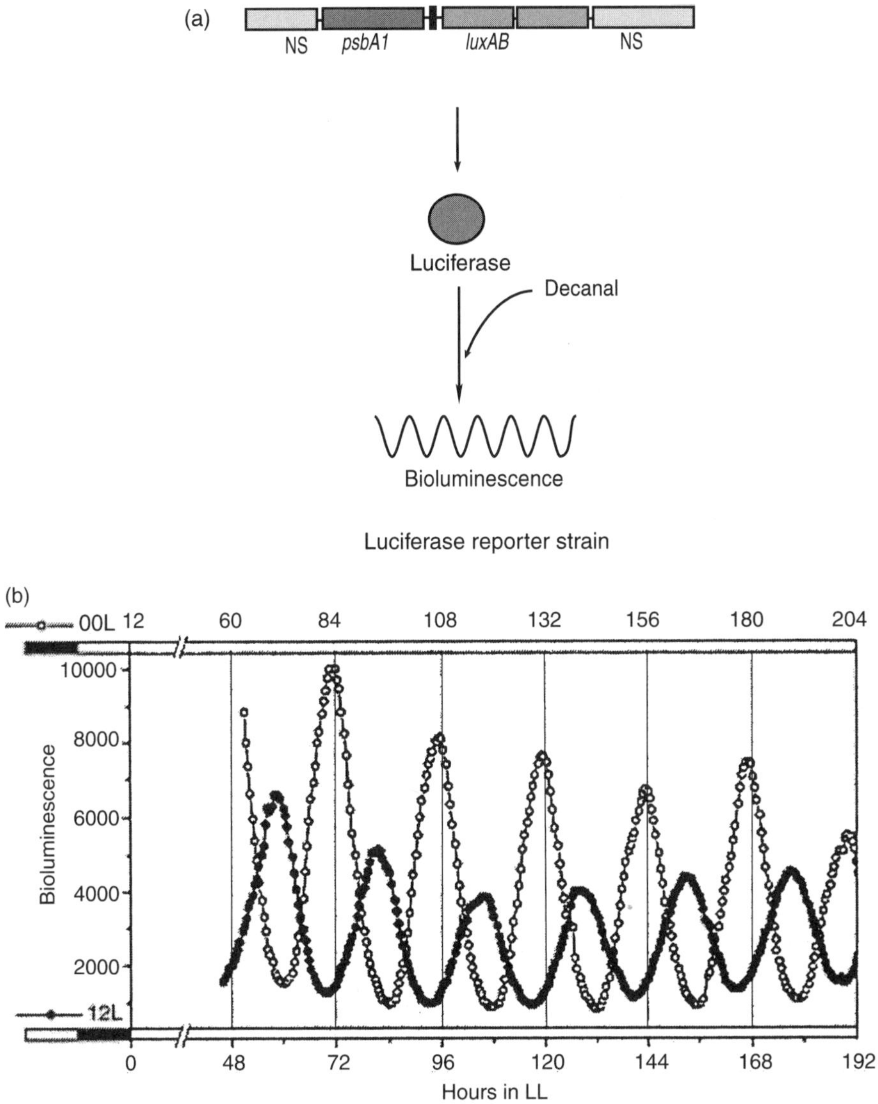

Figure 6.5. The use of a luciferase reporter to measure circadian gene expression in cyanobacteria. (a) Luciferase reporter construct, showing the luciferase gene, *lux AB* downstream of the *psbA1* promoter. (b) Circadian rhythm of bioluminescence from the AMC149 reporter strain in continuous light. The reporter strain was cultured at 30°C under a 12:12 L:D cycle. Shown are two traces from cultures previously entrained to L:D cycles 12 hours out of phase with each other. [Reprinted with permission from Kondo et al. (1993). Copyright 1993 National Academy of Sciences, USA.]

substrate, was provided (Fig. 6.5b). This rhythm satisfied the fundamental criteria for circadian rhythms: (1) persistence of rhythms in constant light, (2) entrainability to light–dark cues, and (3) temperature compensation of the periods. The luciferase reporter has successfully been used to find circadian gene expression in other cyanobacterial species such as *Anabaena* and *Synechocystis*.

Gene Expression

In addition to providing an assay for circadian rhythms in cyanobacteria, the luciferase reporter allowed for automated detection of genes that showed circadian regulation. This was done through a screen of thousands of colonies that represented different insertion sites of the luciferase reporter. The luciferase gene set (*luxAB*) was inserted into the *Synechococcus elongatus* genome so as to achieve random insertions of the *luxAB* gene set throughout the chromosome. Of approximately 30,000 transformed clones, 800 clones whose luminescence was bright enough to be easily monitored were selected. Examination of the luminescence expression patterns of these 800 glowing clones revealed, to the researcher's surprise, that all the clones displayed clear circadian rhythmicity. These rhythms exhibited a range of waveforms, amplitudes, and phase relationships. Two predominant phase relationships were distinguished. Class 1 genes were defined as those whose expression peaks at the end of the day, and class 2 genes were those that peak at the end of the night. The photosynthetic gene *psbAI* is a class 1 gene, peaking at dusk (Fig. 6.6). *PurF*, which encodes glutamine PRPP amidotransferase, a key regulatory enzyme in the de novo purine synthetic pathway, is a class 2 gene. Transcriptional activity peaks at dawn and is minimal at dusk (Fig. 6.6).

On the basis of results from the random promoter trap experiments, a model for the global circadian control of gene expression was formulated (Fig. 6.7). The model incorporates both nonspecific circadian control and circadian regulation by specific *cis* elements and *trans* factors. As such a large number of genes appear to be influenced by the clock in cyanobacteria, it seems unlikely that each of them is controlled by a specific regulatory factor. Clearly, some global factors could be involved as well. Thus, according to this model, there would be one

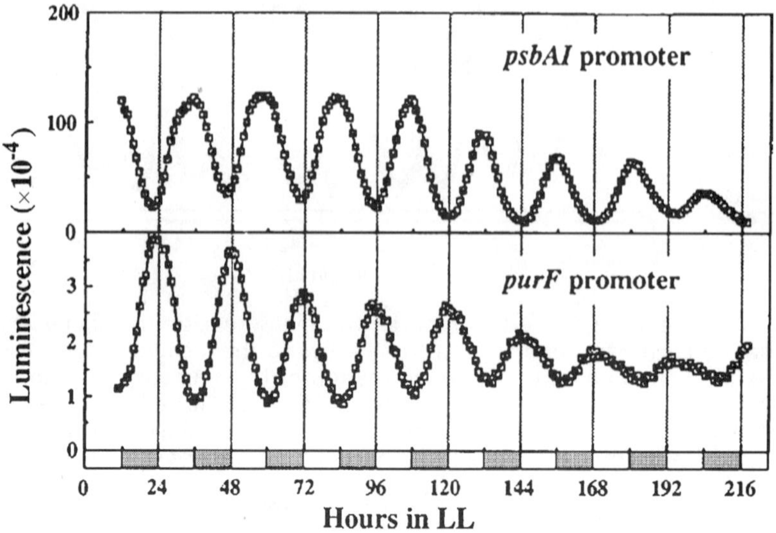

Figure 6.6. Promoter activity of class 1 and class 2 genes as assayed with the *luxAB* luminescent reporter. The upper trace is from the class 1 gene *psbAI*; the bottom trace is from the class 2 gene *purF*. Data were recorded in constant light (LL), but the times of expected or subjective night are shown by the gray bars along the abscissa. [Reprinted from Johnson and Golden (1999) wiht permission from the *Annual Review of Microbiology*, Volume 53. © 1999 by Annual Reviews, www.annualreviews.org.]

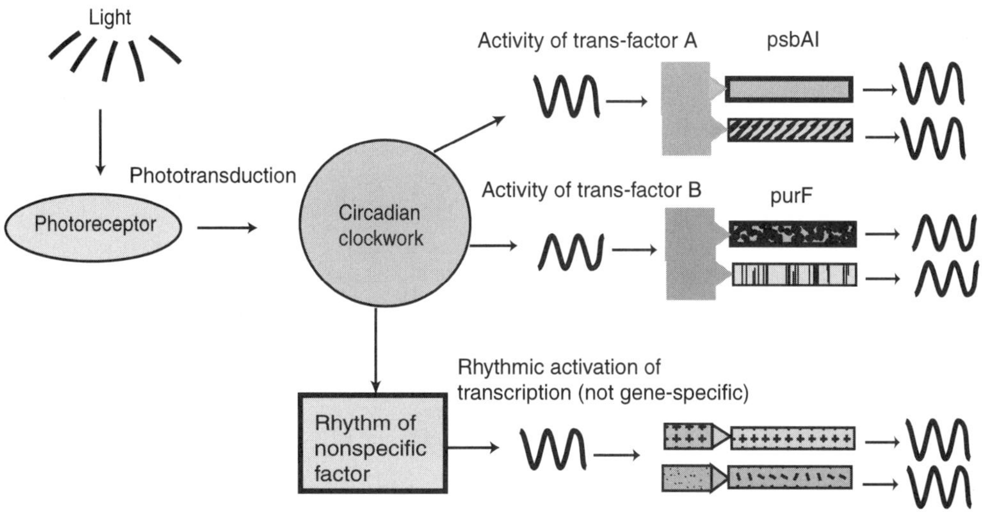

Figure 6.7. Model for global circadian control of gene expression. Coding regions of clock-controlled genes are depicted as rectangles with adjacent promoter regions. Some ensembles of genes may be controlled by rhythmic activity of specific *trans* factors that coordinately regulate groups of genes. These *trans* factors could be activated at different phases. Other genes may be regulated by global but nonspecific factors that oscillate. [Modified from Johnson et al. (1996).]

or more types of class 1–specific *cis* elements turned on during the day by a class 1–specific *trans* factor (*trans* factor A in Fig. 6.7). A different set of class 2–specific *cis* elements would be turned on at night by a class 2–specific *trans* factor (*trans* factor B), and so on. The model is supported by the following observation. The altered expression of the *rpoD2* gene significantly lowers the amplitude of the luminescence rhythm driven by some promoters (e.g., *psbAI*) but not of luminescence rhythms driven by others (e.g., *purF*). The *rpoD2* gene is a member of the sigma factor gene family and encodes a sigma70-like transcription factor.

kai Gene Cluster in Cyanobacterial Clock

Using an exhaustive ethylmethanesulfonate (EMS) mutagenesis screen and the luciferase reporter strain, about 100 mutants displaying a wide range of circadian phenotypes were isolated. The phenotypes included arrhythmia, altered waveforms and atypical periods (ranging between 14 and 60 hours). Most of these mutants grow as well as wild type and show no obvious phenotype other than circadian abnormalities. Of these 100 or so mutants about 30 were rescued by the introduction of a wild-type genomic *Synechococcus elongatus* DNA library. A gene cluster composed of 3 novel open reading frames (ORFs) was identified in the rescuing DNA fragments. This gene cluster was named *kai* (cycle or rotation in Japanese), with the ORFs termed *kaiA, kaiB*, and *kaiC* (Fig. 6.8).

Nineteen mutations were initially mapped by genetic rescue and DNA sequencing to the three kai genes (Table 6.1). Fourteen mutations mapped to *kaiC*, three to *kaiA*, and two to *kaiB*. All are missense mutations arising from single-nucleotide exchanges.

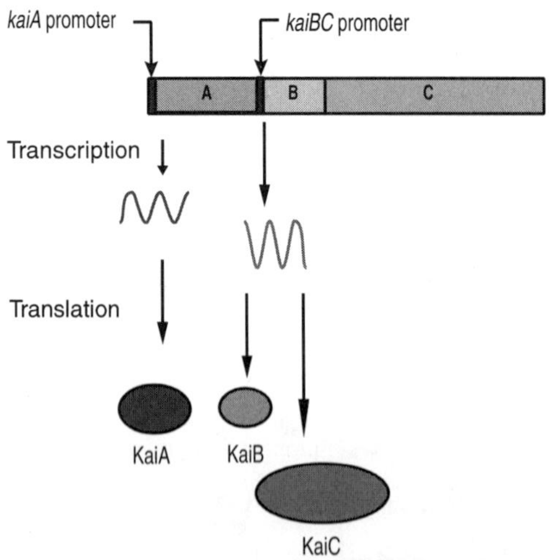

Figure 6.8. Schematic representation of the *kai* gene cluster. The kai gene cluster is transcribed by 2 promoters, *PkaiA* and *PkaiBC*. PkaiA produces monocistronic *kaiA* mRNA, which gives rise to the 284 aa KaiA protein; PkaiBC generates the dicistronic *kaiBC* mRNA, which produces KaiB, a 102 aa protein, and a 519 aa KaiC protein.

TABLE 6.1. Mutations Mapped to the *kai* Genes

Allele Name	Strain Name	Period Phenotype	Mutation
kaiA1	A33a	33 hours	G2280A E103K
kaiA2	A30a	30 hours	G2719A R249H
kaiA3	ALAa	Arrhythmic	G2793A E274K
kaiB1	B21a	21 hours	C2948T L11F
kaiB2	B22a	22 hours	C3137T R74W
kaiC1	C22a	22 hours	C3535T A87V
kaiC2	C29a	29 hours	C3745T S157C
kaiC3	C16a	16 hours	C3918T R215C
kaiC4	C28a	28 hours	C3981T P236S
kaiC5	cLAa	Arrhythmic	C4017T P248L
kaiC6	CLAd	Low amplitude	C4018T P248L
kaiC7	C40a	40 hours	G4033A R253H
kaiC8	C37a	37 hours	G4094A M273I
kaiC9	C21a	21 hours	G4237A R321Q
kaiC10	C27a	27 hours	A4500G T409A
kaiC11	C44a	44 hours	G4536A G421R
kaiC12	C60a	60 hours	T4599C Y442H
kaiC13	CLAb	Arrhythmic	G4654A G460E
kaiC14	CLAc	Arrhythmic	T495A

kai Gene Deletion or Inactivation

Deletion of the entire *kai* gene cluster or inactivation of individual *kai* genes abolishes circadian rhythmicity. Cyanobacteria in which the entire *kai* gene cluster is inactivated grow as well as wild-type cells and emit bioluminescence, but are completely arrhythmic (Fig. 6.9). The cluster is thus essential for circadian rhythmicity but not for growth. Reintroduction of the *kai* gene cluster into another site in the genome of the deletion mutant restores rhythmicity to the cells (Fig. 6.9). Thus, the *kai* gene cluster contains the entire DNA sequence information required for *kai* function. In addition, strains carrying an inactivated *kaiA*, *kaiB*, or *kaiC* are arrhythmic (Fig. 6.9), indicating that each of the three *kai* genes is critical for the circadian rhythm of *Synechococcus*.

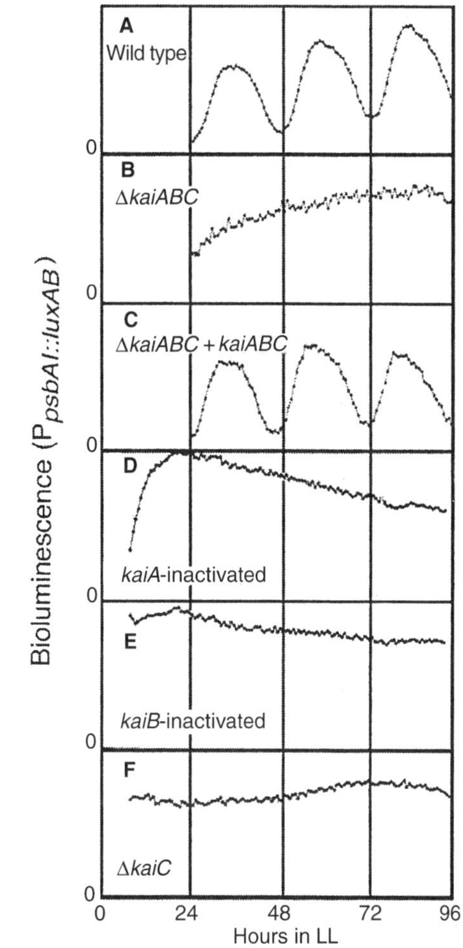

Figure 6.9. Arrhythmia of bioluminescence caused by inactivation of the *kai* genes. (a) WT cells; (b) *kaiABC*-deleted cells (Δ*kaiABC*) in which the entire *kai* gene cluster was deleted by replacement with the Ω fragment; (c) *kaiABC*-transposed cells (Δ*kaiABC* + *kaiABC*) in which the entire WT cluster was transposed to NSII in the genome of Δ*kaiABC* cells; (d) *kaiA*-inactivated cells in which the *kaiA* gene was inactivated by a nonsense mutation; (e) *kaiB*-inactivated cells in which the *kaiB* gene was inactivated by a nonsense mutation; (f) *kaiC*-disrupted cells in which the *kaiC* ORF was replaced with the Ω fragment. [Reprinted with permission from Ishiura et al. (1998). Copyright (1998) American Association for the Advancement of Science.]

kai Clock Proteins

The *kai* gene cluster of *Synechococcus* is transcribed by two promoters, *PkaiA* and *PkaiBC* (Fig. 6.8). The *kaiA* promoter produces a monocistronic *kaiA* mRNA while the *kaiBC* promoter generates a dicistronic *kaiBC* mRNA. Both *kaiA* and *kaiBC* transcripts cycle in abundance. Overexpression of *kaiC* represses the *pkaiBC* promoter acting as a negative-feedback regulator. Thus KaiC functions as the key component of the circadian oscillatory loop. Overexpression of *kaiA* enhances *PkaiBC* transcription, thereby providing the positive, interlocking component described for other systems in previous chapters. Therefore, KaiA functions as the positive element in the kaiBC feedback loop, and KaiC acts as the negative element to inhibit its own transcription. KaiB does not as yet have a specific role assigned to it.

The predicted proteins KaiA, KaiB, and KaiC are composed of 284, 102, and 519 amino acids, respectively. The Kai proteins have neither DNA binding motifs nor any other obvious functional motifs such as the PAS domain. However, the primary structure of KaiC is interesting in that it has a tandem repeat. The first half of the KaiC sequence (1–260) called the CI domain, shares about 42% sequence similarity to the second half, the CII domain (261–519) (Fig. 6.10). KaiC also has two ATP or GTP binding motifs (P loops or Walker's motif), whose consensus is GXXXXGKT/S, and two catalytic carboxylate glutamate residues that are found in ATP binding proteins. In fact, KaiC was shown to possess an ATP binding activity that is important for circadian rhythms. Two DXXG sequences that are highly conserved in GTP binding proteins are also found in the KaiC CI domain. Genetic disruption of a nucleotide binding loop in KaiC strongly impairs the *kaiBC* expression profile. Substitution of a conserved lysine (K52), in P loop 1 by histidine results in complete loss of the *kaiBC* circadian rhythm. A similar substitution of lysine 294 in P loop 2 results in an

KaiA

MLSQIAICIW VESTAILQDC QRALSADRYQ LQVCESGEML LEYAQTHRDQ 50
IDCLILVAAN PSFRAVVQQL CFEGVVVPAI VVGDRDSEDP DEPAKEQLYH 100
SAELHLGIHQ LEQLPYQVDA ALAEFLRLAP VETMADHIML MGANHDPELS 150
SQQRDLAQRL QERLGYLGVY YKRDPDRFLR NLPAYESQKL HQAMQTSYRE 200
IVLSYFSPNS NLNQSIDNFV NMAFFADVPV TKVVEIHMEL MDEFAKKLRV 250
EGRSEDILLD YRLTLIDVIA HLCEMYRRSI PRET

KaiB

MSPRKTYILK LYVAGNTPNS VRALKTLKNI LEVEFQGVYA LKVIDVLKNP 50
QLAEEDKILA TPTLAKVLPL PVRRIIGDLS DREKVLIGLD LLYGELQDSD 100
DF

KaiC

MTSAEMTSPN NNSEHQAIAK MRTMIEGFDD ISHGGLPIGR STLVS**GTSGT** 50
GKTLFSIQFL YNGIIEF**DEP G**VFVTFEETP QDIIKNARSF GWDLAKLVDE 100
GKLFILDASP **DPEG**QEVVGG FDLSALIERI NYAIQKYRAR RVSIDSVTSV 150
FQQYDASSVV RRELFRLVAR LKQIGATTVM TTERIEEYGP IARYGVEEFV 200
SDNVVILRNV LEGERRRRTL EILKLRGTSH MKGEYPFTIT DHGINIFPLG 250
AMRLTQRSSN VRVSSGVVRL DEMCGGGFFK DSIILAT**GAT GTGK**TLLVSR 300
FVENACANKE RAILFAYEES RAQLLRNAYS WGMDFEEMER QNLLKIVCAY 350
PESAGLEDHL QIIKSEINDF KPARIAIDSL SALARGVSNN AFRQFVIGVT 400
GYAKQEEITG LFTNTSDQFM GAHSITDSHI STITDTIILL QYVEIRGEMS 450
RAINVFKMRG SWHDKAIREF MISDKGPDIK DSFRNFERII SGSPTRITVD 500
EKSELSRIVR GVQEKGPES

Figure 6.10. Amino acid sequences of the Kai proteins. KaiA consists of 284 amino acids with a molecular mass of 32.6 kD and a calculated isoelectric point (pI) of 4.69. KaiB consists of 102 amino acids, is 11.4 kD and has a pI of 7.11. KaiC consists of 519 residues, is 58 kD and has a pI of 5.74. The Walker's motif A in KaiC is in bold, imperfect motif Bs are underlined, DXXG motifs are bold and underlined, and C_{KABD} domains are highlighted in gray. Amino acid abbreviations are as follows: A, Alanine; C, Cysteine; D, Aspartate; E, Glutamate; F, Phenylalanine; G, Glycine; H, Histidine; I, Isoleucine; K, Lysine; M, Methionine; N, Asparagine; p, Proline; Q, Glutamine; R, Arginine; S, Serine; T, Threonine; V, Valine; W, Tryptophan; and Y, Tyrosine.

extremely long period phenotype with lowered amplitude. The K52H substitution also significantly reduces the ATP binding activity of KaiC. In addition, replacement of threonines in both loops by alanine residues disrupts the rhythm completely.

Like their mRNAs, KaiB and KaiC proteins display circadian oscillations under constant conditions (LL). The KaiA protein shows little, if any, circadian variation.

KaiC: The Central Clock Component

KaiC is clearly the central component of the cyanobacterial clock. Most of the mutants displaying altered circadian phenotypes map to *kaiC*. The KaiC protein continues to be rhythmically expressed under conditions of constant darkness and metabolic shutdown. The rhythm of KaiC abundance appears to be essential for circadian timekeeping since induced expression of KaiC within the physiological range results in phase resetting. Together with the activity of KaiC in repressing the *kaiBC* promoter, these findings indicate that KaiC is the canonical clock protein that negatively regulates itself and whose oscillations drive overt rhythms.

Interaction among Kai Proteins

A clue to the mechanism of the cyanobacterial clock comes from the finding that the Kai proteins interact. In vitro binding assays, yeast two hybrid assays and a resonance energy transfer assay indicate that KaiA, KaiB, and KaiC proteins interact both homotypically and heterotypically. The tandem similar domains of KaiC, CI, and CII, interact individually with KaiA, KaiB, and KaiC in vitro. A long-period mutation in *kaiA* results in an altered heterotypic interaction between KaiA and KaiB (a dramatically enhanced KaiA–KaiB interaction in vitro), suggesting that protein–protein contact is important to the clock mechanism. Bioluminescence resonance energy transfer (BRET), a technique that assays protein-protein interactions, has also indicated that KaiB polypeptides interact.

BRET is a naturally occurring phenomenon resulting from the nonradiative energy transfer between a bioluminescent donor and a fluorescent acceptor that are in close molecular proximity. For instance, in the sea pansy *Renilla reniformis*, the luminescence that results from the catalysis of coelenterazine by *Renilla* luciferase (Rluc) is transferred to *Renilla* green fluorescent protein (GFP), which in turn fluoresces when the two proteins dimerize. The BRET assay has been developed using the *Renilla* luciferase and enhanced yellow fluorescent protein (EYFP) tagged to the proteins of interest. *Escherichia coli* cells expressing constructs of *kaiB-Rluc* and *kaiB-Eyfp* genes, produce a luminescence that results from the interaction of the tagged gene products, indicating that KaiB homodimerizes.

KaiC enhances the interaction of KaiA with KaiB in vitro and in yeast cells, suggesting the ability of these three Kai proteins to form a multimeric complex. Yeast two hybrid studies and in vitro interaction assays indicate that KaiC possesses two KaiA binding domains, C_{KABD1} and C_{KABD2}. The interaction between KaiA and KaiC appears to be crucial for circadian rhythmicity as known rhythm disrupting mutations in the respective binding domains alter the C_{KABD}–KaiA interactions. Relatively new data provide further evidence of the key interaction between KaiA and KaiC—biochemical and genetic data strongly suggest that the *C*-terminal domain of KaiA stimulates KaiC phosphorylation, which appears to be important for circadian timing. The *N*-terminal domain of KaiA is a pseudoreceiver domain and is thought to be a timing input device that regulates KaiA stimulation of KaiC autophosphorylation. It has been proposed that KaiA and KaiC function cooperatively

```
                                                       50
PCC7942 ---------- ---------- ---------SHGGLPIGR STLVSGTSGT GKTLFSIQFL
PCC 7421 ---------- ---------- --------SHGGLPVGR TTLVSGTSGT GKTLFSLQFL
PCC6803 ---------- ---------- ---------THGGLPIGR STLVSGTSGT GKTLLAVQFL
PCC7203 ---------- ---------- ---------SHGGLPLGR TTLFSGTSGT GKTLFSVQFL
PCC6712 ---------- ---------- ---------AHGGLPIGR STLVSGTSGT GKTLLAIQFL
PCC 7305 ---------- ---------- --------THGGLPEGR TTLVSGTSGT GKTLLALQFL
PCC 7105 ---------- ---------- --------SHGGLPRGR TTLVSGTSGT GKTLFAIQFL
PCC 7417 ---------- ---------- --------SHGGLPIGR TTLVSGTSGT GKTLLSLQFL
PCC73104 ---------- ---------- --------SHGGLPVGR TTLVSGTSGT GKTLFSLQFL
PCC 7906 ---------- ---------- --------SHGGLPIGR TTLVSGTSGT GKTLLSLQFL
PCC 7110 ---------- ---------- --------SHGGLPVGR TTLVSGTSGT GKTLFSLQFL
DCC D550 ---------- ---------- --------THGGLPEGR TTLVSGTSGT GKTLLALQFL
PCC 6718 ---------- ---------- --------SHGGLPIGR TTLVSGTSGT GKTLFSLQFL

                                                                150
PCC7942 GKLFILDASP DPEGQEVVGG FDLSALIERI NYAIQKYRAR RVSIDSVTSV FQQYDASSVV
PCC7421 GKLFILDASP DPEGQDVVGN FDLSALIERI QYAIRKYKAR RVSIDSITAV FQQYETVGVV
PCC6803 GKLFILDASP DPEGQEVVGI FDLSALIERI QYAVRKYKAK LVSIDSVTAV FQQYDAASVV
PCC7203 GKLFILDASP DPEGQEVIGN FDLSALIERL QYAIHKYKAR RVAIDSITAV FQQYEAASVV
PCC6712 GKLFILDASP DPEGQEVVGN FDFSALIERI NYAINKYRAK LVSIDSVTSI FQQYDAASVV
PCC7305 GKLFILDASP DPEGQEVVGN FDLSALIERI QYAIRKYKAK LVSIDSVTAV FQQYDAASVV
PCC7105 GKLFILDASP DPEGQDVVGN FDLSALIERI QYGIRKYKAK RVSIDSVTAV FQQYDAAPVV
PCC7417 GKLFILDASP DPEGQDIVGN FDLSALIERI QYAIRKYKAK RVSIDSVTAV FQQYEAMGVV
PC73104 GKLFILDASP DPEGQDIVGN FDLSALIERI QYAIRKYKAQ RVSIDSITAV FQQYEAMGVV
PCC7906 GKLFILDASP DPEGQDIVGN FDLSALIERI QYAIRKYKAK RVSIDSITAV FQQYEAMGVV
PCC7110 GKLFILDASP DPEGQDVVGN FDLSALIERI QYAIRKYKAK RVSIDSITAV FQQYEAVGVV
DCCD550 GKLFILDASP DPEGQEVVGN FDLSALIERI QYAINKYKAK LVSIDSVTAV FQQYDAAGVV
PCC6718 GKLFILDASP DPEGQDIVGN FDLSALIERI QYAIRKYKAK RVSIDSITAV FQQYEAVGVV
```

Figure 6.11. Alignment of 11 new deduced KaiC homologs with the KaiC sequence of *Synechococcus* PCC7942 and *Synechocystis* PCC688. The completely conserved P loop is underlined.

in both positive- and negative-feedback processes in the cyanobacterial clock by controlling the state of KaiC phosphorylation in a circadian manner.

How the interactions among the Kai proteins contribute to the cyanobacterial circadian clock is not as clearly understood as in Drosophila, where interactions between CLK, CYC, PER, and TIM are central to the circadian clock. Formation of the CLK–CYC complex activates transcription of *per* and *tim*. PER and TIM proteins are synthesized in the cytoplasm, heterodimerize, and move into the nucleus. The PER-TIM complex negatively regulates *per* and *tim* transcription by inhibiting CLK-CYC activity. While the process may not be identical in cyanobacteria, it appears that physical associations between the Kai proteins contribute to circadian rhythmicity.

kaiC Gene Distribution

Proteins of the *kai* system show no sequence similarity to the clock proteins of eukaryotic model organisms. However, a homolog to *kaiABC* is present in the completed genome sequences of *Synechocystis* and other cyanobacteria. In addition, Miao and co-workers demonstrated that *kaiC*-related sequences are widely distributed among phylogenetically diverse strains of cyanobacteria. A PCR-based survey, using degenerate primers designed to identify genes of the *kai* cluster, was used. Over 40 strains of cyanobacteria, representing the five different sections of the phylum, were determined to possess *kaiC* homologous sequences. Deduced sequences of putative *kaiC* genes from the 40 strains showed 64% identity with a 198-residue region of the *Synechococcus elongatus* KaiC protein.

```
                                                                 100
YNGIIEFDEP GVFVTFEETP QDIIKNARSF GWDLAKLVDE
YNGISYFDEA GVFVTFEESP SDIIKNACIF GWNLQRLIDD
YQGIHHFDYP GLFITFEESP SDIIENAYSF GWDLQQLIDD
YNGILFFDEP GVFVTFEESP SDIIKNAYSF GWNLPKLIEE
YHGIKEFDYP GVFVTFEETP KDIITNARSF GWDLQSLIDQ
YHGIKYFDYA GVFVTFEESP QDVIKNAYSF GWDLQNLVER
YNGITYFDEA GVFVTFEESP TDLIKNAQSF GWDLQGLVDE
YNGITYFDEA GVFVTFEESP SDIIKNANIF GWNLHRLIEE
YNGITYFDEA GVFVTFEESP SDIIKNAQIF GWNLQRLIDE
YNGITYFDEA GVFVTFEESP SDIIKNAHIF GWNLPRLLEE
YNGISYFDEA GVFVTFEESP SDIIKNACIF GWNLQRLIDD
YHGIKYFDYA GVFVTFEESP QDVIKNAYSF GWDLQNLVER
YNGISYFDEA GVFVTFEESP NDIIKNAHIF GWDLQRLIEE

                                                  200
RRELFRLVAR LKQIGATTVM TTERIEEYGP IARYGVEEFV PCC7942 SDNVVILRNV LEGERRRRTL EILKLRGTS
RREIFRLVAR LKQLSVTTLI TTERTDEYGP VACVGVEEFV PCC7421 SDNVVIVRNV LEGERRRRTI EILKLRGTT
RREIFRLVAR LKQLQVTSIM TTERVEEYGP IARFGVEEFV PCC6803 SDNVVVLRNV LEGERRRRTV EILKLRGTT
RREIFRLVAR LKQIGVTTII TTERTEEYGA IASFGVEEFV PCC7203 ADNVAIVRNV LEGERRRRTM EILKLRGTT
RREIFRLVAR LKLLNVTSIM TTERVEEYGP IARFGVEEFV PCC6712 SDNVIIIRNV LEGERRRRTI EILKLRGTT
RREIFRLVAR LKQLKVTSIM TTERIEEYGP VARFGVEEFV PCC7305 SDNVVIVRNV LEGERRRRTI EILKLRGTT
RREIFRLVAR LKQLGATTVM TTERIEEYGP VARFGVEEFV PCC7105 SDNVVIVRNV LEGERRRRTI EILKLRGTT
RREIFRLVAR LKQLSVTTII TTERSEEYGP VASFGVEEFV PCC7417 SDNVVIVRNV LEGERRRRTI EILKLRGTT
RREIFRLVAR LKQLYVTTII TTERXEEYGP VASFGVEEFV PC73104 SDNVVIVRNV LEGERRRRTI EILKLRGTT
RREIFRLVAR LKQLNVTTVI TTERSEEYGP VASFGVEEFV PCC7906 SDNVVIVRNV LEGERRRRTI EILKLRGTT
RREIFRLVAR LKQLSVTTLI TTERTDEYGP VACFGVEEFV PCC7110 SDNVVIVRNV LEGERRRRTI EILKLRGTT
RREIFRLVAR LKQLKVTSIM TTERIEEYGP VARFGVEEFV DCCD550 SDNVVIVRNV LEGERRRRTI EILKLRGTT
RREIFRLVAR LKQLNVTTII TTERTEEYGP VACYGVEEFV PCC6718 SDNVVIVRNV LEGERRRRTI EILKLRGTT
```

Figure 6.11. *(Continued)*

Most importantly, the P-loop motif that possesses ATP binding activity essential for circadian oscillatory function is completely conserved (100% identity) in all strains examined (Fig. 6.11). That the *kaiC* gene is conserved over a wide range of cyanobacterial strains suggests that the clock system may be universal in this prokaryotic organism.

Circadian Organization in *Synechococcus* as Model for Cyanobacterial Clock

A transcription/translation feedback model has been proposed for the circadian oscillator of the cyanobacterium, *Synechococcus* (Fig. 6.12). The *kai* genes are rhythmically expressed and the proteins KaiB and KaiC oscillate. Continuous overexpression of KaiC represses the *pkaiBC* promoter, implying a negative feedback mechanism. Overexpression of KaiC for a few hours resets the phase of the rhythm. The level of KaiC expression is clearly linked to the phase of the oscillation. KaiA overexpression enhances the *pkaiBC* promoter activity, suggesting a positive-feedback mechanism. This transcription/translation feedback model for the cyanobacterial clock is schematically similar to those proposed for Drosophila, Neurospora, and mammals. While this model shares many features with the eukaryotic clock models, it has fewer known players. It is likely that, as in Drosophila and mammals, there are transcription factors involved in feedback regulation. The role and identity of these transcription factors are currently not known.

SasA as a Circadian Amplifier

A histidine kinase protein gene, *sasA*, has been identified in a yeast two-hybrid screen for KaiC associating proteins. *sasA* encodes a *Synechococcus* sensory histidine kinase. The sensory histidine kinase and its cognate response regulator constitute

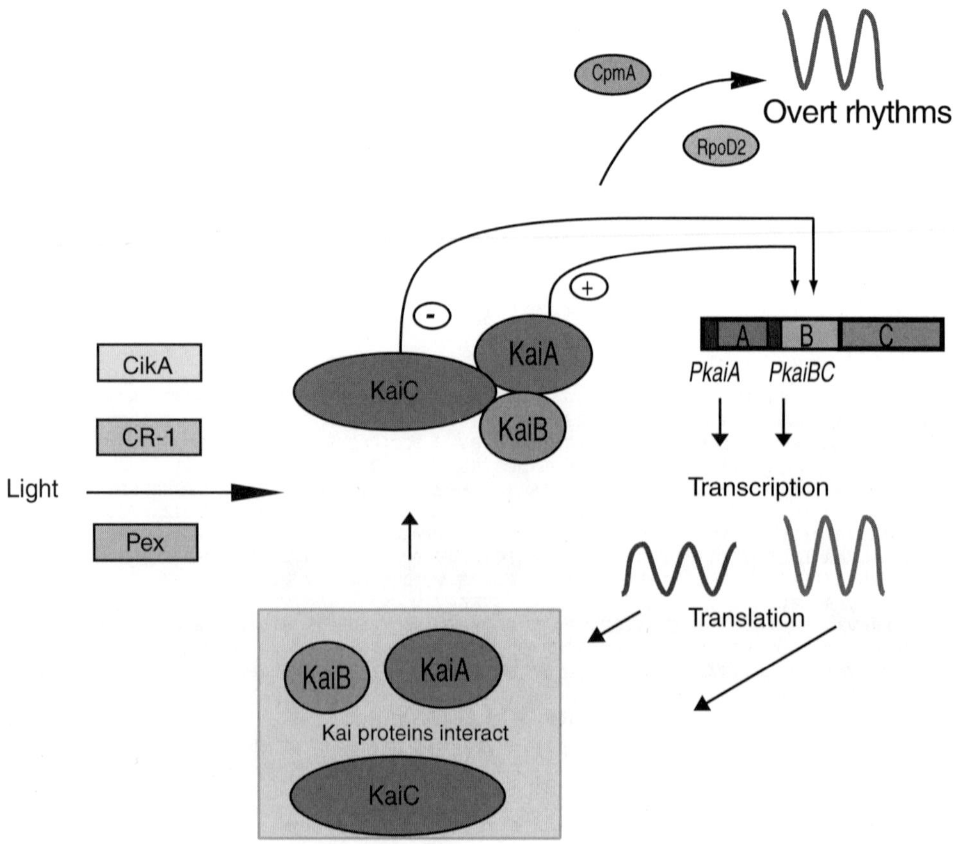

Figure 6.12. A transcription/translation-based molecular feedback model for the cyanobacterial circadian clock. KaiC negatively regulates its own (*kaiBC*) expression to generate a molecular feedback loop, while KaiA activates *kaiBC* expression as a positive element to make the loop oscillate. CikA is a novel phytochrome related histidine kinase that may be a circadian photoreceptor. CR1 and PEX proteins may be associated with input pathways, while CpmA and RpoD2 are components of the output pathway.

phosphorylation-based signal transduction proteins that make up a two-component regulatory system. The simplest two-component signal transduction systems have two protein components: a "sensor" often located in the cytoplasmic membrane, that monitors some environmental parameter; and a cytoplasmic "response regulator" that mediates changes in gene expression or enzymatic activity in response to sensor signals. This system uses the phosphorylation state of a "receiver" (in the response regulator) to propagate a signal. The phosphorylation state of the receiver is controlled by "transmitter" domains of one or more histidine kinases, which may act to bring about its phosphorylation or dephosphorylation through their kinase or phosphatase activities respectively. The signal constituted by the phosphorylation of the receiver may be translated several ways depending on the particular system and, as mentioned above, may include regulation of gene transcription or enzymatic activity. Transmitters have an autokinase activity that transfers phosphoryl groups from ATP to a histidine residue, from which they are transferred to the receiver. The amino terminal end of the sensory kinase protein is thought to be the signal input domain. It is this domain of SasA (residues 1–97) that interacts with KaiC and shares sequence similarity to KaiB (Fig. 6.13). The two polypeptides share 60% similarity, and each is able to bind to KaiC. The binding site on KaiC appears to be the tandem repeat composed of CI and CII domains. Either CI or CII domain of KaiC is sufficient for the interaction with SasA.

Effect of *sasA* Disruption on Circadian Regulation

The functional relevance of SasA to the cyanobacterial circadian clock was examined by analyzing cyanobacterial transformants in which the *sasA* gene was overexpressed or disrupted. Genetic disruption of the *sasA* gene dramatically reduced the magnitude of *kaiA* and *kaiBC* expression and shortened the period length of the *kaiBC* oscillation (Figs. 6.14a–d). Histidine 162, a putative autophosphorylation site on SasA, appears to be necessary for *kaiBC* expression. Replacement of this conserved residue with a glutamine reduced *kaiBC* promoter activity and lowered the amplitude of the rhythm, indicating that normal robust rhythmicity in *Synechococcus* requires autophosphorylation of histidine 162 on SasA.

A role for SasA was supported by studies in which it was constitutively overexpressed. Continuous overexpression of *sasA* completely suppresses the *kaiBC* promoter, eliminating circadian rhythms. However, the effect is not immediate. After

(a)

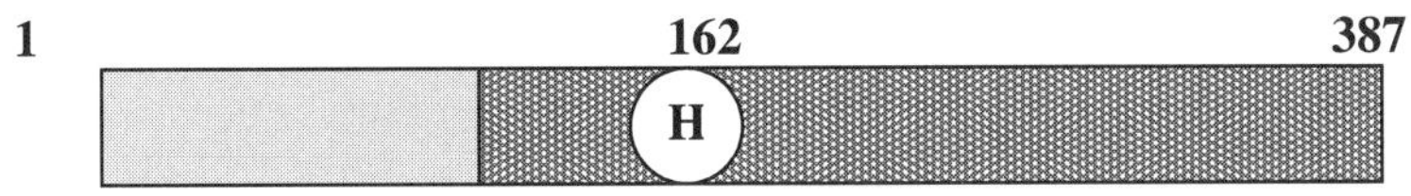

(b)

```
        10        20          30        40        50         60
SasA    QALAQPLLLQLFV-DTRPLSQHIVQRVKNILAAVEATVPISLQVINVADQPQLVEYYRL
        ..  .  .:.:.:  . : : . .. .:::::  ::  :   ..::.:  .:::.:  ..
KaiB    MSPRKTYILKLYVAGNTPNSVRALKTLKNILAEVEQGVYA-LKVIDVLKNPQLAEEDKI
              10        20        30        40           50

           70          80       90        100
SasA    VVTPALVKIGP-GSRQVLSGIDLTDQLANQLPQWLV
        ..::.:.:. :   :.... ..   ...   :
KaiB    LATPTLAKVLPLPVRRIIGDLSDREKVLIGLDLLYG
       60        70        80        90
```

Figure 6.13. (a) Schematic representation of SasA. Putative histidyl residue for autophosphorylation (His 162) and conserved *C*-terminal kinase domain are indicated by shadowing on the SasA molecule. (b) Alignment of SasA with KaiB showing sequence similarity. Conserved residues are shaded. [Reprinted from Iwasaki et al. (2000) with permission from Elsevier Science.]

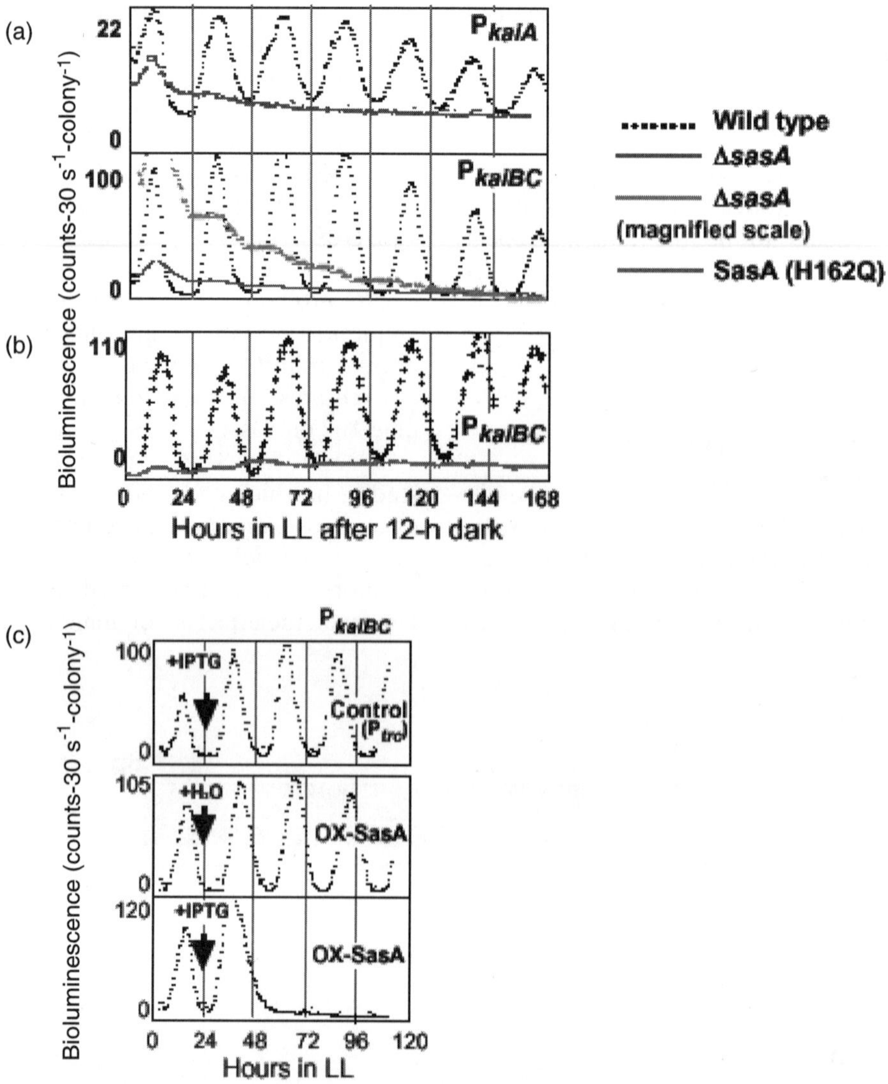

Figure 6.14. Disruption and overexpression of *sasA* affects circadian *kai* gene expression. (*a*) Low amplitude bioluminescence rhythms seen in the absence of *sasA*. Bioluminescence rhythms of the P*kaiA* and P*kaiBC* reporters, in both wild type and *sasA* lacking backgrounds, are shown. For clarity, the bioluminescence profile of the P*kaiBC* reporter in the strain lacking *sasA* is also shown in magnified scale (both solid lines in this panel correspond to the profile in a *sasA* null. (*b*) The *kaiBC* expression profile in the H162Q mutant, in which the putative autophosphorylation residue in *sasA* was substituted with Gln. Bioluminescence from wild-type (. . .) and mutant lines (solid line) was monitored in LL. (*c*) Continuous overexpression of *sasA* disrupts circadian rhythms. Bioluminescence of a P*kaiBC* reporter in a P*trc* vector or a P*trc*::*sasA* (for overexpression of SasA) construct was monitored. Cells were grown on agar plates in LL, exposed to darkness to synchronize their clocks, returned to LL and treated with IPTG or water at hour 24 after dark exposure (arrows indicate addition of water or IPTG).

SasA is overexpressed, a robust *kaiBC* expression cycle occurs before expression shuts down (Fig. 6.14e). In contrast, *kaiC* overexpression immediately suppresses *kaiBC* expression. Transient overexpression of *sasA* advances or delays the phase of *kaiBC* expression rhythm.

Together these studies indicate that SasA is not an essential component of the basic timing loop but acts as an amplifier of the oscillator to generate robust rhythms. Since the expression of most genes in *Synechococcus* is under circadian clock control, the effect of *sasA* disruption on the expression pattern of several other genes was also examined. Disruption of *sasA* attenuated expression of all genes tested (Table 6.2 and Fig. 6.15). It should be noted that growth in *sasA*-disrupted strains is also severely hindered in light–dark cycles (Fig. 6.16), although there are no obvious morphological differences between the wild-type and *sasA* mutant strains under these conditions. This suggests that under natural diurnal conditions, *sasA* is essential for normal growth.

Sensory kinases function in a phosphoryl relay system that begins with the autophosphorylation of the kinase and the transfer of a phosphoryl group. The relay terminates by modifying function of a DNA binding, transcription regulating, response regulator, or other output domain on a response regulator. A SasA-mediated

TABLE 6.2. Altered Gene Expression Patterns in a *sasA* Null Background

Gene/Promoter	Expression
kaiA; kaiBC	Very low amplitude, short period bioluminescence
psbAII; cikA	Arrhythmic, constitutive expression
rpoD2; ndhD	Period, amplitude or phase angle of expression rhythm altered

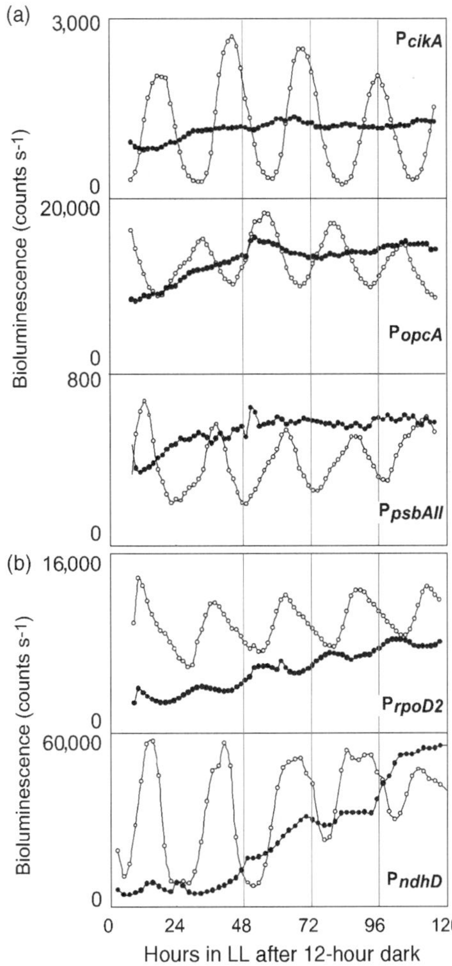

Figure 6.15. Effect of *sasA* disruption on the expression of other circadian genes. Bioluminescence rhythms comparing wild-type (open circles) to *sasA*-disrupted (filled circles) reporter strains. The promoter driving *luxAB* is indicated for each panel. (a) Three reporter strains with an arrhythmic circadian phenotype in a *sasA*-disrupted background are shown. Note that the P_{opcA} rhythm has a class 2 (180° offset) phase angle. (b) Two additional reporter strains that maintain detectable rhythmic expression patterns are shown. [Reprinted from Iwasaki et al. (2000) with permission from Elsevier Science.]

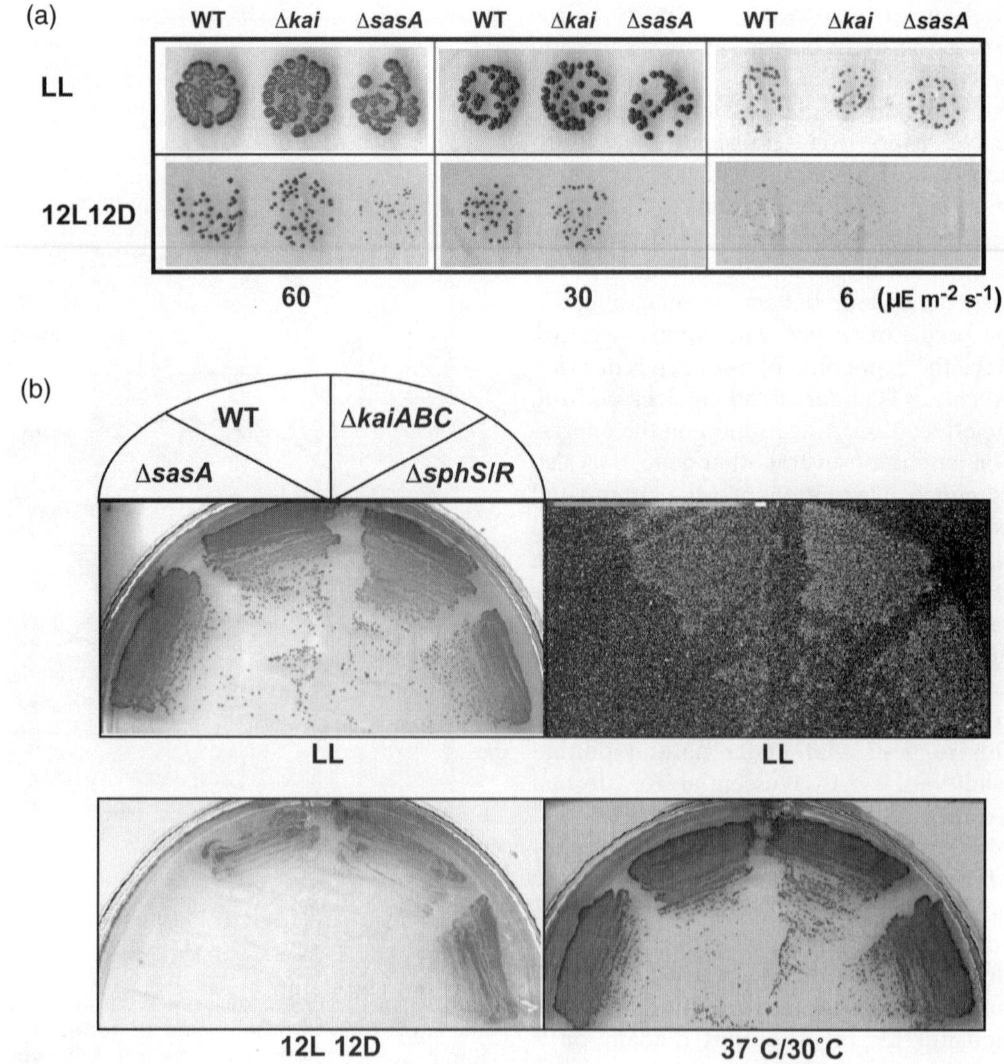

Figure 6.16. *sasA* strains show slow growth in LD but not in LL or temperature cycle conditions. (a) Colonies of wild-type, *kaiABC*-deficient, or *sasA*-disrupted strains were allowed to form on solid media for 10 days under continuous illumination (LL) or 12 hours light/12 hours dark (12L/12D) conditions at the indicated light intensity. (b) Growth phenotypes of wild type and strains lacking *sasA*, *kaiABC*, or *sphSR* on agar plates grown under LL (50 μE m^{-2}s^{-1}) and 12L (50 μE m^{-2}s^{-1})/12D at 30°C, or LL (50 μE m^{-2}s^{-1}) with 12-hour 30°C/12-hour 37°C cycles. These strains carried a P_{kaiBC}::*luxAB* reporter. [Reprinted from Iwasaki et al. (2000) with permission from Elsevier Science.]

phosphoryl relay would therefore include a SasA cognate response regulator, which may plausibly function to activate *kaiBC* gene transcription. While such a cognate response regulator has not yet been identified for SasA, all the data support the hypothesis that a two-component signaling system may be involved. The role for this postulated response regulator in the cyanobacterial circadian oscillator would be similar to that described for the PAS domain containing transcription factors, CLK/CYC, and CLOCK/BMAL1 in Drosophila and mammals, respectively, and WC1/WC2 in neurospora.

Model for Role of *SasA* in *Synechococcus* Circadian System

On the basis of their experimental data, Iwasaki and co-workers have proposed a model for SasA-mediated modulation of *kaiBC* expression (Fig. 6.17). SasA and KaiC function as both positive and negative elements in feedback control of *kai* gene expression. During the subjective day, total KaiC concentration is low, but considerable amounts of SasA-KaiC complex exist in the cell. This complex functions as a positive element by stimulating SasA autophosphorylation and subsequent activation of a cognate response regulator that directly or indirectly enhances *kaiBC* expression. SasA may act as a negative element when it inactivates its cognate response regulator with its phosphoryl-protein phosphatase activity. This would occur when SasA does not bind KaiC or when KaiB binds the SasA-KaiC complex.

Additionally, two alternatives for SasA function in cyanobacterial circadian organization have also been proposed (Fig. 6.17). The first proposes that SasA functions as the first output from the circadian timing mechanism, transmitting timing signals to all clock-controlled processes. This would encompass feedback on *kai* expression and on light sensitivity of the oscillator and metabolic adaptation to natural growth conditions. The second alternative suggests that SasA may act as a circadian amplifier to modulate a photic input pathway into the oscillator. The amplitude of the circadian oscillation in a *sasA* mutant is more sensitive to light fluence than the wild-type strain, indicating that SasA senses light directly or indirectly. Binding of SasA to KaiC could presumably alter its light sensitivity to further regulate SasA's activity.

INPUT/OUTPUT PATHWAYS OF CYANOBACTERIAL CIRCADIAN SYSTEM

Input

The circadian clock is heuristically considered to consist of three components: input pathways, which relay environmental information to a circadian pacemaker; a circadian pacemaker, which generates the daily oscillations; and output pathways through which the pacemaker regulates various observable rhythms. While circadian organization in reality is more of a feedback web/network, this model is still useful to assign roles to molecular and cellular components.

Circadian clocks are entrained to environmental changes in light and temperature. Until relatively recently, very little information existed on the molecular biology of input pathways to the cyanobacterial clock. In *Synechococcus* sp. RF1, experiments suggest that a break in photosynthetic activity by darkness is responsible for photic entrainment. A *Synechococcus* sp. RF1 rhythm mutant, designated CR1, is not fully entrainable to light–dark cycles but is to temperature cycles. CR1 may be the first cyanobacterial photic–input mutant. Other candidates for input factors are *pex* and *cikA*. *Period extender* (*pex*), is a newly identified gene that is involved in adjusting the period but does not appear necessary for robust oscillations. Disruption of the *pex* gene in wild-type cells shortens the period by 1 hour, while overexpression of *pex* lengthens the period by 3 hours, with a concomitant decrease in amplitude. Overexpression of *pex* in various cyanobacterial clock mutants causes arrhythmicity. *Pex* encodes a 148–amino acid protein and is thought to function as a modifier of the circadian clock in *Synechococcus*.

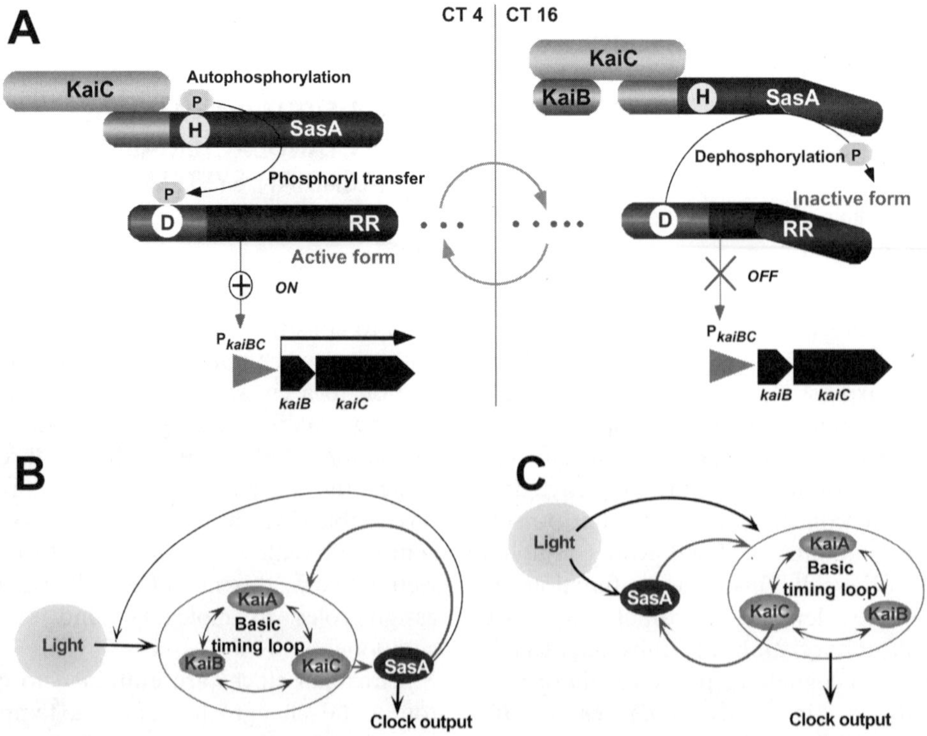

Figure 6.17. Models for SasA function in the circadian system. (a) A model for the function of the SasA-KaiC complex in *kaiBC* expression. In the early morning, when total KaiC abundance is low, KaiC may bind SasA, stimulating its autophosphorylation and thereby activating its cognate response regulator (RR) through phosphoryl transfer from His (H) to Asp (D). Activated RR is proposed to directly or indirectly activate expression of *kaiBC* genes from P_{kaiBC} (left). As the KaiBC concentration rises in the evening, the KaiC-SasA complex may bind additional KaiC or KaiB to form a larger complex(es), which reduces SasA autophosphorylation. SasA may then act as a phosphoryl protein phosphatase, inhibiting RR-mediated stimulation of P_{kaiBC} activity (right). (b,c) SasA may function as the first output of the clock to regulate downstream clock-controlled processes (b) or as an element of redundant input pathways (c). In both models, SasA and clock protein KaiC form an outer feedback loop to amplify the KaiABC-based basic timing process by modulating *kai* gene expression as depicted in (a). In model (b), KaiC-SasA complex controls its cognate response regulator and thereby directly controls downstream gene expression. In model (c), SasA controls the robustness of the circadian oscillation in a light-dependent manner, although SasA is not essential to entrain the clock. SasA may sense light signals directly or indirectly, and the binding of SasA to KaiC would alter SasA's enzymatic activity. [Reprinted from Iwasaki et al. (2000) with permission from Elsevier Science.]

cikA

cikA, circadian input kinase, is a clock associated gene that lies on the input pathway and provides phase-setting information to the *Synechococcus* clock. Persistence of robust rhythms in the *cikA* genetic background indicates that the product of this gene is not necessary for circadian oscillator function. Inactivation of *cikA* shortens the circadian period of gene expression rhythms by approximately 2 hours and changes the phase of a subset of rhythms. Wild-type *Synechococcus* strains maintained in continuous light respond to a dark pulse by changing the phase of subsequent peaks by 10–12 hours. *cikA*-inactivated reporter strains are unable to reset the phase of the clock in response to a similar 5-hour dark pulse, suggesting that *cikA* functions in an input pathway to the circadian oscillator.

CikA belongs to the family of bacteriophytochromes, has a conserved histidine kinase motif, and has a carboxyl terminus similar to the receiver domains of response regulators. The similarity of the CikA amino-terminal sequence to those of *Arabidopsis* and *Deinococcus* photoreceptors raises the possibility that CikA is a photoreceptor (Fig. 6.18).

Output

Currently, two genes, *rpoD2* and *cpmA*, have been identified as components of the output pathway. A mutation in the *rpoD2* gene lowers the amplitude of circadian oscillations of *psbAI* and *ndhD* gene expression but has no effects on that of many others, including *purF*. The *RpoD2* gene is a member of the sigma factor family and encodes a sigma70-like transcription factor. Disruption of the *circadian phase modifier A* (*cpmA*) gene alters the phase of the expression rhythm of certain genes, including *psbAII*, and *kaiA*, but has no effect on *psbAI*, *purF*, *kaiBC*, *rpoD2*, or *ndhD*. *CpmA* was identified from a trans-

(a)

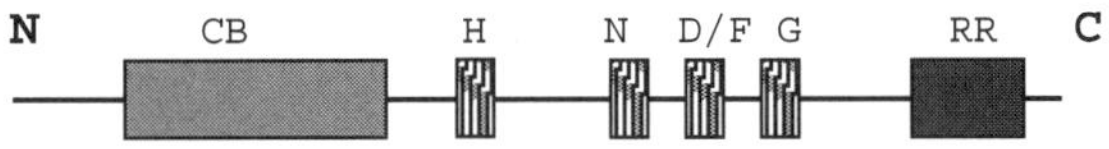

(b)

```
CikA   184   DLSEILTTTIREIRQLLNADRAIIYQFKPQCLDAGLDQRW--PLYIPSQSY--ITYEDRRNE-A
PhyE   216   DIGALCDTVVEDVQRLTGYDRVMVYQFHE-------------------DDHGEVVSEIRRSD-L
RcaE   113   QLKEILHTTVTEVQRILQADRVLIYHVLP-------------------DGTGKTISESVLPDY-
Cph1   152   NLRDFYDVIVEEVRRMTGFDRVMLYRFDE-------------------NNHGDVIAEDKRDD-M

CikA   243   LLSVIDPLVQPGLLITTEEWQRFQQGETLLIDSVG-----------------------FYKERL
PhyE   261   EPYLGLHYPATDIPQA--ARFLFKQNRVRMICDCNATPVKVVQSEELKRP--LCLVNSTLRAPH
RcaE   157   PTLMDLEFPQEVFPQ--EYQQLYAQGRVRAIADVHDP----------------------TAGLA
Cph1   196   EPYLGLHYPESDIPQP--ARRLFIHNPIRVIPDVYGVAVPLTPAVNPSTNRAVDLTESILRSAY

CikA   284   PEQYSFYERVQVRSVCKIPILVQGR------IWGLLVAHQCQQDHRWQPRERDILQHLAEH
PhyE   321   GCHTQYMANMGSVASLALAIVVKGK--DSSKLWGLVVGHHC---SPRYVPFPLRYACEFLMQ
RcaE   197   ECLVEFVDQFHIKAKLIVPIVQNLNANSQNQLWGLLIAHQCDS-VRQWVDFELELMQQLADQ
Cph1   258   HCHLTYLKNMGVGASLTISLIKDGH------LWGLIACHHQ---TPKVIPFELRKACEFFGR
```

Figure 6.18. (a) Graphic representation of the 754 amino acid CikA protein indicating the chromophore binding (CB) domain of phytochromes at the amino terminus, the H, N, D/F, and G boxes of histidine protein kinases and the receiver domain (RR) of response regulators at the carboxyl end. (b) Comparison of chromophore binding domains of CikA, PhyE from *Arabidopsis thaliana*, *Synechocystis* Cph1, and RcaE from *Fremyella diplosiphon*. Residues conserved in all sequences are in boldface type, residues conserved in some sequences are highlighted in gray. Numbers at the beginning of each line indicate position in the respective protein sequence. [Modified from Schmitz et al. (2000): *Science* **289**, 765–768.]

poson-generated mutant of *Synechococcus PCC7942* that maps to a 780-bp (base pair) open reading frame. *CpmA* homologs have been identified in *Synechocystis* sp. strain PCC6803 and in three species of *Archaea*. All these organisms also have a putative *kaiC* homolog. This suggests that *kaiC* may also encode a central clock component in other prokaryotes and that *cpmA* acts as a circadian output factor in those organisms as well.

Control of Cell Division by the Circadian Clock

Until fairly recently, it was widely accepted that circadian rhythms were exclusively found in eukaryotes. This was a plausible idea, since it was assumed that an endogenous timekeeper with a period close to 24 hours would not be very useful to organisms that divide more rapidly than once very 24 hours, as do many prokaryotes. The discovery of the circadian oscillation of nitrogen fixation in 1986 dispelled this idea. The subsequent development of molecular markers allowed researchers to observe the relationship between the cell division cycle and the circadian cycle in actively dividing cultures. It was found that cell division in *Synechococcus* is actually gated by the circadian clock; thus it is a circadian output.

Cultures of *Synechococcus elongatus* grow with doubling times as rapid as one division per 6–10 hours. In order to investigate circadian rhythms in these cells two research groups concurrently examined gene expression in *Synechococcus elongatus* using the AMC149 strain that carries a luciferase reporter. As described earlier, AMC149 contains a *PpsbAI:luxAB* translational fusion integrated into a neutral site of the wild-type strain. The *psbAI* gene encodes the photosynthetic protein D1. When the luciferase gene *luxAB* is correctly oriented to the promoter *PpsbAI*, luciferase is expressed and the cyanobacterial colony glows. Kondo and Golden demonstrated that cultures of *Synechococcus elongatus* that divided twice or more per day displayed circadian rhythms of bioluminescence and levels of mRNA. Mori and Johnson found that cell division was prohibited for several hours in the circadian cycle. In actively growing continuous cultures in light–dark, cells divided in the light but not in the dark. In constant light a circadian pattern of division continued even though the doubling time of the culture was faster than once every 24 hours (Fig. 6.19). Cell division occurred in the late subjective night and the subjective day. Cells slowed or stopped dividing early in the subjective night. What this meant was that although cells were dividing more than twice every 24 hours, the timing of division was under the control of a circadian oscillator. *Synechococcus* cells are therefore able to keep track of two temporal circuits independent of each other.

The independence of the circadian timing circuit from the cell division cycle has also been demonstrated by overexpressing the cell septum protein FtsZ in *S. elongatus*. When the FtsZ protein is overexpressed, cell division but not cell growth is suppressed. This results in green filamentous cells, and not unicells, being formed (Fig. 6.19). This inhibition of cell division does not affect gene expression rhythms (Fig. 6.19), implying that the periodicity of the circadian oscillations is stable and independent of the division state.

■ ADAPTIVE SIGNIFICANCE OF CYANOBACTERIAL CIRCADIAN CLOCK

Cyanobacteria Fitness

Resonating circadian clocks confer enhanced fitness in cyanobacteria. From an evolutionary perspective, it is believed that

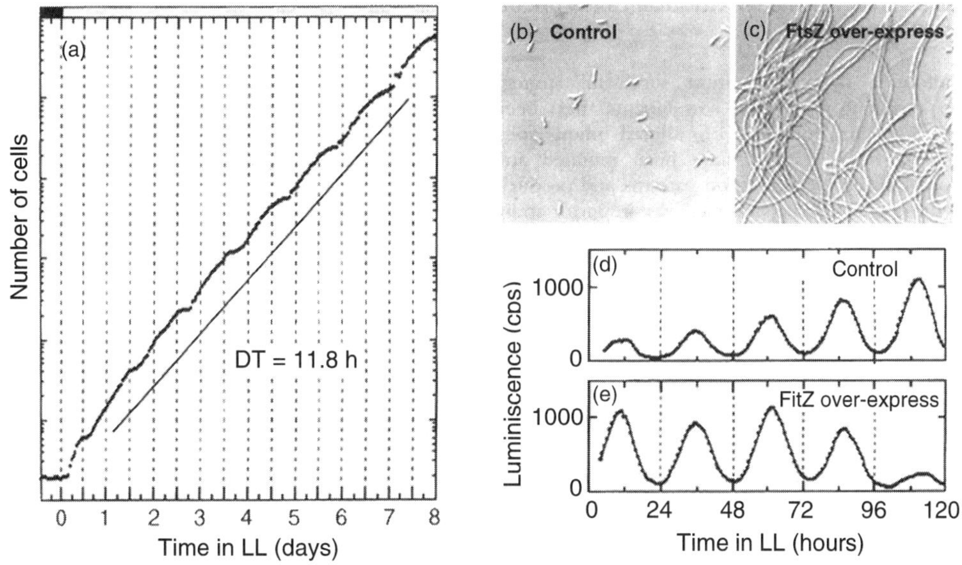

Figure 6.19. Circadian rhythm of cell division in continuous cultures of rapidly growing *S. elongatus* cells and luminescence rhythms in dividing versus nondividing cells. (a) Growth curve of log-phase culture. (b–e) Cell division of growing bacteria is stopped by overexpression of FtsZ. Morphology of control cells (b), cells in which FtsZ is overexpressed (c). Luminescence rhythms in control cells (d) and in FtsZ overexpressing cells (e). [Reprinted from Mori and Johnson (2001) with permission from Elsevier Science.]

circadian clocks improve survival by aiding adaptation to the environment. While it is argued that circadian regulation enables a more predictive control, the merit of circadian control is not obvious, as clock mutants in several species often survive as well as wild-type organisms in the laboratory.

The value of circadian programming to reproductive fitness in cyanobacteria was tested by using wild-type and mutant *Synechococcus* strains that exhibited different periods under constant conditions. The strains tested all grew in pure culture at essentially the same rate under light–dark conditions and under conditions of constant light, so there did not appear to be any advantage to having a particular circadian period. However when the different strains were mixed together and grown in competition with each other a pattern emerged that depended on the endogenous period of each strain and on the period of the light–dark cycle. Light–dark cycles that provided equal amounts of light and dark but with varying periodicity were tested. In all cases examined, the strain whose period most closely matched the period of the light–dark cycle outgrew its competitors (Fig. 6.20). It has been suggested that there are two possible reasons for this observation: (1) clocks allow optimal utilization of limiting resources such as CO_2, light and nutrients—clearly the strain whose period matches that of the environment can best exploit these resources; and (2) cyanobacteria rhythmically secrete inhibitory factors that dampen the growth of other strains. While there are no physiologic data to support either possibility, it is clear that the phase relationships between biological rhythms and environmental cycles are important and having an endogenous period that approximates that of the external cycle confers a selective advantage.

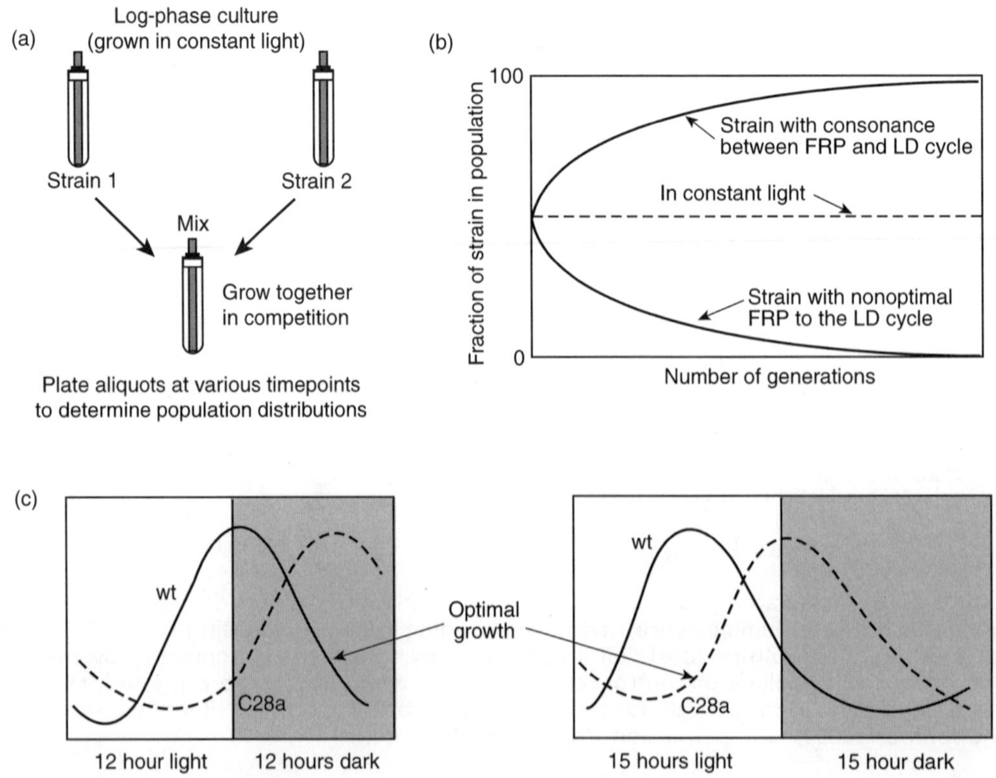

Figure 6.20. Competition of circadian mutant strains in different light–dark cycles. (a) Different strains of cyanobacteria were mixed together in batch cultures and grown in competition under different light–dark cycles. At 8-day intervals, the cultures were diluted with fresh medium. At various times during the competition, aliquots were plated as single colonies, and the luminescence rhythms of individual colonies were monitored to determine the frequency distribution of the different circadian phenotypes. (b) The strain whose endogenous free-running period most closely matched that of the environmental light–dark cycle was able to outcompete strains with a nonoptimal period. In constant light (nonselective conditions), all the strains were able to maintain their initial fraction in the population. (c) Phase of a luminescence rhythm for wild-type (= wt) and the mutant exhibiting a 30-hour period (= C28a) in a light–dark cycle of 12 hours light/12 hours dark (left) versus 15 hours light/15 hours dark (right). [From Johnson (2001) with permission from the *Annual Review of Physiology*, volume 63. © 2001 by Annual Reviews, www.annualreviews.org.]

Circadian Rhythms in Other Prokaryotes

As discussed earlier, it was previously believed that prokaryotic organization and rapid growth lifestyle were not compatible with circadian rhythms. It has since been established that the circadian clock is an integral part of the cyanobacterial organization. Further, circadian programming clearly has an adaptive value for the photosynthetic cyanobacteria. This raises the question of whether a circadian clock may provide similar value in nonphotosynthetic prokaryotes. There had been a report that *E. coli* might have a circadian oscillator, but that report was based on an old study with poor temperature control, and it appears that the daily trends observed were a result of diurnal temperature cycles. Nevertheless, researchers think that there is a good possibility that circadian clocks that enhance adaptation and fitness will be found in other eubacteria.

Evolution of the Circadian Clock

The physiological properties of circadian rhythms are so remarkably similar in a wide range of genera that it was generally accepted that the circadian oscillatory mechanism was developed once and was evolutionarily conserved. Clock models of similar structure have been proposed for cyanobacteria, fungi, and animals. However, following the cloning of the clock genes, it has become apparent that key proteins, encoded by these genes, do not share sequence similarity. The clock systems in cyanobacteria, fungi, animals, and probably plants appear to have developed independently. Natural selection by the alternating environment has led varied organisms to evolve similar clock mechanisms like the negative feedback of gene expression. In each group of organisms, the feedback loop might have developed a mechanism for temperature-compensated 24-hour periods and evolved into the circadian clock.

The cyanobacterial *kai* genes are interesting from an evolutionary perspective for several reasons. First, the *kai* genes are generally well conserved among cyanobacterial species and potential homologs have also been identified in Archaea. Genes similar to *kaiB* and *kaiC* have been found in Archaea and in at least two other lineages of photosynthetic Eubacteria. The amino acid identity between the cyanobacterial and Archaeabacterial proteins range between 25.7 and 34.2%. More importantly, the P loop motif, GXXXXGK(T/S), a GTP/ATP binding region, is highly conserved. To date a circadian clock has not been identified in Archaea. Next, since cyanobacteria are believed to be ancestral to chloroplasts there is the distinct possibility that there may be a phylogenetic link between cyanobacterial and plant circadian clocks. The fact that many chloroplast encoded genes in *Chlamydomonas* and in higher plants are expressed in a circadian manner raises the likelihood that *kai* genes could have been introduced into ancestral plant cells together with photosynthesis genes. However, no homolog of *kai* has been found in plant genomes thus far sequenced. Finally, a genomic survey asserts that *kaiC* is a member of the bacterial RecA/DnaB family. RecA is an ATP-dependent DNA recombinase, and DnaB is the replication fork helicase in bacteria. It is hypothesized that the ancestral *kaiC* was a single-domain RecA-like protein that originated in eubacteria and was laterally transferred to Archaea, where gene duplication and fusion occurred. The double-domain gene was subsequently transferred from the Archaea to cyanobacteria after the main eubacterial lineages had been established.

The discovery of genes encoding proteins similar to eukaryotic photoreceptors (cryptochromes and phytochromes) in *Synechocystis* and the existence of CikA, a member of the bacteriophytochrome family, in *Synechococcus* further suggests an evolutionary link between the cyanobacterial clock and those of higher organisms.

■ CONCLUSIONS

Although developed recently (as of 2003), the cyanobacteria model for circadian rhythms has already made important contributions to our understanding of circadian biology. While components of this molecular clock are not conserved in animal systems, the mechanisms certainly are, and it is likely that the components will be similar to plant circadian proteins. In addition, cyanobacteria provide a powerful tool to study the relationship between the circadian cycle and basic cellular processes such as the cell cycle. Many other aspects of the circadian control of cell metabolism are likely to emerge from studies of the cyanobacteria clock.

FURTHER READING

Ishiura, M, Kutsuna, S, Aoki, S, Iwasaki, H, Andersson, CR, Tanabe, A, Golden, SS, Johnson, CH, Kondo, T (1998): Expression of a gene cluster *kaiABC* as a circadian feedback process in cyanobacteria. *Science* **281**: 1519–1523.

Iwasaki, H, Williams, SB, Kitayama, Y, Ishiura, M, Golden, SS, Kondo, T (2000): A KaiC-interacting sensory histidine kinase, SasA, necessary to sustain robust circadian oscillation in cyanobacteria. *Cell* **101**: 223–233.

Iwasaki, H, Kondo, T (2000): The current state and problems of circadian clock studies in cyanobacteria. *Plant Cell Physiol* **41**: 1013–1020.

Johnson, CH, Golden, SS, Ishiura, M, Kondo, T (1996): Circadian clocks in prokaryotes. *Mol Microbiol* **21**: 5–11.

Johnson, CH, Golden, SS (1999): Circadian programs in cyanobacteria, adaptiveness and mechanism. *Annu Rev Micrbiol* **53**: 389–409.

Johnson, CH (2001): Endogenous timekeepers in photosynthetic organisms. *Annu Rev Physiol* **63**: 695–728.

Kondo, T, Mori, T, Lebedeva, NV, Aoki, S, Ishiura, M, Golden, SS (1997): Circadian rhythms in rapidly dividing cyanobacteria. *Science* **275**: 224–227.

Kondo, T, Strayer, CA, Kulkarni, RD, Taylor, W, Ishiura, M (1993): Circadian rhythms in prokaryotes: Luciferase as a reporter of circadian gene expression in cyanobacteria. *Proc Natl Acad Sci USA* **90**: 5672–5676.

Lakin-Thomas, PL (2000): Circadian rhythms: New functions for old clock genes. *Trends Genet* **16**: 135–142.

Liu, Y, Golden, SS, Kondo, T, Ishiura, M, Johnson, CH (1995): Bacterial luciferase as a reporter of circadian gene expression in cyanobacteria. *J Bacteriol* **177**: 2080–2086.

Lorne, J, Scheffer, J, Lee, A, Painter, M, Miao, VP (2000): Genes controlling circadian rhythms are widely distributed in cyanobacteria FEMS. *Microbiol Lett* **189**: 129–133.

Mori, T, Binder, B, Johnson, H (1996): Circadian gating of cell division in cyanobacteria growing with average doubling times less than 24 hours. *Proc Natl Acad Sci USA* **93**: 10183–10188.

Mori, T, Johnson, CH (2001): Circadian programming in cyanobacteria. *Semin Cell Dev Biol* **12**: 271–278.

Schmitz, O, Katayama, M, Williams, SB, Kondo, T, Golden, SS (2000): CikA, a bacteriophytochrome that resets the cyanobacterial circadian clock. *Science* **289**: 765–768.

Schneegurt, MA, Sherman, DM, Nayar, S, Sherman, LA, Mitchell, CA (1994): Oscillating behavior of carbohydrate granule formation and dinitrogen fixation in the cyanobacterium Cyanothece sp. strain ATCC 51142. *J Bacteriol* **176**: 1586–1597.

7

MOLECULAR ANALYSIS OF CIRCADIAN RHYTHMS IN NEUROSPORA

Amita Sehgal

Neurospora crassa, commonly known as the "bread mold," has been exploited as a genetic system to study many different aspects of biology. Its potential use for the study of clocks was first noted in 1953 with the description of a rhythmic growth pattern in a Neurospora strain lacking the amino acid proline (thereby called a *proline auxotroph*). A few years later, Pittendrigh, also known for his contributions to Drosophila circadian biology, reported that the rhythm was circadian. He also found that the period was unchanged at different temperatures, thereby satisfying the criterion of temperature compensation expected of a circadian rhythm (see Chapter 1). Sargent and colleagues went on to develop assays for the Neurospora rhythm and to characterize it at many levels including determining its response to light and temperature. Since then, there has been no looking back as researchers have used first genetics and then molecular tools to uncover the complexities of the Neurospora circadian system.

■ OVERT RHYTHMS IN NEUROSPORA

Production of Asexual Spores (Conidia)

Like other filamentous fungi, Neurospora spread over the surface of the medium in the form of undifferentiated filaments (mycelia). A developmental switch, that is under circadian control, results in the pro-

Molecular Biology of Circadian Rhythms. Edited by Amita Sehgal
ISBN 0-471-41824-2

duction of aerial hyphae (mycelia growing perpendicular to the medium) that elaborate asexual spores (conidia). The switch usually occurs toward the end of the subjective night, and the differentiation process ends around the middle of the subjective day, at which point surface filaments dominate the culture again. This rhythm of conidiation, as it is called, is easily monitored in race tubes, which are basically glass tubes containing a solid medium on which the fungus grows (Fig. 7.1). The position of the growth front is marked at regular intervals and a constant linear growth rate is assumed. At 25°C these rhythmic growth patterns occur with a periodicity of ~21.6 hours.

Through molecular analysis, it is now becoming clear that aspects of Neurospora metabolism and function other than conidiation are under circadian control. Thus, a number of output cycling genes have been isolated that are not required for conidiation, but play an important role in other processes. One such process, that is most likely regulated by the clock, is the response to stress. The identification of other rhythms, including those of gene expression, provides researchers with new assays that could either replace or supplement the race tube.

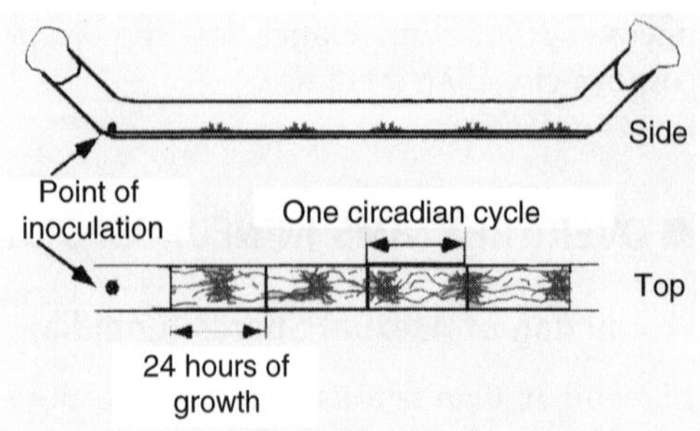

Figure 7.1. The race tube assay for measuring circadian rhythms in Neurospora. Conidiation, characterized by the production of a cluster of aerial hyphae, can be directly visualized in the race tube. The endogenous rhythm is <24 hours, as is evident from the pattern in the tube. [Reprinted from Bell-Pedersen et al. (1996) with permission from the Indian Academy of Sciences.]

■ MOLECULAR BASIS OF THE NEUROSPORA CLOCK

The *frq* Gene as a Clock Component

As in Drosophila, progress in the Neurospora circadian field has relied largely on the use of genetics. The first circadian rhythm mutations in Neurospora were found by Jerry Feldman in 1971, about the same time that Ron Konopka reported the Drosophila *period* mutants. Many of the original mutant alleles mapped to the *frequency* (*frq*) gene, which thus became to Neurospora circadian biology what *per* was to Drosophila. *frq* mutants affect circadian rhythms in Neurospora in many different ways, ranging from altered periodicity to arrhythmicity under normal, race tube assay conditions (Fig. 7.2). Many of them also affect temperature compensation.

The *frq* gene was isolated in Jay Dunlap's laboratory and shown to encode a protein of unknown function. Although *frq* shares a glycine-threonine/glycine-serine repeat region with *per*, this is not indicative of a common origin of these two genes, which are otherwise very dissimilar. Nor did this repeat provide any clues to the biochemical function of *frq* (note that the presence of this repeat in *per* was initially thought to indicate a proteoglycanlike nature of the protein, but that notion was subsequently dispelled). The molecular lesions underlying mutant *frq* phenotypes were identified and found to be largely missense mutations resulting in single amino acid subsitutions (Fig. 7.3).

As in the case of *per*, it was the regulation of *frq* that turned out to be the most intriguing aspect of this gene. Thus, *frq* RNA and protein levels also cycle with a circadian rhythm, and the protein nega-

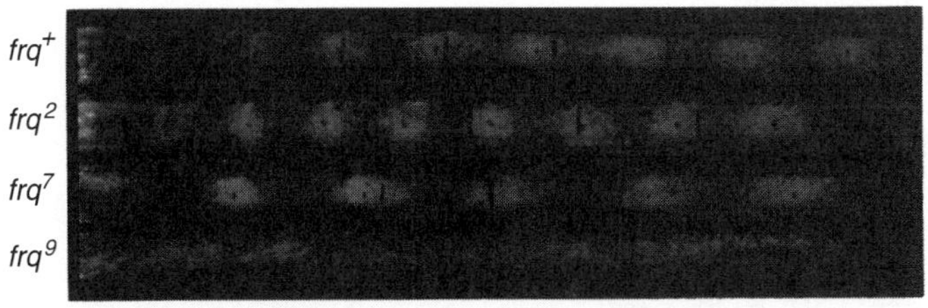

Figure 7.2. Effects of *frq* mutants on the conidiation rhythm. *frq*^{2} is a short-period mutant as seen by the advancing of its phase relative to wild type each day. *frq*^{7} lengthens period, as demonstrated by the progressive delay in its phase relative to wild type. *frq*^{9} is a null that lacks a circadian rhythm of conidiation. [Reprinted from Bell-Pedersen et al. (1996) with permission from the Indian Academy of Sciences.]

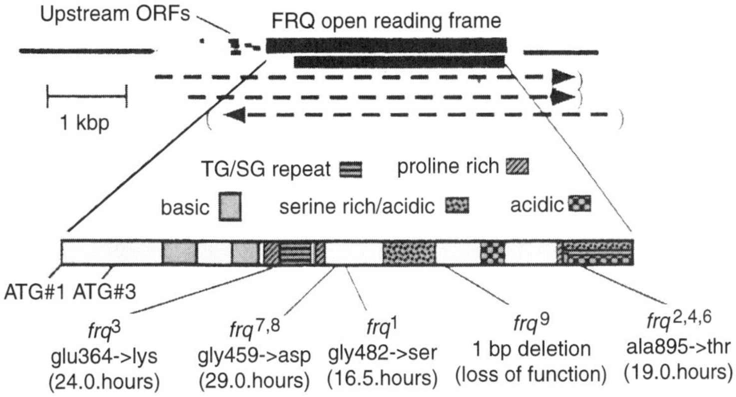

Figure 7.3. Molecular basis of different *frq* mutants. Major domains and translation initiation sites in *frq* are depicted schematically. The locations as well as amino acid changes associated with mutations that shorten and lengthen period are indicated. The frameshift mutation in *frq*^{9} truncates the protein. (ORF—open reading frame; TG—threonine glycine; SG—serine glycine.) [Reprinted with permission from the *Annual Review of Genetics*, Volume 30, 1996, by Annual Reviews (www.annualreviews.org).]

tively regulates the synthesis of its own mRNA, resulting in a feedback loop analogous to the ones we have read about in the preceding chapters. Constitutive expression of high levels of *frq* eliminates the overt rhythm and acute induction of *frq* resets the clock. Together, these data demonstrate the importance of *frq* cycling to the Neurospora circadian system.

Post-translational Regulation of FRQ

The ability to manipulate gene expression together with the use of pharmacological agents has allowed dissection of the Neurospora molecular cycle at a level that is difficult in other organisms. To examine each aspect of the feedback loop independently, a *frq* transgene driven by a quinic acid inducible promoter, was introduced into a *frq*^{9} (*frq*-null) background (Fig. 7.4). Treatment of this strain with quinic acid resulted in rapid induction of *frq* RNA followed by the accumulation of FRQ protein. To address the time course of feedback, levels of endogenous *frq*^{9} RNA were assayed. It was found that feedback, as assayed by a decrease in *frq*^{9} levels, was achieved in 3–6

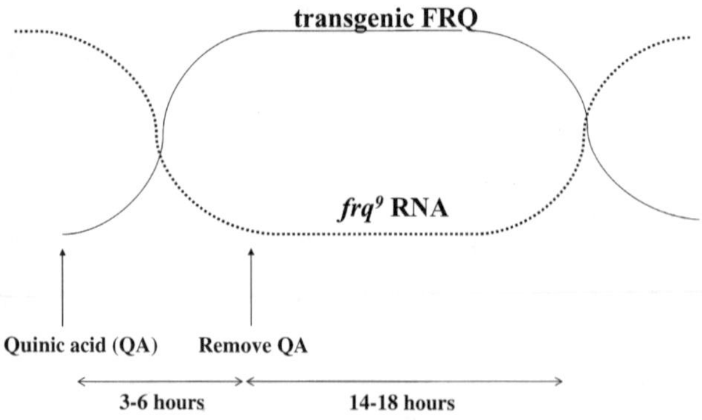

Figure 7.4. Time course of feedback inhibition. *frq* was expressed under control of a quinic acid–inducible promoter in a frq^9 background. The time course of feedback on frq^9 RNA was examined by inducing expression of transgenic *frq* with quinic acid and monitoring levels of the endogenous frq^9 transcript (dotted line). As indicated, negative feedback is achieved rapidly and stays in effect for several hours after the quinic acid is removed. The extended period of feedback correlates with the amount of time the transgenic FRQ protein (solid line) stays around. [Adapted from Merrow et al. (1997).]

hours and could be effected by as few as 10 molecules of FRQ per nucleus. Interestingly, the peak of FRQ protein normally lags behind the peak of its mRNA by 4–6 hours, which corresponds to the time taken to achieve negative feedback in this experiment. Recovery from feedback, again as indicated by levels of frq^9 RNA, takes much longer. Thus, *frq9* RNA levels do not return to normal for 14–18 hours. The FRQ protein stays around for ~14 hours and, although levels have decreased by this point, they are still sufficient to mediate repression.

As described earlier for the Drosophila proteins, FRQ is progressively phosphorylated as it accumulates. Phosphorylation regulates turnover and is also important for other aspects of FRQ function. Blocking phosphorylation with kinase inhibitors results in decreased FRQ turnover and lengthens circadian period. Biochemical experiments in Y. Liu's laboratory led to the identification of a calcium/calmodulin-dependent kinase, CaMK1, as being responsible for half of the FRQ kinase activity. Strains lacking this kinase have very modest effects on phase and period, but do appear to alter light responses of the conidiation rhythm.

A second kinase that phosphorylates FRQ and was found to copurify with it is casein kinase II (CKII). Null mutants of this kinase affect general growth and development of Neurospora, thereby precluding analysis of the conidiation rhythm. However, in the absence of this kinase, *frq* RNA and protein fail to cycle. Both are expressed at high levels, supporting the idea that hypophosphorylated FRQ is resistant to degradation and, in addition, indicating that phosphorylation is important for negative feedback, specifically, for reducing levels of *frq* mRNA.

FRQ can also be phosphorylated by casein kinase I α and β. Recall that in Drosophila and mammals, CKIε plays an important role in the central feedback loop. In Neurospora, mutants of CK1 show some effects on rhythms, but its function in the clock is not clear yet.

wc-1 and *wc-2* Genes as Positive Regulators of *frq*

The WC-1 and WC-2 proteins were first identified as clock-related components on the basis of their requirement for the induction of *frq* by light (further discussed below). Work done in the Macino laboratory, where the two *wc* genes were cloned, showed that these proteins mediate all photic responses in Neurospora, in particular responses to blue light. The Loros and Dunlap laboratories subsequently demonstrated that not only are both genes required for the *frq* light response, they also operate in constant darkness as integral components of the circadian feedback loop.

The two proteins, WC-1 and WC-2, are PAS-domain-containing transcription factors that act together as a complex to activate expression of *frq*. Thus, they are the Neurospora equivalents of the CLK/BMAL1 complex found in Drosophila and mammals and they function similarly to activate expression of the autoregulating, so-called "negative" component. In Drosophila, these negative components are encoded by the *per* and *tim* genes, in mammals they are PER and CRY and in Neurospora it is *frq*. In addition to PAS domains, each of the two WC proteins contains a transcriptional activation domain and a zinc finger DNA binding domain (Fig. 7.5). In this respect, WC-1 and WC-2 differ from their animal counterparts, which fall in the category of basic helix–loop helix (bHLH) transcription factors rather than zinc finger factors. In addition, WC-1 contains an LOV (for "light," "oxygen," and "voltage") domain, which is similar to a PAS domain and is found in molecules implicated in photoreception, most notably in the plant photoreceptor NPH1.

FRQ dimers interact directly with WC-2 and WC-1 and, in doing so, inhibit the transcriptional activity of the WC complex (WCC). Analysis of these interactions in the CKII mutant background have shed some light on the role of phosphorylation in this process. It turns out that the defect in negative feedback in this mutant is not due to reduced association of FRQ with the WCC. Increased amounts of the WC proteins coimmunoprecipitate with FRQ in CKII mutant Neurospora, suggesting that it is the hypophosphorylated form of FRQ that preferentially associates with the WCC. Thus, it appears that association with the WCC and repression of its activity are two separable functions of FRQ—the former involves predominantly hypophosphorylated FRQ, while the latter requires phosphorylation by CKII. Degradation of FRQ in the latter part of the night allows WCC-mediated transcription of *frq* mRNA to resume.

As one might predict, loss-of-function mutations in any of these three genes have dramatic effects on overt rhythmicity. While the null mutations lead to loss of rhythmicity under standard conditions, some alleles of *frq* and *wc-2* alter circadian period. Period altering mutations have not yet been reported for *wc-1*.

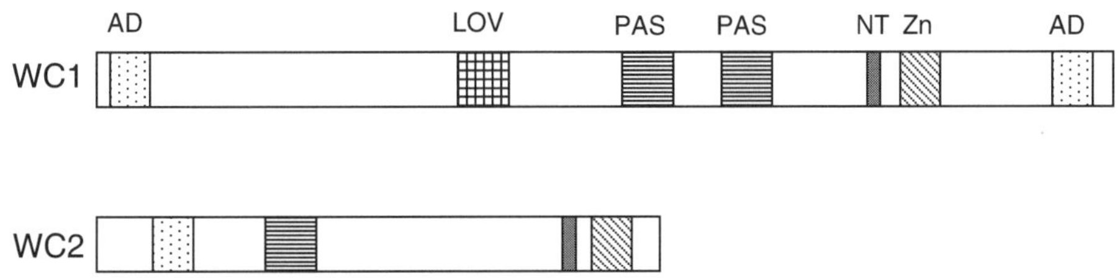

Figure 7.5. Schematic representation of the WC-1 and WC-2 proteins (AD—acidic domain; NT—nuclear transport signal; Zn—zinc finger). WC-1 contains two PAS repeats and a LOV domain, while WC-2 has only one PAS domain. [Adapted from Ballario et al. (1998): Roles in dimerization and blue light photoresponse of the PAS and LOV domains of *Neurospora crassa* white collar proteins. *Mol Microbiol* **29**: 719–729 and from Lee et al. (2000).]

Function of WC-1

WC-1 expression cycles and is regulated by FRQ. The WC-1 protein shows 48% similarity to the mammalian BMAL1 protein. This degree of homology in proteins from two such distant organisms is striking and attests to the evolutionary conservation of circadian mechanisms. *wc-1* RNA is induced by light, but it does not cycle under freerunning conditions. However, the protein cycles, although with a phase that is different from what is observed for FRQ. Thus, levels of WC-1 peak at night while FRQ peaks during the day (Fig. 7.6). The phase is consistent with WC-1's role as a transcriptional activator of *frq*, given that the expression of *frq* RNA precedes the expression of FRQ by several hours. WC-1 is also regulated by posttranslational mechanisms. In response to light, it is phosphorylated and degraded, suggesting a causal relationship between the two processes.

In *frq*-null mutants, levels of WC-1 are low suggesting that FRQ is a positive regulator of WC-1. In fact, induction of FRQ, using an inducible promoter, results in upregulation of WC-1 levels without altering *wc-1* transcript levels. The expression of WC-1 under these conditions lags behind FRQ expression by ~8 hours, consistent with the relative profiles of these two proteins under normal wild-type conditions. This cross-regulation of FRQ and WC-1 indicates that in Neurospora, as in the other systems we have read about in previous chapters, there are interlocked molecular loops at the heart of the clock mechanism (Fig. 7.7).

To determine whether FRQ regulates WC-1 at the level of synthesis or degradation, Lee et al. grew Neurospora in medium containing cycloheximide, which blocks new protein synthesis, in the presence of light as well as in the dark. They found that, under these conditions, levels of WC-1 were independent of FRQ levels. FRQ levels were substantially higher in the light but were degraded at the same rate in light as well as dark. WC-1, on the other hand, showed more rapid turnover in the light as compared to the dark, indicating that its turnover is independently controlled. The overall levels also did not support a role for FRQ in WC-1 turnover. Thus, FRQ controls WC-1 synthesis through an unknown posttranscriptional mechanism.

Function of WC-2

Levels of WC-2 do not cycle. The protein is expressed constitutively in nuclei in abundant nonlimiting quantities. The primary role of WC-2 in freerunning clock function is apparently to mediate interactions between WC-1 and FRQ. FRQ and WC-1 do not interact in the absence of WC-2.

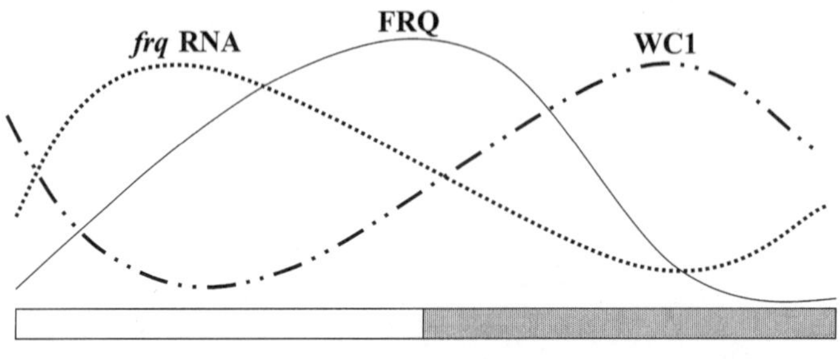

Figure 7.6. Molecular oscillations of *frq* and WC. WC-1 and *frq* are the two known oscillating components of the Neurospora clock. While *frq* RNA as well as protein cycle, WC-1 cycles only at the level of the protein with a phase different from that of FRQ. The oscillations are shown relative to the light–dark cycle (depicted by the bars at the bottom).

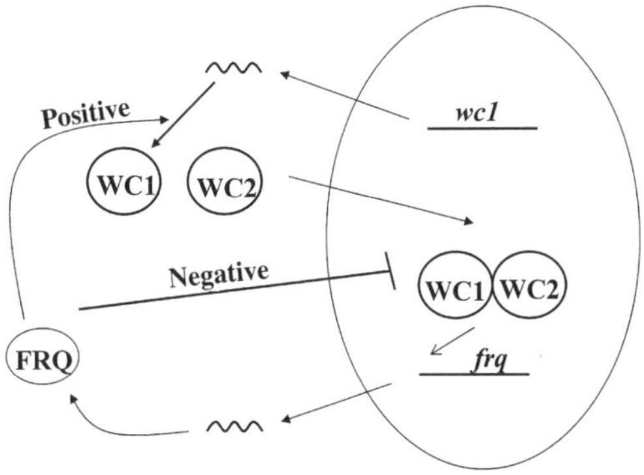

Figure 7.7. The molecular clock in neurospora. FRQ negatively regulates its transcription by blocking activity of the WC complex in the nucleus. At the same time it positively regulates expression of the WC-1 protein, resulting in an interlocked loop. [Adapted from Lee et al. (2000).]

Thus, nuclear entry of FRQ leads to its WC-2-dependent interaction with WC-1, thereby inhibiting the activity of the WC complex. Degradation of FRQ relieves this repression, allowing the WC-1/WC-2 heterodimer to resume transcription of the *frq* gene.

Rhythms in *frq*-null Revealed by Metabolic Defects

Analysis of circadian rhythms in Neurospora is complicated by aspects of development and physiology that either interfere with circadian rhythms or else display rhythms themselves. The presence of these rhythms in *frq*-null strains has also challenged the notion that *frq* is the central oscillator.

Some of the metabolic mutations that affect circadian periodicity will likely turn out to affect functioning of clock proteins. For instance, the *cys-9* mutation, which occurs in the gene encoding thioredoxin reductase, has dramatic effects on period length. The phenotype is initially a period that is 5 hours shorter than the wild type, but after a few days in freerunning conditions irregular periodicity, that can be long or short, is observed. Given reports postulating a role for redox activity in clock entrainment and perhaps also in clock output, it is easy to see how thioredoxin reductase might be involved in the circadian system. Other metabolic mutants have small effects on period that probably do not result from direct effects on the clock. For instance, the *arg-13* mutation affects an amino acid carrier in the mitochondrial membrane. It is likely that the overall effect on cellular metabolism produces a small change in circadian periodicity.

The most intriguing interactions between metabolic processes and clock function are seen with lipid-deficient strains *cel-1* and *chol-1* (Table 7.1). The *cel-1* mutant is defective in fatty acid synthesis and so displays a reduced growth rate unless exogenous fatty acids are provided. In the absence of saturated fatty acids or in the presence of unsaturated fatty acids, the conidiation rhythm in *cel-1* strains is poorly temperature compensated, such that the periods are longer at lower temperatures. In addition, the presence of unsaturated fatty acids also lengthens the period at ambient temperature (22°C). Interestingly, the period lengthening defects are also observed when *cel-1* mutants are assayed in a *frq*-null background (Table 7.1). However, the *frq;cel-1* rhythms do not entrain to light.

The *chol-1* strain requires choline for its growth. In the absence of choline it displays abnormal, swollen morphology, reduced growth rate, and altered circadian rhythms that range from arrhythmia to extremely long periodicity. The addition of choline to the medium increases the growth rate and also reduces the severity of the circadian phenotype. As in the case of *cel-1*, long-period rhythms displayed by *chol-1* in the absence of the nutritional supplement (choline) are also observed in *frq*-null backgrounds, but are not entrained by light (Table 7.1).

Given that the lipid mutations affect growth in general, it is difficult to dissociate

TABLE 7.1. Experiments Using Null Alleles (Either *frq*^{9} or *frq*^{10}) as *frq* Mutants

Mutant	Lipid-Deficient Conditions	Lipid Supplemented Conditions	Light Entrainability
cel-1	Loss of temperature compensation; long periods in the presence of unsaturated fatty acids	Wild type	Yes
chol-1	Long period	Wild type	Yes
frq; cel-1	Long period	Arrhythmic	No
frq; chol-1	Long period, loss of temperature compensation	Arrhythmic	No

Source: Based on data from Lakin-Thomas and Brody (2000).

the effect on circadian rhythms from the general effect on growth. It is possible that they unmask an underlying low-amplitude rhythm in growth rate that is present in all strains. The reduction of growth rate increases the amplitude of the rhythm. In support of this idea, choline-starved wild-type strains will phenocopy the *chol-1* strain to produce a long-period rhythm. Since the background in this case is *frq*+, this rhythm can be entrained by light. The oscillator activity observed under choline starvation conditions may correspond to the oscillator revealed by temperature cycles in *frq*^{9} strains. It is important to remember that the growth rhythms, whose periods can vary greatly, do not meet the criteria for "circadian rhythms" (entrainability by light, temperature compensation, etc.), nor are they known to be controlled by the kind of intracellular loops that lie at the heart of circadian control.

RESPONSE OF NEUROSPORA CLOCK TO LIGHT AND TEMPERATURE

Photic Resetting of the Clock by *frq*

The molecular events that reset the clock in response to light were first described in Neurospora. As in the other organisms described earlier, light effects changes in the levels of a clock component, namely, *frq*. Levels of *frq* RNA and thereby that of FRQ protein are increased by light exposure in both the early part of the night as well as toward the end of the night. As *frq* levels are normally high during the day, what this does is to reset the clock to a "day" timepoint, with the early night exposure resetting the clock to dusk and the late-night exposure resetting to dawn; thus, the phase is delayed or advanced, respectively (Fig. 7.8). In this model, Neurospora is reset primarily by a dawnlike cue (lights on) and not by a dusk-like cue. The discovery, in 2003, of an antisense transcript encoded by the *frq* locus has added to our understanding of the mechanisms involved. The antisense transcript is also induced by light and serves to modulate resetting of the clock.

The increase in *frq* (sense mRNA) levels in response to light occurs at the level of transcription. In fact, through deletion analysis, two light-response elements (LREs) have been identified in the *frq* promoter. Each LRE by itself can mediate a limited light response. Thus, deletion of the distal LRE reduces the response by ~50%, while deletion of the proximal one (i.e., the one closer to the transcription start site)

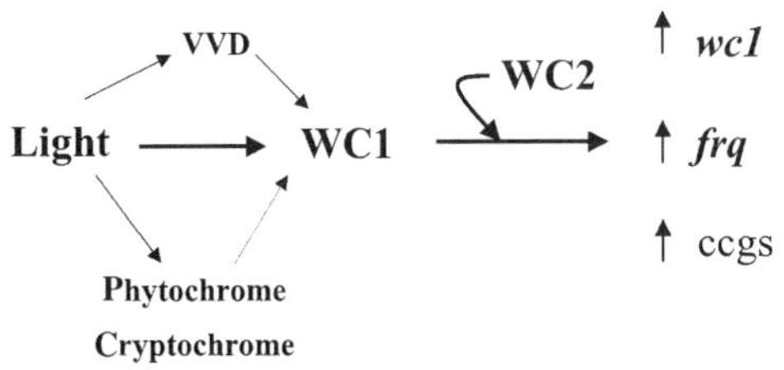

Figure 7.8. Resetting of the clock in response to light. The circadian photoreceptor is the product of the *wc-1* gene, although VVD, and possibly phytochrome and cryptochrome, play a modulatory role. WC-1 recruits its partner, WC-2, and activates transcription of *frq*. [Adapted from Froehlich et al. (2002).] Since WC-1 itself and other clock-controlled genes (ccgs) are also upregulated by light, it is shown to activate transcription of these genes also.

reduces light-induced *frq* induction by ~70%. However, deletion of both completely eliminates the light response. In addition, the LREs can confer light inducibility onto a reporter gene, establishing their role in driving light-induced gene expression.

Mediation of *frq* Light Response by the WC-1 Photoreceptor

As mentioned above, WC-1 contains a LOV domain, which had been found in photoreceptors in other organisms and was known to bind flavin mononucleotide. On the basis of the sensitivity of Neurospora rhythms to blue light, flavin-based photoreceptors had been predicted. Consistent with this idea, mutations in the LOV domain of WC-1 were shown, by the Macino laboratory, to render Neurospora blind to blue light. More recently, the Loros and Dunlap laboratories demonstrated, through an elegant series of experiments, that WC-1 is a blue-light photoreceptor that mediates circadian responses to light.

The WC proteins were found to bind the *frq* LREs in electrophoretic mobility shift assays (EMSAs). While this binding occurred in both dark and light, the DNA-protein complex displayed an altered mobility in the presence of light. More importantly, the formation of the light complex occurred at doses and wavelengths of light required for circadian responses. This light response was reconstituted with in vitro–translated proteins to demonstrate that the EMSA complex was composed only of the WC proteins and that the shift from the "dark" to the "light" form required flavin adenine dinucleotide (FAD). Finally, treatment of WC-1 alone with light produced the "light" complex, indicating that this was the photoreceptor responsible for the shift.

Sequence analysis of the two LREs reveals an imperfect repeat of the consensus sequence, CGATN-CCGCT. The GATN sequence is necessary for the binding of WC, so the presence of two of these sequences in each LRE supports the idea that each LRE is bound by each of the two WC proteins. Mutants lacking WC-1 activity are circadian-blind, indicating that this is the major photoreceptor. The VVD protein (see text below) also binds a flavin via a LOV domain and uses light information to modulate the primary action of WC-1. Finally, two phytochromes and a crypotochrome have also been found in the Neurospora genome, but the role of these in the circadian light response is unclear at present.

The *vvd* Gene

The *vivid* (*vvd*) gene product gates the circadian response to light. Work done in several species indicates that the circadian response to light is gated such that it is restricted to definitive periods of time. One of the most striking examples of this is the "dead zone" that occurs in the phase response curve of many organisms. This dead zone, which is usually found in PRCs generated with low intensity light pulses, corresponds to a time in the middle of the day when light exposure does not produce a shift in the rhythm. In Neurospora, gating is also indicated by other phenomena. Light-induced processes occur transiently, even under constant light conditions, suggesting a mechanism that serves to limit these responses.

The light response is regulated, at least in part, by the product of the *vvd* gene. *vvd* encodes a PAS-domain-containing protein that regulates its own expression through negative feedback (Fig. 7.9). Since VVD is a small protein, the PAS domain actually encompasses almost the entire protein. *vvd* RNA is induced by light in a WCC-dependent manner and then downregulated through a process that requires the VVD protein. While this light response drives rhythmic expression of *vvd* in a light–dark cycle, the oscillations dampen by the second day of constant darkness. Nevertheless, *vvd* is clock-controlled because its expression on the first day of constant darkness is affected by *frq* mutants. A long period *frq* allele delays the *vvd* peak while the null mutant results in reduced *vvd* levels on the very first day of constant darkness. *frq* expression and cycling, on the other hand, are unaffected in a *vvd* null indicating that *vvd* is clock-controlled, but does not itself affect the clock.

vvd does, however, affect input to the clock. In wild-type Neurospora induction of the *frq* and *vvd* RNA is gated such that it is greater at certain times of day. In the absence of the VVD protein, the amplitude of the gating is reduced, especially for *frq* and particularly in response to low-fluence light. In addition, VVD affects the phase-response curve. The dead zone is reduced, the overall light response is greater and the shift from delays to advances is delayed in *vvd* null mutants. This effect of VVD on photic input most likely involves interactions with the WC proteins.

Temperature Regulation by FRQ

The FRQ protein affects the range of temperature over which the clock functions. As described in earlier chapters, temperature interacts with the clock in multiple ways. While the period is compensated against effects of changes in ambient temperature, the phase is reset by temperature signals, and the very occurrence of rhythmicity is restricted to a specific temperature range. *frq* appears to participate in mediating all these different effects of temperature. Although the mechanisms underlying temperature compensation are not known,

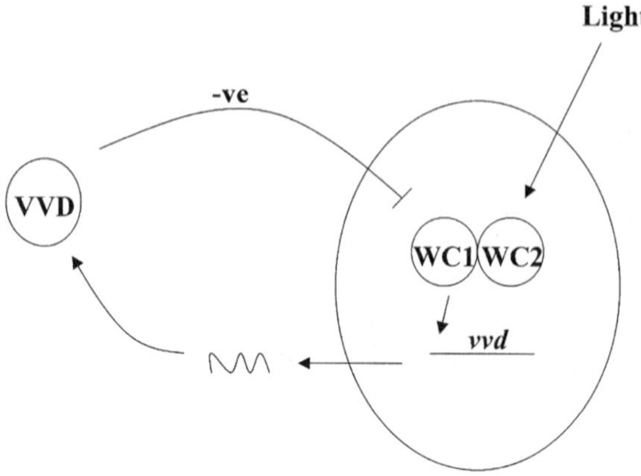

Figure 7.9. Regulation of photic input by the *vvd* gene. *vvd* is induced by light in a WC-dependent fashion. The protein product then reduces activity of the WC complex, thereby regulating the intensity of the light response. [Adapted from Heintzen et al. (2001).]

some of the *frq* mutants are deficient in this property, indicating a role for *frq* in this process.

At higher temperatures levels of FRQ are higher at all times of day. Thus, the protein basically oscillates around a higher baseline level. An increase of temperature automatically resets the clock without any change in the levels of *frq* RNA or protein because the FRQ level corresponds to a different time of day at the new temperature (Fig. 7.10). Since the trough level at 28°C is higher than the peak level at 21°C, Liu et al. predicted that a stepup from 21 to 28°C would always reset the clock to CT0 (the low point of FRQ expression). They did, in fact, find this to be the case. Conversely, a stepdown of temperature reset the phase to CT12, although this resetting was not as strong as that produced by the stepup, most likely because the stepdown required a decrease of FRQ levels and the decrease was somewhat dependent on the starting point, that is the prior phase. While the *frq* RNA levels showed an acute response to temperature such that they increased with the shift to the higher temperature and decreased at the lower temperature, these effects did not last longer than a few hours. Thus, the adjustment to different temperatures involves regulation of FRQ largely at a posttranscriptional level.

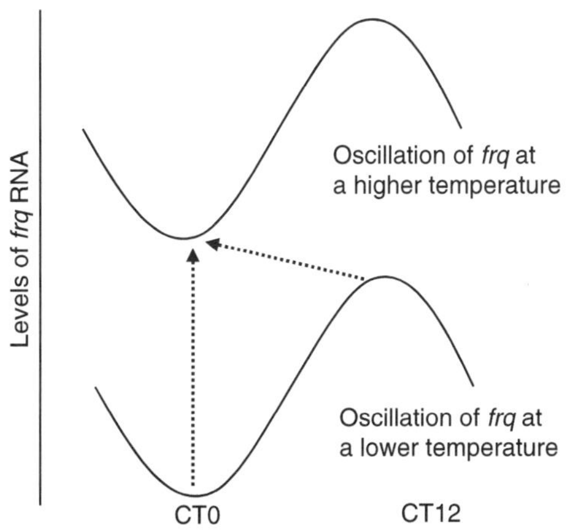

Figure 7.10. Resetting of the circadian clock by temperature. At higher temperature FRQ oscillates around a higher level, such that even the trough levels at 28°C are higher than peak levels at 21°C. It follows then that a temperature shift up at all times of day resets the clock to the trough timepoint (CT0) A temperature shift down (not shown) always resets to the peak timepoint (CT12). [Adapted from Loros and Dunlap (2001).]

Interestingly, temperature can dominate over light in the entrainment of the Neurospora clock: (1) temperature cycles produce greater synchrony of the conidiation rhythm, and (2) when Neurospora were placed under conditions of competing light and temperature cycles, they entrained to temperature, provided the amplitude of the temperature cycle was great enough. Normally, conidiation occurs during the dark phase of a light–dark cycle and the low-temperature phase of a temperature cycle. When light and temperature cycles were superimposed such that the low-temperature phase coincided with the light phase, conidiation was observed during the light phase. Thus, at least in Neurospora, light may not be the dominant entraining signal at all times in nature (Fig. 7.11).

Finally, regulation of *frq* defines the limits of rhythmicity for the Neurospora oscillator. The use of two different translation initiation sites, the choice of which is controlled at a posttranscriptional level, results in two forms of FRQ that are differentially regulated at different temperatures. At higher temperatures the longer form is preferentially made while at lower temperatures, the shorter form dominates. Each is required for rhythmicity at its respective end of the temperature range. While either one, by itself, will rescue rhythms of a *frq*-null strain, the restored rhythms are restricted to a narrower range of temperature. Normally, Neurospora clock functions can be observed from 18 to 34°C. Strains carrying only the short form

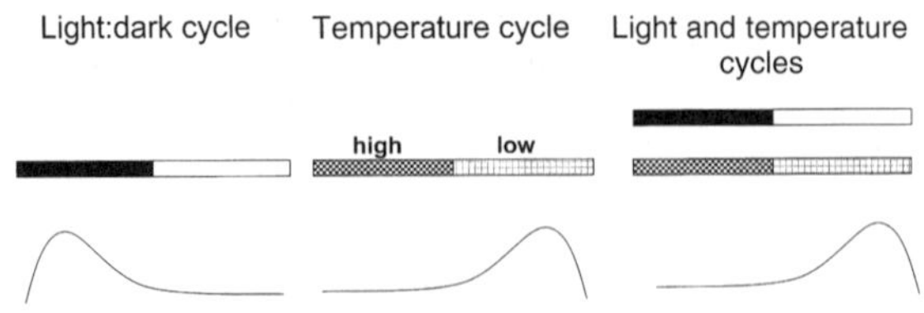

Figure 7.11. Temperature dominates over light in entrainment of the conidiation rhythm. In a light–dark cycle conidiation occurs during the dark phase. In the temperature cycle shown here it occurs during the low-temperature phase. When both cycles are imposed simultaneously such that the light phase coincides with the low-temperature phase, conidiation occurs in the light/low temperature phase, indicating that temperature is a more potent entraining stimulus.

are rhythmic at lower temperatures, but become arrhythmic above 27°C while those carrying the long form retain rhythmicity at higher temperatures, but are arrhythmic below 20°C (Fig. 7.12).

■ OUTPUTS CONTROLLED BY THE NEUROSPORA CLOCK

As discussed in previous chapters, outputs of the clock are molecular, cellular, physiological, or behavioral rhythms that are manifested only when a functional clock is present. In most cases the molecular and cellular rhythms drive physiological processes to occur at specific times of day. In Neurospora the dominant and best studied overt rhythm is that of conidiation. However, it is now apparent that many other aspects of fungal physiology are under circadian control: lipid metabolism, activity of various enzymes, and possibly also the response to environmental stresses.

The classical approach toward dissecting output pathways in Neurospora has consisted of molecular screens to identify genes that cycle under control of the clock. The term *clock-controlled-gene* (CCG) was, in fact, coined by Neurospora researchers, Loros and Dunlap in 1989, and the first such screens were done in this organism. Morning- and evening-specific RNAs were isolated first through subtractive hybridization technology and then by differential screening of libraries made from "morning" or "evening" RNA. More recently, with the advent of genomics technology, efforts have focused on high-throughput screens using microarrays.

As a result of these methods, several CCGs have been identified, and many of them turn out to be regulated not only by the circadian clock but also by light. Clearly, many of these will turn out to be involved in the remodeling process required for the production of aerial hyphae and conidia. For instance, the *eas* gene encodes hydrophobin, a hydrophobic protein that coats conidia and prevents dehydration (note that the formation of these conidia involves exposure of fungal filaments to the air after contact with a wet medium). Other CCGs encode enzymes, many of which are involved in metabolism. Levels of glyceraldehyde-3-phosphate dehydrogenase and of a carotenoid biosynthetic enzyme, geranylgeranyl pyrophosphate synthase, cycle under constant conditions. The cycling of trehalose synthase appears to be required for the conidiation rhythm, but may also play a role in conferring circadian regulation to the stress response.

Promoter Elements in CCGs

Careful dissection of promoter regions of *frq* as well as of some of the CCGs, in par-

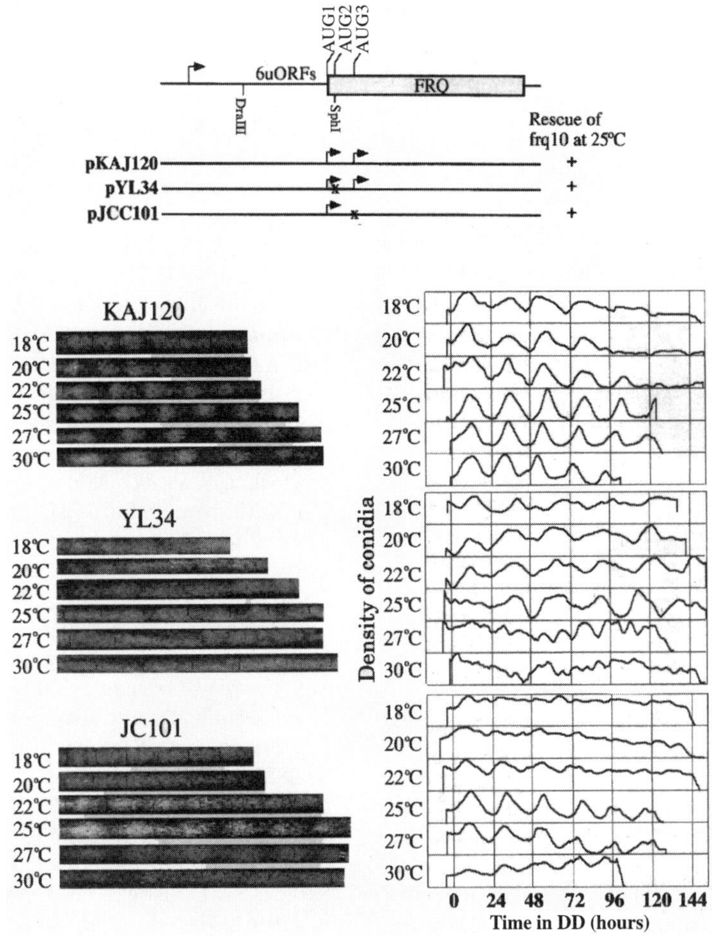

Figure 7.12. Different forms of FRQ mediate rhythmicity at different temperatures. The different forms of FRQ are produced through the use of alternate initiation codons that are ~100 amino acids apart (AUG1 or AUG3). The choice of the codon is made at a posttranscriptional level. While both can restore rhythmicity in *frq*-null mutants at 25°C, only the short form (YL34) restores rhythms at low temperatures and only the long form (JCC101) restores rhythms at higher temperatures. [Reprinted from Liu et al. (1997), copyright 1997, with permission from Elsevier Science.]

ticular *eas*, have defined elements required for clock control and distinguished these from others required for developmental or light control. As discussed above, in case of *frq*, two regions (LREs) are required for the light response. Both regions are bound by the WCC in EMSA assays, but only one, the distal LRE, is required to confer clock control. In addition, a 68-bp region in the *eas* gene is required for clock control and can also confer circadian regulation on heterologous genes.

■ CONCLUSIONS

Research in Neurospora has clearly been invaluable in testing classical hypotheses about circadian rhythms, identifying molecular mechanisms of clock function, and unraveling the importance of circadian regulation in basic cellular physiology. The discovery of *frq*, second only to that of *per* in Drosophila, was a major landmark in the circadian field. More recent data that question the all-important role of *frq* only serve to highlight the use of Neurospora for probing the generation of rhythmicity at a deeper level. While the feedback loops reported here and in other chapters are an integral part of the circadian timekeeping mechanism, rhythmicity is so ingrained in organisms that aspects must persist even when these loops fail.

FURTHER READING

Ballario, P, Macini, G (1997): White collar proteins: PASsing the light signal in Neurospora crassa. *Trends Microbiol* **5**: 458–462.

Bell-Pedersen, D, Garceau, N, Loros, JJ (1996): Circadian rhythms in fungi. *J Genet* **75**: 387–401.

Dunlap, JC (1996): Genetics and molecular analysis of circadian rhythms. *Annu Rev Genet* **30**: 579–601.

Feldman, JF, Hoyle, M (1971): Isolation of circadian clock mutants of Neurospora crassa. *Genetics* **75**: 605–613.

Froehlich, AC, Liu, Y, Loros, JJ, Dunlap, JC (2002): White Collar-1, a circadian blue light photoreceptor, binding to the frequency promoter. *Science* **297**: 815–819.

Heintzen, C, Loros, JJ, Dunlap, JC (2001): The PAS protein VIVID defines a clock-associated feedback loop that represses light input, modulates gating, and regulates clock resetting. *Cell* **104**: 453–464.

Lakin-Thomas, PL, Brody, S (2000): Circadian rhythms in *Neurospora crassa*: Lipid deficiecies restore robust rhythmicity to null *frequency* and *white-collar* mutants. *Proc Natl Acad Sci (USA)* **97**: 256–261.

Lee, K, Loros, JJ, Dunlap, JC (2000): Interconnected feedback loops in the neurospora circadian system. *Science* **289**: 107–110.

Liu, Y, Garceau, NY, Loros, JJ, Dunlap, JC (1997): Thermally regulated translational control of FRQ mediates aspects of temperature responses in the neurospora circadian clock. *Cell* **89**: 477–486.

Liu, Y, Merrow, M, Loros, JJ, Dunlap, JC (1998): How temperature changes reset a circadian oscillator. *Science* **281**: 825–829.

Loros, JJ, Dunlap, JC (2001): Genetic and molecular analysis of circadian rhythms in Neurospora. *Annu Rev Physiol* **63**: 757–794.

Merrow, MW, Garceau, NY, Dunlap, JC (1997): Dissection of a circadian oscillation into discrete domains. *Proc Natl Acad Sci (USA)* **94**: 3877–3882.

Pittendrigh, CS, Bruce, VG, Rosenzweig, NS, Rubin, ML (1959): A biological clock in Neurospora. *Nature* **184**: 169–170.

Sargent, ML, Briggs, WR, Woodward, DO (1966): The circadian nature of a rhythm expressed by an invertase strain of *Neurospora crassa*. *Plant Physiol* **41**: 1343–1349.

Yang, Y, Cheng, P, Liu, Y (2002): Regulation of the Neurospora circadian clock by Casein kinase II. *Genes Dev* **16**: 994–1006.

8

PHYSIOLOGICAL AND MOLECULAR CHARACTERISTICS OF PLANT CIRCADIAN CLOCKS

Jose A. Jarillo, Juan Capel, and Anthony R. Cashmore

■ INTRODUCTION

The circadian clock represents a fascinating adaptation to life on Earth. Organisms have endogenous biological clocks enabling them to tell the time of day. Such circadian timing mechanisms were first described in photosynthetic organisms. These organisms depend on sunlight for their energy, so the requirement to adapt to daily and seasonal fluctuations in light provides strong selective pressure for a capacity to determine time. Whereas many physiological and biochemical processes in plants have been found to be regulated in a circadian manner, a detailed mechanism underlying the circadian oscillator remains to be elucidated.

Advances in the identification and characterization of components of the plant circadian system have been made largely through genetic studies in Arabidopsis. Approximately 20 genes have been shown to affect the function of the Arabidopsis circadian clock, contrasting with the smaller number of genes known to regulate the circadian clock of insects, mammals or fungi. As in the mammalian circadian clock, several clock-associated genes from Arabidopsis have overlapping functions. The complexity of phototransduction pathways in plants may contribute to the large number of genes implicated in clock function.

In this chapter, we review the most recent progress in the field, highlighting

Molecular Biology of Circadian Rhythms. Edited by Amita Sehgal
ISBN 0-471-41824-2

work on the central clock mechanism itself and on the photopigments involved in the entrainment of the central clock.

BEGINNING OF PLANT CIRCADIAN BIOLOGY

Plants played a fundamental role in the discovery that organisms possess an accurate timing mechanism allowing them to synchronize their physiology with the daily environmental cycle. As early as 1729, the French astronomer de Mairan observed that the daily leaf movements of *Mimosa pudica* persisted for several days after the plants were placed in constant darkness. Plant circadian rhythms continued to intrigue scientists from Linnaeus to Darwin, but little progress ensued until two centuries later when Erwin Bunning identified the first plant clock mutants in bean. Bunning proposed that plants relied on their endogenous circadian clock to monitor the duration of day and night.

In 1920, Garner and Allard found that a tobacco mutant, Maryland Mammoth, grew profusely to ~5 m in height, but failed to flower in the prevailing summer conditions. However, the plants flowered in the greenhouse during the winter. By covering plants grown during the long days of summer with a light-tight tent late in the afternoon, artificial short days were then provided; these short days also caused plants to flower. Garner and Allard coined the term "photoperiodism" for this phenomenon, and plant researchers established the involvement of the circadian clock in controlling these timed events.

The pervasive influence of the circadian clock in plants is reflected in the variety of processes employed as circadian markers by researchers. Over the years, overt rhythms have been measured in processes such as stem elongation, root pressure, stomatal opening, cell membrane potential, chloroplast movements, and CO_2 exchange. Presumably, control of these processes by the clock allows plants to anticipate and prepare for changes in the environment that occur at dusk and dawn. The effect of disrupting the clock on photoperiodism indicates that the circadian clock provides a timing mechanism for the measurement of daylength. This allows plants to follow the changing of the seasons as the Earth orbits the Sun, and as such the clock is involved in the regulation of processes sensitive to photoperiod like the timing of flowering.

PLANT PHOTORECEPTORS

As in other organisms, the circadian system in plants consists of input pathways that provide temporal information from the environment to the clock, the central oscillator mechanism itself, and a set of pathways through which the temporal information provided by the clock is used to generate overt rhythms in several processes (Fig. 8.1). In addition to its role in entraining the clock, light is a crucial developmental regulator throughout the life history of plants, from the regulation of seed germination through seedling establishment to the timing of flowering. Plants have consequently evolved an array of photoreceptors capable of detecting light over a large range of fluence rates and wavelengths. Plant photoreceptors fall into three main classes: the phytochromes, which absorb primarily in the red and far-red region of the spectrum, and the cryptochromes and phototropins, both of which absorb in the blue and ultraviolet A (UVA) region of the spectrum. The majority of the most recent work on plant photoreception has been carried out in *Arabidopsis thaliana*.

Mediation of Response to Red/Far-Red Light by Phytochromes

The phytochrome photoreceptors are homodimeric chromoproteins; each subunit

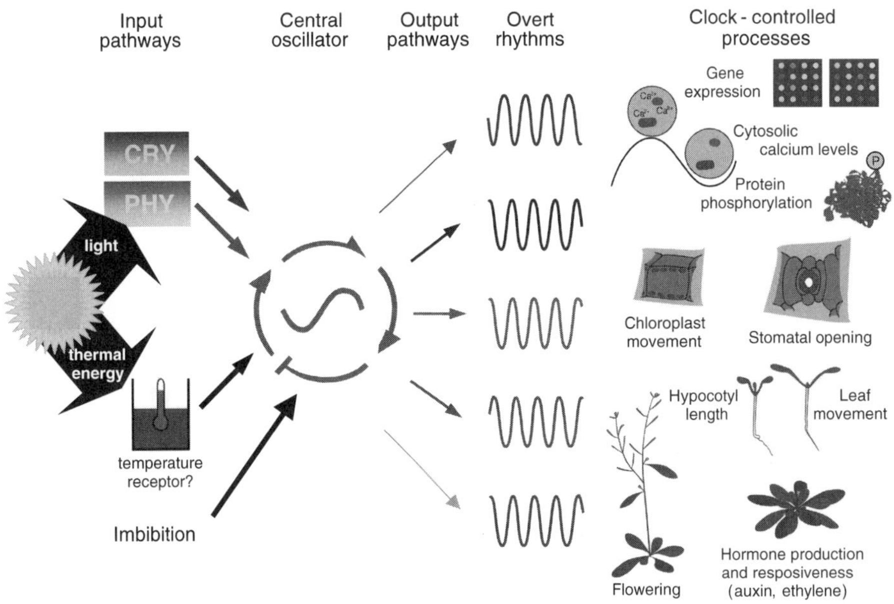

Figure 8.1. Model of a simple plant circadian system. The system consists of a set of entrainment pathways (inputs), a putative central oscillator, and sets of output pathways. Entraining stimuli include red and blue light perceived by phytochromes and cryptochromes, temperature (by a putative temperature receptor), and possibly imbibition. The central oscillator is illustrated as a loop including positive and negative elements that give a self-sustaining oscillation with a period close to 24 hours. Multiple output pathways are drawn as each regulating an overt rhythm with a distinct phase. Plant clocks control different biological processes, including the expression of several genes, cytosolic Ca^{2+} concentration, phosphorylation of some proteins, chloroplast movement, stomatal opening, hypocotyl elongation, cotyledon and leaf movement, and hormone production and responsiveness. The clock is also crucial for synchronizing developmental processes such as flowering time. Indeed, mutations in almost all the putative clock-associated genes cause altered photoperiodic control of flowering.

of the dimer is composed of a linear tetrapyrrole chromophore that is covalently bound to an apoprotein (Fig. 8.2). Phytochromes can exist as one of two spectrally distinct forms, a red-light absorbing form (Pr), and a far-red-light absorbing form (Pfr) that are interconvertible by the absorption of red (R) light and far-red (FR) light, respectively. In higher plants, the Phy apoproteins are encoded by a small gene family, with the holoprotein derived from each isoform having both distinct and overlapping functions in light perception. Arabidopsis contains five phytochromes (A–E) classified in two groups: (1) type I (PHYA) is light-labile and (2) type II (PHYB-PHYE) are light-stable. At the molecular and cellular levels Phy responses include development of the chloroplast, inhibition or promotion of cell growth, ion fluxes, and gene expression responses. The carboxy-terminal moiety of all phytochromes contains two PAS repeats and a histidine-kinase-related (HKR) domain (Fig. 8.2). PAS domains are generally involved in protein–protein interaction and ligand binding and have been found in many central clock proteins in other systems.

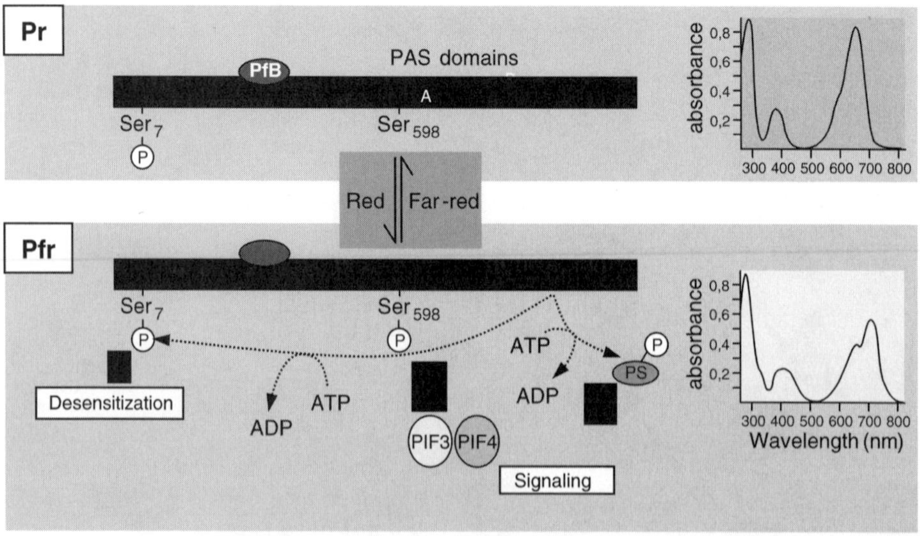

Figure 8.2. A phytochrome signaling model. Phytochromes exist in two different forms, the red-light absorbing (Pr) and the far-red-light (Pfr) absorbing forms, which have different absorption spectra. The red-light-stimulated phytochrome kinase activity may initiate light signaling by phosphorylating different phytochrome substrates (PSs), such as PKS1 and/or by specific interactions with downstream elements of the phytochrome signaling pathway like PIF3 and PIF4. PHYA and PHYB regulate multiple aspects of seedling deetiolation, including cell expansion and plastid development via PIF3. PIF4, in contrast, is proposed to act specifically in the PHYB pathway to regulate a subset of genes involved in cell expansion. The phytochrome response might also be downregulated by phosphorylation of the serine-rich amino-terminal region of phytochrome. PϕB phytochromobilin, the phytochrome chromophore.

Mediation of Responses to Blue Light by Cryptochromes and Phototropins

Two cryptochromes are present in Arabidopsis: CRY1 and CRY2 (Fig. 8.3). They have been identified in several plant species, including ferns and algae, and appear to be ubiquitous throughout the plant kingdom. The *N*-terminal domain of plant cryptochromes shares strong homology with the chromophore binding domain of DNA photolyases. Thus, they bind FAD and pterin chromophores; however, they lack photolyase activity. In contrast to CRY1, CRY2 is rapidly degraded in response to high-intensity blue light. The major photoreceptor mediating blue-light inhibition of the elongation of hypocotyl (the stem of the seedling between the cotyledons and the radical), is CRY1. Conversely, CRY2 performs an important role in the regulation of flowering time. Expression of the *C*-terminal domain of Arabidopsis CRY in transgenic seedlings results in a constitutive light response. This latent activity of the *C*-terminal domain is apparently repressed in the dark in the intact molecule. Irradiation with blue light relieves this repression, presumably through a redox reaction involving the flavin bound to the *N*-terminal photolyase-like domain. A hypothetical model of CRY signaling is shown in Fig. 8.4.

The phototropin family of blue-light photoreceptors in Arabidopsis has two members: PHOT1 and PHOT2 (Fig. 8.3). Mutants deficient in PHOT1 fail to bend and grow toward unidirectional low fluence blue light. PHOT2 controls chloroplast relocation induced by blue light. Recombinant pho-

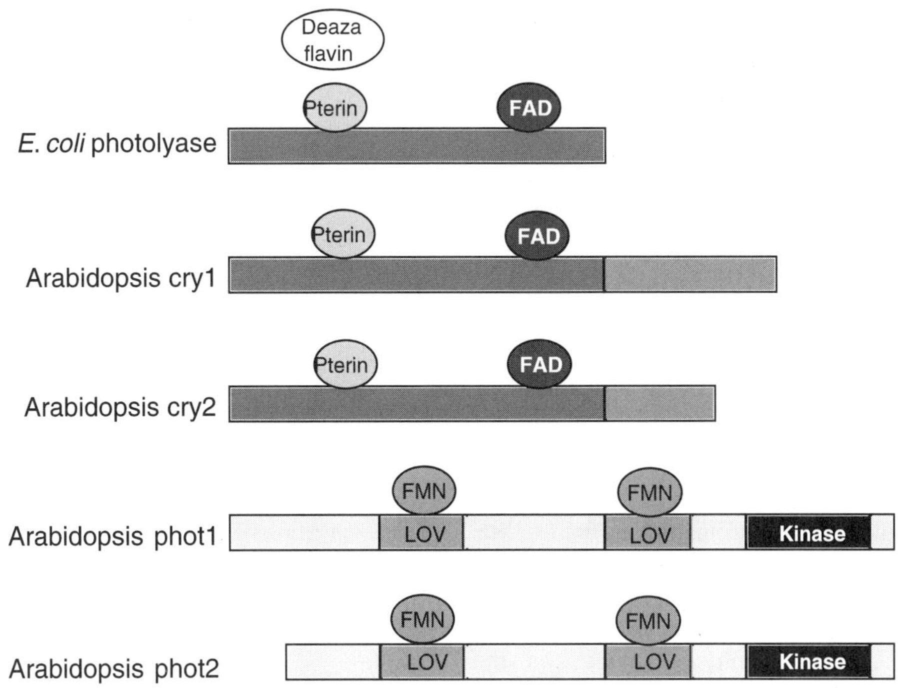

Figure 8.3. Protein structures of *E. coli* type I photolyase, and Arabidopsis cryptochromes and phototropins.

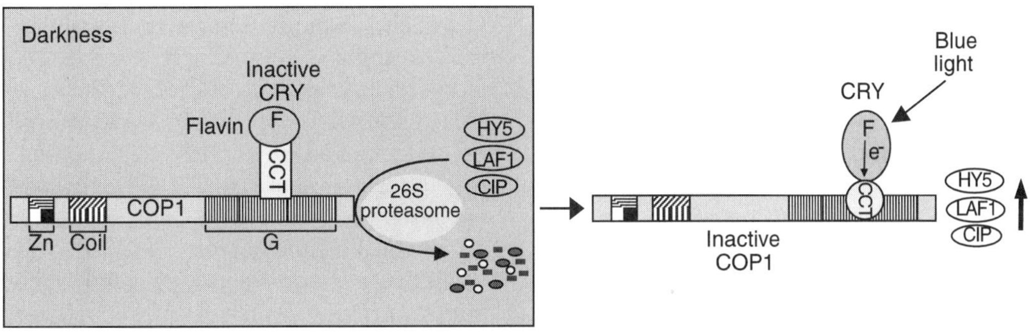

Figure 8.4. A hypothetical model of cryptochrome signaling. Arabidopsis cryptochromes contain a conserved *N*-terminal flavin-binding domain and a distinct *C*-terminal extension (CCT). The CCT interacts with the *C*-terminal WD-40 repeat domain of COP1. In darkness, COP1 is active and promotes the degradation of different interacting proteins via a 26*S* proteasome-mediated process. CRY blue-light activation may cause a change in the redox state and/or the conformation of the CCT. The activated CCT interacts with COP1, abrogating the COP1 degradation of different substrates (HY5, LAF1, CIP).

totropins noncovalently bind the chromophore flavin mononucleotide (FMN) and undergo autophosphorylation in response to blue-light irradiation. Hypocotyl phototropism, light-induced chloroplast relocation, and blue-light regulation of stomatal opening are severely impaired in a *phot1 phot2* double mutant. All three of these physiological responses serve to improve the efficiency of photosynthesis.

PHYTOCHROME–CRYPTOCHROME INTERACTION IN A CIRCADIAN INPUT PATHWAY

The identification of timing mutants and the related clock-associated genes was facilitated by the development of technologies that exploited the circadian oscillation of transcription of specific genes; these promoters were used to drive reporter genes whose expression is amenable to automated quantitation. The *chlorophyll A/B binding protein 2* (*CAB2*) gene encodes a component of the photosynthetic machinery and shows a strong rhythm of transcription, peaking during the early morning. In an assay that measured the rhythm in bioluminiscence from the luciferase (luc) reporter gene linked to the *CAB2* gene promoter, both the phytochromes and the cryptochromes were demonstrated to be involved in light input to the clock in plants. However, there was no perturbation of circadian function in a *phot1* mutant, possibly indicating that phototropins do not mediate input to the circadian clock. In addition, both red and blue light were effective in shortening *CAB2* expression period length, further implicating Phy and CRY plant photoreceptors. Mutants lacking *phyB* show a deficiency in responses to high fluence rates of red light for shortening of period length, whereas mutants lacking in *phyA* are deficient in the response to both red and blue light of low fluence rates.

Analysis of mutants deficient in one or both cryptochrome photoreceptors reveals a redundancy in the action of CRY1 and CRY2 in blue light. At intermediate fluence rates of blue light, CRY1 and CRY2 redundantly act to maintain circadian period length, whereas at higher fluence rates of blue light that would degrade CRY2, CRY1 acts alone in circadian photoperception. Significantly, the *cry1* mutant is also deficient in responses to low fluence rates of blue light in the range in which PHYA is known to act. Despite the fact that PHYA and CRY1 act in this range, they fail to compensate for loss of each other, suggesting that they are both required to mediate low-fluence-rate blue light to the clock. Indeed, observations indicate that both PHYA and CRY1 are also required for low-fluence-rate red-light input to the clock, implicating CRY1 as a signal transduction component downstream of PHYA in light input to the clock. The interaction between PHY and CRY is supported by functional interactions between PHYB and CRY2 and between PHYD and CRY1. PHYB interacts directly with CRY2 as observed in coimmunoprecipitation experiments using transgenic plants overexpressing CRY2. Through fluorescence resonance energy transfer microscopy, it was shown that PHYB and CRY2 interact in nuclear speckles that are formed in a light-dependent fashion.

Arabidopsis *cry1cry2* double mutants still show robust rhythmicity, indicating that cryptochromes do not form a part of the central circadian oscillator in plants as they do in mammals. Interestingly, a quadruple photoreceptor mutant of *cry1*, *cry2*, *phyA*, and *phyB* still keeps track of time and retains circadian rhythmicity. The mutant was nearly blind for developmental responses but perceived light cues for entraining the circadian clock.

All five *PHY* and two *CRY* genes show diurnal variation, with peaks of expression at different timepoints during the light period. On transfer to constant conditions (light or dark), circadian regulation is maintained for all genes, with the exception of *PHYC*; nevertheless, mRNA accumulation follows a circadian rhythm for all seven genes under constant light. Similarly, both a *suppressor of phyA 1* (*spa1*) and a component of the phototropic response, *root phototropic 2* (*rpt2*), are also clock-regulated. *PHOT1* gene expression is also clock-controlled; however, circadian function is unaffected in *phot1* mutants, suggesting that *PHOT1* is solely an output component. Because phytochromes and cryptochromes mediate light input to the clock, and are themselves regulated by the clock, they are both clock input and output components creating feedback loops. Clock control of these light signaling components may account for the circadian gating of photic signals.

■ RHYTHM ENTRAINMENT IN PLANTS BY OTHER INPUTS

Temperature, as well as light, is an effective entraining stimulus and has been demonstrated to entrain rhythms in CO2 assimilation in crassulacean acid metabolism (CAM) plants. Cyclic temperature treatments of dark-grown pea seedlings induce a rise in specific transcript levels of light-regulated genes related to photomorphogenesis. Rhythms in *light harvesting complex B* (LHCB) and *catalase 3* (CAT3) transcription are also entrained by temperatures cycles in Arabidopsis and low temperatures reset the phase of LHCB and *cold-circadian rhythm-RNA binding 2* (CCR2) oscillations. In tomato, a chilling-sensitive species, cold pulses stop the clock. Several studies indicate that a circadian clock also affects seed germination. A circadian oscillator is running in etiolated dark-grown seedlings that have not seen the light or have not been exposed to temperature cycles. The amplitude of the acute induction of both *LHCB* and *CAT2* mRNA abundance by light varies according to the timing of the onset of illumination, which indicates that a circadian clock running in etiolated seedlings gates the induction of *LHCB* and *CAT2* by light. The timing of hydration of the dry seed may also serve as a entraining stimulus. The acute induction of *CAT2* mRNA varies with time after imbibition, indicating that this event provides a signal capable of resetting the circadian clock.

A role for redox potential in resetting the clock mechanism has been proposed in mammals. DNA binding activity of both NPAS2::BMAL1 and CLK::BMAL1, the components utilized by the cellular clock in mammals, are exquisitely sensitive to the cell's redox state (see Chapter 4). A similar role for redox potential as an input to the clock may also exist in plants as it is known that the cellular redox state changes over circadian time in plants.

■ OTHER COMPONENTS OF LIGHT SIGNALING IN PLANTS

Two Arabidopsis mutants that undergo photomorphogenic seedling development in darkness, *constitutive photomorphogenic 1* (COP1) and *deetiolated 1* (DET1), display altered circadian rhythms. Both mutants have a shortened circadian period of *CAB2::luc* in continuous dark and light conditions. Since DET1 and COP1 are thought to mediate signaling from both phytochromes and cryptochromes, this aberrantly short period may reflect an imbalance between these different signaling pathways. Alternatively, and because these mutations cause inappropriate expression of several gene sets, their effects may not be specific to light. COP1 has three different protein–protein interaction motifs, a RING finger, a coiled coil domain, and a seven WD-40 motif domain. Several RING finger-containing proteins have recently been implicated in the ubiquitination of specific substrates, and the subsequent degradation by the proteasome. COP1 interacts directly with HY5, a b-ZIP factor that binds to promoters and regulates expression of light-regulated genes. COP1 targets HY5 for degradation by the proteasome in the dark. A direct physical interaction between COP1 and SPA1 has also been demonstrated suggesting that SPA1 may function to link the phytochrome A–specific branch of the light signaling to COP1. Moreover, both cryptochromes CRY1 and CRY2 bind to COP1 and apparently repress COP1 activity through direct protein–protein interaction. COP1 also binds to the *C*-terminal domain of PHYB in yeast two-hybrid assays, although the relevance of this interaction to phytochrome signaling remains to be established; as in contrast to the cryptochromes, the isolated *C*-terminal domains of the phytochromes do not appear to be active in vivo.

COMPONENTS OF THE CENTRAL OSCILLATOR IN PLANTS

Much is known about how plants perceive light, providing many tools for the exploration of light interactions with the plant clock. Components of the plant central oscillator are being identified, although no homologs of clock genes described in other organisms have been found in arabidopsis.

Timing of CAB1 (*TOC1*)

Toc1 mutation affects different clock-controlled processes. Using the *cab2*::*luc* luminiscence reporter system mentioned above, mutants were identified in which the peak of free-running cycling bioluminiscence was altered. More than 20 mutants were recovered displaying period lengths ranging from 21 to 27 hours (wild type = 24.5 hours). The best characterized of these mutants, *toc1-1*, runs with a period of 21 hours in continuous white light. The morphology of the mutant is wild type, consistent with the normal phenotypes observed for period length mutants in other organisms. The simplest observation suggests that as in other organisms the clock is not based on metabolic processes that are fundamental to the maintenance of the organism but rather, arises from interactions between processes that are at least in part specific to the clockwork itself.

The *toc1-1* mutation affects all clock-controlled processes so far examined. The period of cycling in transcription and mRNA abundance of two differently phased genes (*CAB2* and *CCR2*) are both shorter than wild type by 2–3 hours in the *toc1* background. *CAB2* expression peaks about 4–6 hours after dawn, whereas peak *CCR2* mRNA abundance occurs 5–6 hours later. This different phasing suggests that different signaling pathways lead from the oscillator to control of each of these two genes, and that *TOC1* acts upstream of the divergence.

Toc1-1 also shortens the *cab2*::*luc* rhythm induced by a red-light pulse in etiolated seedlings and causes shortened circadian rhythms in the absence of light input to the clock. Moreover, leaf movement and stomatal conductance rhythms in *toc1-1* are proportionately shorter than in the wild type. TOC1 also controls photoperiodic flowering responses through its effects on the clock, and the mutation results in early flowering.

The *TOC1* gene encodes a nuclear protein containing an atypical response regulator receiver domain and two motifs that suggest a role in transcriptional regulation: a basic motif conserved within the *CONSTANS* family of transcription factors (the CCT motif) and an acidic domain. In the motif that is similar to the receiver domain of the response regulators from two component signal transduction systems (see Chapter 6), the conserved Asp residue that undergoes phosphorylation in other characterized response regulators is substituted in TOC1. Consistent with this finding, TOC1 and other pseudoreceivers are unable to undergo phosphorylation in vitro, and the function of these putative domains remains unclear. *TOC1* is itself circadianly regulated and participates in a feedback loop to control its own expression. In fact, *TOC1* mRNA cycles with a shortened period in the *toc1-1* mutant. *TOC1* mRNA oscillations dampen rapidly in darkness, whereas for other genes (*CCR2* and *CAT3*), expression rhythms persist in the dark. In transgenic *TOC1*-overexpressing plants, the free-running robust circadian rhythm of *CAB2* was dampened and the circadian profiles of potential clock-associated genes *CCA1*, *LHY*, *GI*, and *CCR2* were all markedly altered, implicating TOC1 as a player within, or close, to the central oscillator.

It is of interest that in *Synechococcus elongatus*, the bacteriophytochrome CikA (circadian input kinase), which provides light input to the cyanobacterial clock and is important for resetting the clock (see Chapter 6), also carries an unusual receiver motif that, like the motif in TOC1, lacks the expected aspartyl residue needed for phosphoryltransfer in response regulator proteins. Whether this similarity between TOC1 and CikA is indicative of a biochemical function that cyanobacterial and plant mechanisms share has yet to be determined. CikA also lacks the conserved Cys residue expected as a bilin ligand for phytochromes, although the similarity of CikA to phytochromes may provide a potential evolutionary parallel between cyanobacterial and eukaryotic circadian systems.

The APRR1/TOC1 Family

Arabidopsis possesses at least five members belonging to the APRR family of pseudoresponse regulators. One of the members, APRR1, is identical to TOC1. All these APRR proteins are transcribed with a circadian rhythm. Furthermore, in a given 24-hour period, the APRR mRNAs accumulate sequentially after dawn, separated by 2–3-hour intervals, suggesting that the APRR family of proteins may all be associated in some way with the Arabidopsis circadian clock. Certain light stimuli are crucial determinants to induce the robust circadian waves, and the first light-induced *APRR9* appears to be primarily responsible for this light response of the circadian waves. A red-light pulse perceived by phytochromes appears to be sufficient to synchronize the APRR circadian waves.

Function of LHY and CCA1 in Autoregulatory Loops

Late elongated hypocotyl (LHY) and *circadian clock-associated 1* (CCA1), two other molecular components controlling the plant circadian system, possess a Myb-related DNA-binding domain. *LHY* and *CCA1* fulfill some of the required criteria for oscillator components. Both genes show robust circadian oscillations of transcript and protein levels in plants kept in continuous light. They also fulfill the criterion that the component controls its own levels by feedback inhibition of its synthesis. In constant light, overexpression of either *CCA1* or *LHY* results in repression of each of the two genes' endogenous expression. In addition, overexpression of *CCA1* and *LHY* stops all overt rhythmicity that has been measured, including leaf movement, hypocotyl elongation, and the expression of genes that peak with different phases. However, in a *cca1* null mutant, robust rhythmicity of output gene expression is maintained, albeit with a shorter period than in wild-type plants. These findings indicate that although *CCA1* is clearly an important element of clock function, *LHY* oscillations might be able to compensate for the loss of *CCA1*. Indeed, by constructing *lhy cca1* double mutants, it has been demonstrated that LHY and CCA1 are partially redundant genes required to maintain circadian rhythms. For the first 24–48 hours in continuous light, the *lhy cca1* mutant showed further shortening of the period of rhythmicity of several clock-controlled genes. However, by 2 days in constant light these rhythms and those of leaf movements damped out completely. Thus circadian rhythms cannot be sustained without both LHY and CCA1, and the results are consistent with them being components of the central oscillator.

But how can the plant show any oscillations in the absence of LHY and CCA1? A family of several DNA binding proteins, the REVEILLE (RVE) proteins, which have MYB-like domains similar to those of LHY and CCA1, has been identified. Their mRNAs show rhythmic expression, and the constitutive expression of *RVE1* and *RVE2* leads to arrhythmic *CAB2::luc* expression in continuous light. The homology with *LHY* and *CCA1* is particularly striking in the MYB domain, suggesting that they may bind similar DNA sequences. These proteins might partially substitute LHY and CCA1, but not be able to sustain the rhythms.

A reciprocal regulation between *TOC1* and *LHY/CCA1* within the Arabidopsis circadian clock has been described. LHY and CCA1 negatively regulate *TOC1*, such that its expression is strongly reduced in plants that constitutively overexpress either *LHY* or *CCA1*. Indeed, both MYB proteins bind to a region in the *TOC1* promoter that contains a sequence (the evening element) that is critical for its circadian regulation. The genetic interaction between TOC1 and LHY/CCA1 is reciprocal, as TOC1 appears to participate in the positive regulation of *LHY* and *CCA1* expression. Both the *LHY* and *CCA1* messages cycle with a shorter period length in *toc1-1* and *toc1-2* alleles. The interactive regulation between TOC1 and CCA1/LHY may define a basic framework for the clock mechanism in arabidopsis.

CCA1 also acts as a positive regulator of *CAB* gene expression, which suggests that these MYB factors may simultaneously regulate genes that are phased to different times of the day. The specification of circadian phase may entail differential binding by different members of this protein family at distinct circadian phases. Alternatively, phase specification may involve differential phosphorylation of family members at distinct circadian phases or different interacting partners recruited to the promoters that modulate CCA1/LHY function.

OTHER CLOCK-ASSOCIATED GENES

ZGT Gene

A clock- and light-regulated tobacco gene *ZGT*, links the circadian oscillator to *LHCB* expression. *ZGT* transcripts have alternate forms that are differentially expressed in different tissues. *ZGT* is expressed rhythmically in light–dark cycles and in constant light. Constitutive expression of *ZGT* sustains the expression of the clock-controlled light harvesting complex *LHCB1*1* gene in constant darkness, when it would normally dampen, but does not affect *LHCB1*1* expression in constant light. *ZGT* expression is induced rapidly by light and its overexpression increases the sensitivity of the circadian oscillator to brief light pulses. The *ZGT* promoter includes a G-box motif, found in many light-regulated promoters in plants, and evening element motifs that are correlated with circadian control of plant genes.

Phytochrome Interacting Factors (PIF)

The phytochrome interacting factor (PIF) proteins bind to and regulate putative clock components. Light input to the clock via phytochrome is likely to be mediated through an interaction with PIF3. PIF3 is a novel basic helixp–loop helix (bHLH) protein containing a PAS domain, necessary for normal photoinduced signal transduction. PHYA-PIF3 and PHYB-PIF3 interactions are known to occur. A brief irradiation with red light induces the rapid binding of the activated phytochrome to PIF3, whereas a pulse of far-red light releases activated phytochrome from the complex. PIF3 binds specifically to a G-box DNA-sequence motif present in *CCA1*, *LHY*, and *SPA1* promoters and possibly to other targets in the clock. Indeed, the induction of *CCA1* and *LHY* was reduced in transgenic plants expressing *PIF3* antisense RNA. PHYB binds reversibly to

G-box-bound PIF3 on light-triggered conversion of the photoreceptor to its biologically active conformer.

HFR1 is an atypical bHLH protein that acts in the phytochrome A signal transduction pathway. HFR1 did not bind directly to either PHYA or PHYB. However, HFR1 did bind to PIF3, suggesting heterodimerization, and both the HFR1/PIF3 complex and PIF3 homodimer bound preferentially to the Pfr form of both phytochromes. Thus, HFR1 may function to modulate PHYA signaling via heterodimerization with PIF3. On the other hand, PIF4, another bHLH protein, appears to interact preferentially with PHYB in its Pfr form. PIF4 does not contain a canonical PAS domain; however, it has a region similar to the putative PAS-like domain of PIF3 located close to the bHLH region. PIF4 overexpression resulted in red-light hyposensitivity, while loss-of-function mutations at PIF4 resulted in red-light hypersensitivity. PIF4 may function as a nuclear negative regulator of PHYB signaling, sequestering the active form of PHYB in a nonproductive complex, thereby reducing the photoreceptor's effective activity elsewhere.

Early Flowering 3 (*ELF3*)

ELF3 modulates resetting of the circadian clock. *Elf3* mutants are aphotoperiodic and exhibit light-conditional arrhythmia. *Elf3*-7 mutant retains oscillator function in the light but blunts circadian gating of *CAB* gene activation, indicating that deregulated phototransduction may mask rhythmicity. Furthermore, *elf3* mutations confer the resetting pattern of SD photoperiodism, indicating that gating of phototransduction may control resetting. Temperature compensation can bypass the requirement for normal *ELF3* function for the oscillator and partially restore rhythmic *CAB* expression. Therefore, ELF3 specifically affects light input to the oscillator, similar to its function in gating *CAB* activation, allowing oscillator progression past a light-sensitive phase in the subjective evening.

ELF3 encodes a novel nuclear protein that may function as a transcriptional regulator. The *ELF3* transcript level is regulated in a circadian manner, and this cycling requires ELF3 protein. Constitutive expression of *LHY* does not abolish the circadian rhythm of *ELF3* gene expression. Light and the circadian clock regulate ELF3 protein accumulation in nuclei. Furthermore, yeast two-hybrid assays and in vitro binding experiments demonstrate that ELF3 interacts with PHYB. Genetic analyses suggest that *ELF3* requires *PHYB* function in early morphogenesis but not for the regulation of flowering time. In addition, ELF3 modulates resetting of the clock. Plants overexpressing *ELF3* have an increased period length in both constant blue and red light; furthermore, etiolated *ELF3*-overexpressing seedlings exhibit a decreased acute *CAB2* response after a red-light pulse, whereas the null mutant is hypersensitive to acute induction. These findings suggest that ELF3 negatively regulates light input to the clock and its outputs. The action of ELF3 is phase-dependent. Absence of ELF3 activity causes a significant alteration of the phase response curve during subjective night, and constitutive overexpresssion of ELF3 results in decreased sensitivity to the resetting stimulus. Indeed, the phase of ELF3 function correlates with its peak expression levels in the subjective night. It is possible, then, that ELF3 action represents a mechanism by which the oscillator modulates light resetting.

Plant Genes Affecting Clock Function and Flowering Time

GIGANTEA (*GI*) is a circadian clock-controlled gene that plays a role in the photoperiodic control of flowering in Arabidopsis. *GI* transcripts levels show circadian oscillations in both light and dark conditions, with peak expression in the evening. Moreover, in plants overexpressing either *CCA1* or *LHY*, oscillations of *GI* mRNA are disrupted. Thus, *GI* expression is under control of the clock. Consistent with this idea, *GI* mRNA rhythms are also perturbed in *elf3* mutants, suggesting that GI acts downstream of ELF3. The absence of ELF3 causes arrhythmic and upregulated expression of *GI* plants under continuous light and, although to different extents, in long and short photoperiods. Complicating this scenario, however, is the finding that *gi* mutants exhibit an altered period and reduced amplitude of *CCA1* and *LHY* expression, suggesting that GI functions in input to the clock. GI is nucleus-localized, consistent with its proposed role in PHYB signaling, given evidence that early PHY signaling events are nuclear localized.

ZEITLUPE (ZTL)/ADAGIO 1 (ADO1) and *FLAVIN-BINDING KELCH REPEAT F-BOX 1 (FKF1)/ADO3* genes have also been implicated in both circadian clock function and the regulation of flowering time in response to daylength. Mutations in *ZTL* or *FKF1* delay flowering under long but not short days and affect clock-controlled gene expression. In the *ZTL* mutant, the circadian-clock-controlled peaks in gene expression occur approximately 3 hours later than in the wild type, and the circadian rhythmicity of cotyledon movement is also altered (Fig. 8.5). The *fkf1* mutant has a weaker effect on circadian clock regulation, reducing the level of expression of clock-controlled genes,

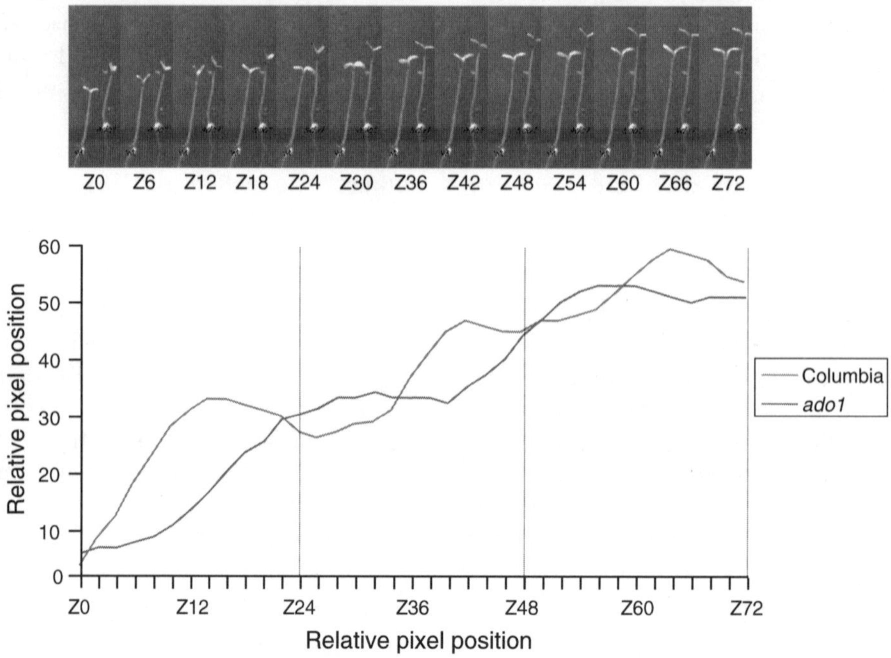

Figure 8.5. Cotyledon tip movement in the *ado1* mutant. Cotyledon movement was recorded for wild-type (Columbia) (left part of each panel) and mutant (*ado1*) (right part of each panel) seedlings grown under a 12-hour photoperiod and then transferred to constant white light. The position in pixels of one cotyledon tip per seedling for 72 hours was recorded. The periodicity of cotyledon movement under continuous white light for the *ado1* mutant was about 6 hours longer than that for wild type.

without altering the level or timing of the peak itself. *FKF1* mRNA is itself clearly circadian clock regulated, peaking toward the end of the day, whereas *ZTL* mRNA is not.

Transgenic plants overexpressing the *ZTL* gene have elongated hypocotyls and petioles with elongated cells and exhibit distinct cotyledon movement during the day. Moreover, these plants flower later than do wild-type plants when they are grown under long-day, but not under short-day conditions. GFP-ZTL fusion protein is observed in nuclei and in the cytosol, indicating that ZTL is a nucleocytoplasmic protein that influences flowering time in the long day pathway of arabidopsis.

A circadian role has also been reported for *LKP2/ADO2*, the other member of the family. Its mRNA expression is not regulated by the circadian clock and is detected in all tissues examined. Overexpression of *LKP2* results in arrhythmic leaf movement and *CAB2::luc* expression in constant light, and arrhythmic *CCR2::luc* expresion in both constant light and constant dark. These findings suggest that LKP2 may function within or close to the circadian clock.

ZTL/ADO1, *LKP2/ADO2*, and *FKF1/ADO3* encode closely related proteins that have a combination of a LOV domain, an F box and six kelch repeats (Fig. 8.6). LOV domains have been implicated in binding flavin chromophores, which suggests that these proteins might act as blue-light photoreceptors. Phenotypic characterization of the *fkf1* and *ztl* mutants suggest roles for these proteins in the response to light. The second domain, the F box, is implicated in targeting proteins for degradation via the ubiquitin system and the kelch domain may serve as a protein–protein interaction domain that recruits specific proteins for degradation. Candidates target proteins for these F-box proteins are circadian clock components and CRY1 and PHYB, which physically interact with ZTL/ADO1 in both yeast two-hybrid and in vitro binding studies.

FLC (*flowering locus C*) encodes a MADS-box protein that functions as an autonomous pathway flowering repressor. Loss-of-function mutations confer early flowering and shorten the circadian period and *FLC* may account for *ANDANTE* (*AND*) QTL (quantitative trait locus), that affects period of circadian leaf movements in continuous light, based on its map position. Overexpression of the *flowering locus M* (*FLM*) gene, which is closely related to *FLC*, produces comparable effects of late flowering to *FLC*, acting as an inhibitor of flowering. Expression of the flowering time

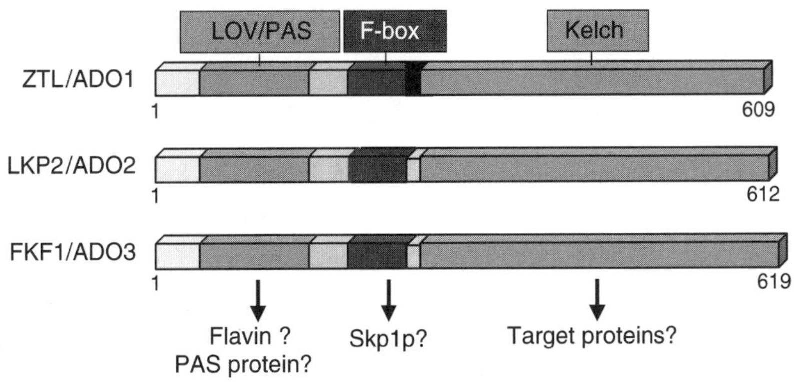

Figure 8.6. Domain structure of the ZTL family of proteins. Schematic alignment of ZTL/ADO1 with paralogs LKP2/ADO2 and FKF1/ADO3. The PAS-like LOV domain, the F box and six kelch repeats are shown, together with their potential interaction targets.

gene *Agamous-like 20* (*AGL20*) is antagonistically regulated by FLC and CONSTANS through a 113-bp promoter sequence. A CarG box (a DNA element that binds a MADS box protein) within the *AGL20* promoter sequence is bound specifically by the FLC protein in vitro.

Other identified QTLs such as *ESPRESSO*, *NON TROPPO*, and *RALENTANDO* may account for new genes that regulate the Arabidopsis circadian system in nature, one of which may be the flowering time *GIGANTEA*, discussed above.

An Arabidopsis Homolog of *TIMELESS* (*TIM*)

TIM is an essential gene for circadian function in drosophila. Mammalian *timeless* genes appear to be not the true orthologs of *dTIM*, but are the likely orthologues of a newly described fly gene, *timeout*. The predicted Arabidopsis TIM protein contains TIM homology domains and a putative bipartite nuclear localization signal. Arabidopsis *tim* mutants appear unaffected in circadian properties.

■ OUTPUT RHYTHMS

One advantage of plants as model clock systems is the myriad rhythmic outputs of plant clocks. The clock gates different overt rhythms to distinct times of the day. In most cases the connections between the clock and the output rhythm are not known; however, oscillations of molecules/processes relevant for various outputs have been reported.

Diverse Mechanisms Underlying Growth Rhythms

The growth rhythms include the pulvinar leaf movement systems, in which the cells in the extensor and flexor regions of the pulvinus swell in antiphase to drive circadian oscillations in leaf position.

The circadian leaf movement of *Mimosa pudica L.* is regulated through the action of leaf opening substances and leaf-closing substances. The balance of concentration between the two leaf movement factors is inverted during the day. A glycoside-type leaf movement factor is hydrolyzed by β-glucosidase, the activity of which is regulated by the biological clock. Circadian oscillations in cotyledon position in Arabidopsis arise from variations in elongation rates of abaxial (dorsal side of the leaf) and adaxial (ventral side of the leaf) cells of petioles. These cotyledon movements are easily monitored by video imaging and have been exploited as the basis for the identification of natural alleles that contribute quantitatively to the circadian period length in Arabidopsis.

There are also rhythms reflecting growth rate, for example, cell elongation. Inflorescence stems of Arabidopsis exhibit a circadian oscillation in elongation rate correlated with indol-3-acetic acid (IAA) levels in rosette leaves, although IAA levels in the inflorescence stem do not oscillate. In *Chenopodium* stems the circadian rhythms in elongation rate were correlated with a rhythm in oleic acid content.

The circadian clock also controls the elongation of the arabidopsis hypocotyl. Light signaling leads to a cessation of hypocotyl elongation. Mutations that reduce photoperception and signaling yield an elongated hypocotyl phenotype; conversely, overexpression of elements of the phototransduction pathways reduces hypocotyl elongation relative to the wild type. The hypocotyl elongation defect may also result from a primary dysfunction in the circadian system with a resulting failure to impose a daily period of growth arrest.

Stomatal Aperture, Gas Exchange, and CO_2 Assimilation

Oscillations of metabolic enzymes contribute to stomatal aperture, gas exchange, and CO_2 assimilation. The stomatal pores of higher plants allow for gaseous exchange in and out of leaves. These pores are situated in the epidermis, where they are surrounded by a pair of guard cells controlling their opening. Circadian rhythms in stomatal aperture are correlated with a rearrangement of the cytoskeleton of the guard cells. On the other hand, Arabidopsis exhibits a circadian rhythm in the rate of CO_2 assimilation and in beans this rhythm includes circadian regulation of the Calvin cycle reactions.

Flattened stems of cactus can survive after detachment from the plant for several months without water. Their stomata are closed all the time and the CO_2 released by respiration is refixed into malate. This process, which has been called *crassulacean acid metabolism* (CAM) idling, allows the plant to survive for prolonged periods of time while losing remarkably little water. Recent observations have provided evidence for the tonoplast functioning as the master switch for the circadian regulation of this process. However, *phosphoenolpyruvate carboxylase (PEPC) kinase* gene transcription is under circadian control and hence the regulation of CAM may also be regulated through PEPC.

Some studies have demonstrated circadian regulation of sucrose phosphate synthase activity in tomato by a protein phophatase and circadian regulation of sucrose metabolism.

Circadian Rhythm in Levels of Calcium

Ca^{2+} is a ubiquitous second messenger implicated in phytochrome-mediated as well as UVA-blue and UVB light signal transduction pathways. Pulses of blue light induce transient spikes in cytosolic but not in chloroplastic Ca^{2+}, and these transients are implicated in signalling downstream of PHOT1 in the phototropic response. Calcium also plays a critical role in guard cell signaling and in the circadian regulation of stomatal aperture and gas exchange. Ca^{2+} may play a role in the entrainment of the circadian oscillator as well as in the regulation of clock-controlled gene expression. Free cytosolic and chloroplastic Ca^{2+} levels, monitored by aequorin luminiscence, oscillate with a circadian rhythm in tobacco and Arabidopsis. The light-to-dark transitions stimulates a spike in chloroplast stromal Ca^{2+} levels, although whether this signals the circadian clock is not known. It has been demonstrated that nuclear calcium does not exhibit circadian rhytmicity and that calcium rhythms of different cell types oscillate with different circadian phases. This may represent different underlying cellular control mechanisms and have significant implications for understanding how Ca^{2+} mediates signal transduction in plant cells.

External applications of either Ca^{2+} or Ca^{2+} ionophores phase-shift the leaflet movement rhythm of *Robinia*. Different circadian oscillators control Ca^{2+} fluxes and *Lhcb* gene expression. Free calcium is responsible for driving the rhythm of *Lhcb* expression. These rhythms free-run with different periods in tobacco seedlings in constant conditions, providing evidence for separate circadian pacemarkers controlling molecular events in plants.

Oscillation of Enzymes Required for Hormone Biosynthesis

Production of ethylene is known to vary in a circadian fashion in several species, including barley, wheat, and rye. In sorghum, this rhythmic ethylene production is correlated with rhythms in the activity of 1-aminocyclopropane-1-carboxylic acid (ACC) oxidase, the enzyme that controls the conversion of ACC to ethylene.

A diurnal oscillation observed in gibberellic acid (GA) levels in sorghum may be circadian. There is also a circadian oscillation in rosette leaf auxin levels in Arabidopsis. The rhythm in inflorescence stem elongation in Arabidopsis in the absence of rhythmic IAA levels might reflect rhythmic responsiveness to auxin. In potato, feedback control and diurnal regulation of gibberellin 20-oxidase, a key regulatory enzyme in the GA-biosynthetic pathway has been reported.

Regulation by light–dark cycles of enzymes implicated in abscisic acid (ABA) biosynthesis in tomato [zeaxanthine epoxidase (ZEP) and 9-*cis*-epoxycarotenoid dioxygenase (NCED)], has been shown. *ZEP* mRNA oscillates with a phase very similar to *LHCII* mRNA, and the oscillations continue in a 48-hour dark period. Genes that resemble response regulators, which are implicated in propagation of phytohormone responses (e.g., ethylene and cytokinin), are also regulated by the circadian clock.

Rhythmic Expression of Genes Involved in Photosynthesis

Genes regulated by circadian output pathways have been shown to peak in expression at different times of the day. Examples include genes involved in photosynthesis, oxidative stress, cold response, and cell wall production. Circadian rhythms in gene expression were first reported in plants with the observation of a circadian oscillation in mRNA abundance of *LHCB/CAB* gene expression.

In most species, the photosynthetic genes *CAB* and *RUBISCO* anticipate dawn by rising to a high level of expression in early morning. There are species-specific differences, such as in the expression of genes encoding the small subunit of RUBISCO (*RBCs*). This is not rhythmic in all species and, where rhythmicity is observed, the circadian regulation can be transcriptional or posttranscriptional.

Other genes related to photosynthesis are also expressed rhythmically. These include the *Rubisco activase* (*RCA*), and in CAM plants, *PEPC kinase*. The peak expression of *PEPC kinase* mRNA occurs at midnight, and correlates with kinase activity, reflecting the requirement of the enzyme at that time. The expression of a single *NADPH-protochlorophyllide oxidoreductase* gene in cucumber, implicated in the light-dependent step in the biosynthesis of chlorophyll, is also positively regulated by diurnal and circadian rhythms.

Circadian Control of Gene Expression

Circadian control of gene expression extends to virtually all aspects of plant metabolism. The expression of two *serine hydroxymethyltransferase* (*SHM*) genes, encoding mitochondrial components of the photorespiratory pathway, is also circadian-regulated. This, together with the circadian regulation of nuclear genes encoding both chloroplastic (RBCS and RCA) and peroxisomal components of the photorespiratory pathway, suggests an integrated temporal regulation of this process. On the other hand, the transcripts of two closely related putative cold-induced glycine-rich RNA binding genes, *CCR1* and *CCR2*, have a high level of expression in the afternoon, with a peak expression 8–12 hours after dawn in the wild type. Overexpression of *CCR2* in arabidopsis strongly depresses the cycling of the endogenous *CCR2* transcript. However, *CCR2* overexpression has no effect on the period or the levels of other unrelated circadian-regulated transcript such as *CAB3* or *CAT3*. *CCR2* gene expression lies most likely downstream of the effects of TOC1 on the clock, and may define a slave oscillator that is still subject to temporal control from the master oscillator, but may itself control an undefined subset of clock-controlled outputs.

ATGER3 germinlike cell wall promoter confers circadian-regulated transcription with peak expression at the beginning of the night. Nonphotosynthetic enzymes under the control of the circadian clock include catalases, which are primarily involved in eliminating toxic H_2O_2 from the cell. *Catalase* genes are expressed at different times; *CAT2* peaks in the morning and *CAT3* peaks in the afternoon, 12 hours out of phase with each other. This difference probably reflects different roles for each in plant metabolism.

Circadian and light regulation of nitrate reductase has been shown in Arabidopsis. Gene expression of *S*-adenosylmethionine decarboxylase (SAMDC), a key enzyme involved in the biosynthesis of polyamines, from *Pharbitis nil* and *Tritordeum*, is under the control of the circadian clock. A circadian and senescence-enhanced expression of a tobacco cysteine protease gene has been identified, and a role in the amino acid remobilization in senescing leaves has been proposed. In Arabidopsis, the expression of *RD19a*, that encodes a drought-inducible cysteine proteinase, is also circadian-clock-regulated.

Identification of Clock-Controlled Genes by Differential Screens

Two clock-controlled genes were isolated from *Pharbitis nil* through a differential screen for mRNAs increased in abundance in photoperiods that induce flowering. One of these is *PnZIP*, a novel mesophyll-specific gene that is regulated by phytochrome and encodes a protein with a leucine zipper motif. The other one, *PnC401*, is also clock-regulated and its transcript accumulates during flower-inductive darkness. Fluctuations in levels of *PnC401* mRNA are regulated by phytochrome and by the circadian clock and are associated with photoperiodic events that include induction of flowering.

Several circadian clock-regulated genes have been identified in Arabidopsis by fluorescent differential display, including *CAB*, *CAB2*, carbonic anhydrase 2 (*CAH2*), *RBCS*, and *CCA1*. Using gene chip arrays representing 8200 different genes, more than 450 genes displayed were found to show circadian changes in steady-state mRNA levels. These included genes involved in photosynthetic light harvesting, synthesis of protective phenolic compounds, lipid modification, carbon metabolism and partitioning, nitrogen and sulfur assimilation, cell elongation and flowering, and genes coding for photoreceptors. Through analysis of the promoter regions of the oscillating genes, a novel promoter element, the evening element (AAAATATC), which confers circadian rhythmicity, was identified. The functional role of this conserved motif was confirmed by mutational analysis. Mutation of the full-length evening element in *CCR2* promoter caused a decrease in rhythmicity, as did deletion of this region.

In another study, 11% of the Arabidopsis genes showed diurnal expression and approximately 2% cycled with a circadian rhythm. By clustering microarray data from additional nonrelated experiments, groups of genes regulated by the circadian clock were identified. Genes within a group were found to be coregulated, such as those involved in photosynthesis and those required for starch metabolism. In Arabidopsis, starch is synthesized in the plastids during the day and broken down during the night, thus providing plants with sugars. These experiments show that amylase is expressed in the leaves late in the day. Another example of genes found in these studies are genes encoding proteins involved in nitrogen metabolism. Nitrate reductase shows strong circadian regulation, whereas nitrite reductase is light-regulated.

The selective advantage that plants gain by having genes under circadian regulation is presumed to be the ability to anticipate diurnal changes in the environment, such as variations in light, UV radiation, and temperature. The fact that several photosynthetic genes, genes involved in the synthesis of phenolic "sunscreens" and in chilling tolerance are under circadian control is consistent with this hypothesis. In addition, the clock may provide a temporal framework for internal coordination of physiological activities. This could account for the circadian rhythmicity of genes involved in carbon partitioning and in the energy-intensive processes of nitrogen and sulfur assimilation.

CONSTANS (CO)

CONSTANS mediates the interaction between the circadian clock and the control of flowering. Flowering time in many species is environmentally-regulated by photoperiod, light quality, and vernalization. This ability to perceive changes in daylength, photoperiodism, is essential for organisms to recognize seasonal changes and is associated with the circadian clock system. In 1935, Erwin Bunning hypothesized that the circadian clock was required for photoperiod sensitivity in flowering and physiological and genetic data from arabidopsis have corroborated his insight.

Some plants, including Arabidopsis, flower earlier when growth conditions mimic the long days of summer. Most of the more recently identified circadian mutants exhibit alterations in flowering time, and some previously identified flowering mutants also exhibit circadian defects. For example, *toc1* and *elf3* show defects in flowering time. Both *LHY* and *CCA1* overexpressing mutants are late flowering. Mutations at the *GI* locus result in late flowering, whereas null mutations in the floral repressor *FLC* confer earliness in flowering.

The Zn finger transcription factor *CONSTANS* (*CO*), is thought to promote flowering in long-day (LD) photoperiod conditions because *co* mutant plants have delayed flowering in LD, but flower at the same time as wild type under short-day (SD) conditions. The abundance of *CO* mRNA undergoes circadian oscillations and the waveform depends on the daylength; under SD, the *CO* peak is largely confined to darkness, whereas under inductive LD, levels of *CO* are high during most of the light period. In the *gi3* mutant, *CO* oscillates with a lower amplitude, and in the *lhy* mutant, *CO* is expressed at a reduced level with a different phase. The *gi* late flowering phenotype is corrected by overexpressing *CO* in the mutant background. Conversely, *CO* oscillates at higher levels in the early flowering *elf3* mutant.

It is possible that the clock exerts its control on photoperiodic response by setting the *CO* rhythm. *CO*, in turn, will promote flowering via downstream genes such as *FLOWERING LOCUS T* (*FT*). It seems likely that at its peak phase, CO activates FT, presumably through its zinc finger DNA binding domain. The photoperiod dependent phase of *CO* cycling is consistent with the view that the *CO* rhythm is the molecular basis for differential light sensitivity. Thus, the perception of the photoperiod may be mediated by adjustments in the phase of the circadian cycle that arise on reentrainment to a different light–dark cycle. The nature of the rhythm underlying the floral response is not known, but candidate molecules are being identified.

A CONSTANS ortholog (*PnCO*) from the short-day plant *Pharbitis nil* also plays a role in flowering. The level of *PnCO* is photoperiodically regulated and shows a circadian pattern of expression. *PnCO* is able to complement the *co* mutant of Arabidopsis. The cloning of *Hd1*, a major QTL controlling heading date (flowing time) in rice, showed that it encodes a GATA1-type protein with high similarity to CONSTANS in the zinc finger domain and in the *C*-terminal region. Thus, similar genes are involved in controlling the daylength response of flowering in short-day (SD) rice and long-day (LD) Arabidopsis plants. Unlike *CO* in Arabidopsis, *HD1* appears to promote heading under short days and to inhibit it under long days. This kind of research on rice may help to elucidate the difference in genetic control of photoperiod sensitivity between SD and LD plants.

In Arabidopsis, CO target genes play distinct roles in reproductive development. Two of them, *AGL20 and FT*, are required for CO to promote flowering; *AGL20* encodes a MADS-box transcription factor, whose expression responds to long photoperiods, consistent with the role that CO plays in the promotion of flowering. *FT* encodes a putative phosphatidylethanolamine binding and nucleotide binding protein, with similarity to Raf kinase inhibitor protein. The *AGL20* and *FT* genes are also regulated by a second flowering time pathway that acts independently of *CO*, and their functions are genetically redundant.

Two Arabidopsis CO-related proteins, CONSTANS-like 1 (COL1) and CONSTANS-like 2 (COL2), are predicted to encode proteins with 67% amino acid identity overall to the CO protein. All three proteins contain two adjacent *N*-terminal zinc finger motifs that are 85% identical between the proteins. They also share high identity in a *C*-terminal basic region containing a putative nuclear localization signal (the CCT motif). *COL1* and *COL2* transcripts levels are regulated by the circadian clock with a transcript peak around dawn. Transgenic plants misexpressing *COL1* and *COL2* show little effect on flowering time. However, overexpression of *COL1* shortens the period of two distinct circadian rhythms: *CAB* expression and leaf movement. The fact that these circadian defects are fluence-rate-dependent may suggest an effect on a light input pathway. Collectively, all these data suggest that there is an intimate interaction between the circadian clock and photoperiodic timing.

All possible molecular interactions described within the plant circadian clock are depicted in Figure 8.7 and listed as follows:

1. Light input to the clock via phytochrome may occur through a complex with PIF3 and PIF4. PIF3 binds to a G-box motif in *CCA1*, *LHY*, and *SPA1* promoters and possibly to other targets in the clock. PIF4 may function as a negative regulator of PHYB signaling. HFR1 seems to modulate PHYA signaling via heterodimerization with PIF3.
2. *ZGT* is a novel clock- and light-regulated tobacco gene that links the circadian oscillator to *LHCB* expression.
3. COP1 may function as a repressor protein that targets different substrates for degradation by the proteasome. SPA1 may link the phytochrome A–specific branch of light signaling to COP1. Photoactivated CRY1 and CRY2 and PHYB also bind to COP1, repressing its activity.
4. At the core of the oscillator, feedback loops generated by reciprocal regulation between the MYB-related transcription factors CCA1 and LHY, and possibly RVEs, and the family of pseudoresponse regulators APRR, which includes TOC1, are responsible for generating circadian rhythms. The MYB factors may function as activators of genes peaking at dawn and repressors of genes peaking at dusk. CCA1 and LHY are phosphorylated by CK2, which may make them substrates for the SCF complex and target them for degradation by the proteasome.
5. ZTL/ADO1 and related proteins might act as novel blue-light photoreceptors, using the LOV domain as a flavin binding site. Through the F-box domain, they may target proteins for degradation via the ubiquitin system,

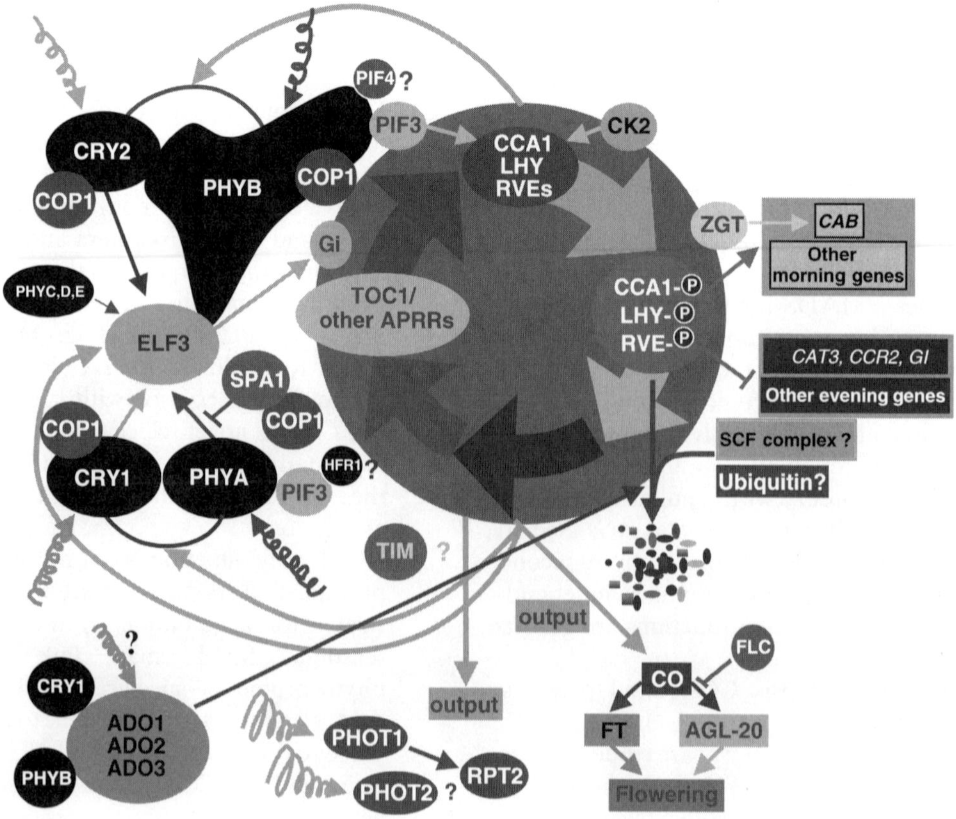

Figure 8.7. Molecular interactions within the plant circadian clock. Both photoreceptors, phytochromes and cryptochromes, mediate the effects of light–dark cycles on the phase and period of the circadian clock. The effects of light on the clock are also modulated by the clock itself through its regulation of both positive and negative light signaling components such as GI and ELF3.

and via the kelch domain they may recruit specific proteins for degradation, playing some role in protein turnover of clock-associated components and interacting with CRY1 and PHYB.

6. Arabidopsis contains a *TIMELESS* homolog gene, although its role in the circadian oscillator remains to be elucidated.
7. Clock-controlled expression of genes involved in the photosynthetic process, cell elongation, light signaling pathways, and flowering-time regulation provides the organism with the ability to anticipate and adapt to periodic changes in the environment. Output pathways come from the oscillator to input components known to be regulated by the clock at transcriptional, mRNA abundance, or protein abundance levels. Because phytochromes and cryptochromes mediate light input to the clock, they are both clock input and output components, creating outer feedback loops.
8. Flowering is triggered by a change in photoperiod. In Arabidopsis, CO promotes flowering in response to long photoperiods, mediating the control of flowering by the circadian clock.

CO can directly increase expression of the *FT* and *AGL-20* genes, promoting the transition to flowering. *FLC*, a floral repressor, negatively regulates *AGL-20*. FLC and CO are proposed to have antagonistic effects on the expression of *AGL-20* and *FT*.

MECHANISMS THAT CONFER CIRCADIAN REGULATION

Both transcriptional regulation and post-translational modifications have been implicated in the control of the plant circadian oscillator.

Promoter Elements

Minimal nuclear promoters able to confer circadian transcription have been identified for several *LHCB* genes in different species, tomato *LHCA* genes, the wheat *psbD* gene and the Arabidopsis *RCA*, *GER3*, *CAT3*, and *TOC1* genes. The circadian clock associated gene *CCA1*, previously implicated in phytochrome regulation, binds to a consensus motif (AAa/cAATCT) widely conserved among *CAB* genes.

A promoter deletion analysis of four *LHC* tomato genes in transgenic tobacco plants led to the identification of a short sequence of 47 nucleotides that is necessary for conferring circadian *Lhc* mRNA oscillations. A novel motif, CAANNNNATC, is conserved in 5′ upstream regions of clock-controlled *Lhc* genes and overlaps with a sequence for relevant phytochrome mediated gene expression.

In vivo analysis of progressively truncated *LHCB 1*1* (*CAB2*) promoter fragments fused to luciferase reporter gene defined a 36-bp region sufficient to confer circadian transcription. In vitro analysis of DNA binding by EMSA and DNA footprinting identified binding sites for multiple complexes in this promoter fragment from −111 to −74. Minimal promoter fragments (278 and 119 bp) have also been defined for two tomato *LHCA* genes. A minimal 330-bp clock-responsive promoter has been defined for the Arabidopsis *RCA* gene, whose expression oscillates in phase with the *LHC* genes. A clock-responsive element sufficient to confer a low-amplitude circadian oscillation lies within 317 bp of the transcription start, but other elements necessary for high-amplitude circadian oscillation lie upstream of −317 and downstream of −970. A CCA1 binding site is found in both *LHCA* and *RCA* functionally defined minimal promoters, although the functional relevance of this motif to the circadian transcription of *LHCA* or *RCA* has not yet been established.

In addition to promoters that confer morning-specific transcription such as *LHCB* and *RCA* promoters, dusk-specific promoters have also been studied. To confer circadian transcription with a dusk-specific phase, 490 bp of the Arabidopsis *GER3* promoter are enough. Clock-response elements contributing to high-amplitude *ATGER3* oscillations largely reside between −967 and −299.

On the other hand, 265 bp of the Arabidopsis *CCR2* promoter confer a robust high-amplitude rhythm with dusk-specific phase, and as few as 56 bp are sufficient to confer a low amplitude oscillation. The smallest *CCR2* promoter fragment contains a CCA1 binding site, as does a minimal promoter (230 bp) of the *CAT3* catalase gene that is sufficient to confer dusk-specific transcription. The functional significance of the CCA1 binding site in the dusk-specific promoters has not been directly tested, but the existence of CCA1 binding sites in promoters that are transcribed nearly 180°C out of phase suggest that the mechanism by which the phase of transcription is determined will be very complex and due not only to phase-specific transcriptional activators. A model in which CCA1 and LHY simultaneously regulate genes that are phased to different times of the day has been proposed. MYB factors act as negative elements that repress *TOC1* expression and probably other genes having evening elements in their promoters, which include, for example, *GIGANTEA* and *CCR2*. Furthermore, given the overall similarity between CCA1 and LHY, it is likely that LHY protein, as CCA1, is also a positive regulator of *CAB* genes.

Phosphorylation

In addition to transcriptional regulation, posttranslational modifications such as phosphorylation and regulation of protein stability have also emerged as pivotal control steps in the plant circadian oscillator. Phosphorylation may regulate multiple properties of clock proteins, including stability and intracellular localization. Interestingly, CCA1 was found to associate with CKB3, a regulatory β subunit of the Arabidopsis casein kinase 2 (CK2) in a yeast two-hybrid screen. CK2 is a serine/threonine kinase that is present in all eukaryotic cells examined to date. It has a heterotetrameric structure α2-β2, consisting of two catalytic (α) and two regulatory (β) subunits. CK2 can phosphorylate several transcription factors that bind to the promoter regions of light-regulated genes, including *CCA1*, *LHY*, and *G-box factors* (*GBFs*) in vitro. CK2 was also proposed to be involved in cyclic phosphorylation of an endosperm-specific transcription factor, *Opaque 2* (*O2*), whose phosphorylation status changes diurnally. O2 activity is downregulated at night by both a reduction in *O2* transcript and hyperphosphorylation of residual O2 protein, suggesting that regulatory gene activity during endosperm development may be acutely sensitive to a diurnal signal directed to the developing seeds.

Overexpression of *CKB3* increases CK2 activity in transgenic Arabidopsis plants and shortens the period of the rhythmic expression of *CCA1*, of *LHY*, and of four other circadian clock-controlled genes and causes early flowering in both LD and SD conditions, suggesting a role for protein phosphorylation mediated by CK2 in the function of the arabidopsis circadian clock. In plants overexpressing *CKB3*, CCA1 may be phosphorylated more rapidly, causing a shorter period length due to an increased rate of degradation of the protein. This might involve F-box proteins and the ubiquitination pathways.

Some studies have revealed an involvement of CK2α in plant phototransduction pathways. In Arabidopsis, *CK2α* antisense RNA affects the expression of some light-regulated genes, although it has little effect on the inhibition of hypocotyl elongation by red light.

In rice, heading date is determined mainly by two factors: duration of the basic vegetative growth and photoperiod sensitivity. *Hd6*, a QTL involved in circadian photoperiod sensitivity, encodes the α subunit of the protein kinase CK2. The Kasalath allele of *CK2α* increases days to heading. This effect is produced by a single nucleotide substitution, which changes the premature stop codon (TAG) in a *japonica* variety (Nipponbare) to a lysine codon (AAG) in an *indica* variety (Kasalath). These findings indicate again that the same genes appear to be involved in photoperiod response in both SD and LD plants.

A well-defined system of post-transcriptional regulation is provided by analysis of PEP carboxylase. In CAM plants, PEP carboxylase activity cycles as a result of a circadian rhythm in the phosphorylation state of the enzyme. Sucrose phosphate synthase activity in tomato is regulated in a circadian manner by a protein phosphatase. The rhythm in nitrate reductase mRNA abundance in Arabidopsis also reflects posttranscriptional control. Cyclic nitrate reductase activity is affected by circadian regulation of its expression, in large part through phosphorylation dependent mechanisms that affect the level of mRNA.

■ MULTIPLE AUTONOMOUS CLOCKS IN PLANTS

Considerable evidence supports the existence of multiple oscillators in multicellular plants. Most of the evidence takes the form of multiple rhythms running with different periods (internal desynchronization), which was demonstrated in *Phaseolus coccineus*. In *Phaseolus vulgaris*, rhythms in CO_2 assimilation and stomatal aperture exhibit a different period from the rhythm in leaf movement. In Arabidopsis, the freerunning periods in leaf movements and *CAB* expression are different, although both are shortened by the *toc1-1* mutation. Similarly, the *gi-2* mutation shortens the period in leaf movement but lengthens the period in gene expression. In the dark, the *Lhcb* transcription period lengthens to 30 hours, whereas the oscillations in *CCR1*, *CCR2*, and *CAT3* mRNA abundance and transcription retain a 24-hour period, suggesting that they are driven by distinct oscillators.

Transcription of the Arabidopsis and wheat *Cab* genes in single tobacco transgenic seedlings exhibit independent rhythms in a developmentally regulated fashion. Indeed, a red-light-insensitive oscillator and a phytochrome-coupled circadian clock coregulate the oscillating expression of individual *CAB* genes at the level of transcription at an early developmental stage. Free calcium is resonsible for driving the rhythm of *Lhcb* expression. Although the rhythms of cytosolic calcium and *Lhcb::luc* peak at the same time, they freerun with different periods in continuous light, suggesting that different circadian pacemakers control these rhythms.

The well-known circadian rhythm of the CAM mode of photosynthesis is expressed as independent dynamic patterns of photosynthetic efficiency over a single leaf. The biological clock controlling this rhythm is a spatiotemporal product of many weakly coupled individual oscillators that operate independently as a consequence of the dynamics of metabolic pools and limitations of CO_2 diffusion between tightly packed cells.

Animal species support multiple peripheral clocks in addition to the neural pacemakers, such as the suprachiasmatic nucleus. Peripheral animal tissues maintain circadian rhythms when isolated in culture, indicating that these tissues contain circadian clocks. Likewise, isolated plant organs and tissues support circadian rhythms, as do the specialized cells that drive rhythmic leaf and stomatal movements. In plants it is unclear whether peripheral circadian clocks are mutually coupled and whether there is a centralized circadian pacemaker. The levels of several plant hormones exhibit circadian rhythms, but it is unknown whether any such coordinated, rhythmic signals can synchronize cellular clocks.

In vivo reporter gene imaging has been used to demonstrate that the circadian systems of plant organs and localized areas of tissues are functionally independent. The rhythmic expression of a single gene can be set to three phases in different anatomic locations of a single plant, by applying light–dark treatments to restricted tissue areas. The rhythms of one organ are unaffected by entrainment of the rest of the plant, indicating that phase resetting signals are also autonomous. Thus, higher plants appear to contain a spatial array of autonomous circadian clocks that regulate gene expression without a localized central pacemaker. The circadian rhythms of cyanobacteria and of unicellular eukaryotes, such as *Chlamydomonas*, also demonstrate that a circadian clock (or clocks) can exist within a single cell and several mammalian cell types retain a functional circadian oscillator in culture. Thus, it is possible that circadian oscillators are distributed throughout the intact plant and function independent of any central pacemaker and possibly of each other. Simultaneous but independent entrainment of these clocks in different anatomic parts may be responsible for synchronizing their circadian rhythms.

■ ADAPTIVE SIGNIFICANCE OF CIRCADIAN CLOCKS

As circadian rhythms are ubiquitous in nature, it appears that an accurate temporal control likely confers some adaptive advantage. This adaptive fitness would be enhanced by the synchronization of an organism's internal clock with the light–dark cycle imposed by the environment, and the advantages in phasing sunlight-sensitive cellular events to the night has been suggested. More recent studies have addressed the adaptive fitness of dawn anticipation in cyanobacteria and green algae. *Chlamydomonas* exhibits circadian rhythms in cell division and in sensitivity to UV radiation. At the end of the day maximal UV sensitivity occurs, and this is coincident with the DNA division period.

■ CONCLUSIONS AND PERSPECTIVES

Study of the plant circadian system has witnessed a great deal of progress with identification of many molecular components. The next and more difficult step will be the assembly of these components in a coherent molecular model. There are no reliable criteria with which to unambiguously assign molecules to roles as input, output, or oscillator components. The data available so far suggest that transcriptional regulation is very important to the function of the plant circadian clock, as it is to other circadian systems. It is becoming increasingly clear that posttranscriptional regulation also plays a central role in the function of the circadian oscillator. The multiple levels of feedback on light input pathways indicate that phototransduction pathways and the circadian oscillator of Arabidopsis are associated in an intricate manner. The distinction between light input and oscillator mechanism is becoming increasingly blurred, supporting the suggestion that the circadian clock may have arisen from phototransduction pathways. Similar regulatory mechanisms are employed in plant and animal circadian clocks. But so far, the putative components of the Arabidopsis oscillator have no orthologs in the circadian systems of cyanobacteria, fungi, or animals.

One common evolutionary characteristic may be the presence of domains related to two-component signaling systems. The PAS domains present in clock-associated genes of animals and fungi show similarities to sensor domains and are found in several response genes to environmental, developmental, or cellular signals. In Arabidopsis several putative components of the circadian clock (phytochromes, PIF proteins, and ZTL and ZTL-like proteins) contain PAS-like domains. Phytochromes contain two PAS repeats at their *C*-termini, and a histidine kinase–related domain similar to that of sensor kinases of two component signaling systems. However, phytochromes seem to function as serine kinases. Phytochrome-mediated phosphorylation would be a mode of action for very rapidly induced phytochrome responses. PIF proteins containing a PAS domain are necessary for normal photoinduced signal transduction, and they bind to G-box sequence motifs found in various light- and circadian-regulated promoters. ZTL/ADO1 and the related LKP2/ADO2 and FKF1/ADO3 have a degenerate form of a PAS domain more similar to that of the phototropins PHOT1 and PHOT2 and to neurospora WC1 and VIVID proteins. Such LOV motifs are found in proteins that bind flavins (FMN in the phototropin case and FAD for WC1) and are involved in redox reactions. An argument for the common ancestry of clocks perhaps is the existence of PAS-containing proteins in the circadian clock from distantly related organisms. Furthermore, the G-box element, found in the promoters of many light- and circadian-regulated plant genes, is the same sequence as the E-box promoter element that plays an essential role in the circadian regulation of animal *PER* genes. An alternative point of view is that different clocks may have evolved convergently from related molecules whose original function was to mediate light perception. This is consistent with the idea that circadian clocks evolved under selective pressure to set some cellular events to the light or dark period of the day as we mentioned earlier. The elucidation of the circadian clock mechanisms of plants will shed more light on the evolutionary origins of the circadian rhythms.

■ ACKNOWLEDGMENTS

Work in the laboratory of ARC was supported by grants from NIH and USDA.

FURTHER READING

Barak, S, Tobin, EM, Andronis, C, Sugano, S, Green, RM (2000): All in good time: The *Arabidopsis* circadian clock. *Trends Plant Sci* **5**: 517–522.

Carre, I (2001): Day-length perception and the photoperiodic regulation of flowering in Arabidopsis. *J Biol Rhythms* **16**: 415–423.

Devlin, P, Kay, S (2001): Circadian photoperception. *Annu Rev Physiol* **63**: 677–694.

Devlin, PF (2002): Signs of the time: Environmental input to the circadian clock. *J Exp Bot* **53**: 1535–1550.

Harmer, SL, Panda, S, Kay, SA (2001): Molecular bases of circadian rhythms. *Annu Rev Cell Dev Biol* **17**: 215–253.

McClung, C (2001): Circadian rhythms in plants. *Annu Rev Plant Physiol Plant Mol Biol* **52**: 139–162.

Millar, A (1999): Biological clocks in *Arabidopsis thaliana*. *New Phytol* **141**: 175–197.

Piechulla, B (1999): Circadian expression of the light-harvesting complex protein genes in plants. *Chronobiol Int* **16**: 115–128.

Roden, LC, Carre, IA (2001): The molecular genetics of circadian rhythms in Arabidopsis. *Semin Cell Dev Biol* **12**: 305–315.

Somers, D (1999): The physiology and molecular bases of the plant circadian clock. *Plant Physiol* **121**: 9–19.

Staiger, D, Heintzen, C (1999): The circadian system of *Arabidopsis thaliana*: Forward and reverse genetic approaches. *Chronobiol Int* **16**: 1–16.

Yanovsky, MJ, Kay, SA (2001): Signaling networks in the plant circadian system. *Curr Opin Plant Biol* **4**: 429–435.

Young, MW, Kay, SA (2001): Time zones: A comparative genetics of circadian clocks. *Nat Rev Genet* **2**: 702–715.

Part IV

CIRCADIAN ORGANIZATION IN COMPLEX ORGANISMS

9

MULTIPLE OSCILLATORS

Jadwiga M. Giebultowicz

■ INTRODUCTION

In the previous chapters we learned that molecular mechanisms underlying the circadian clock are governed by similar principles in all organisms from the simplest (bacteria) to the most complicated (mammals). We also found that, no matter how complex the organism, circadian oscillations are generated at the level of a single cell. In this chapter, we discuss how cellular oscillators are organized into coherent circadian systems. Several lines of evidence demonstrate that animals contain multiple circadian oscillators. Circadian systems in mammals seem to be organized hierarchically with a central, or master, clock, which coordinates many rhythmic phenomena in the body. It is increasingly evident, however, that circadian systems also include many peripheral oscillators coordinating tissue-specific functions. The relative contributions of central and peripheral oscillators in the organization of the timing system vary substantially among different groups of animals.

■ PHYSIOLOGICAL EVIDENCE FOR MULTIOSCILLATORY SYSTEMS

Individual organisms display a multitude of behavioral, physiological, and molecular rhythms in a highly specific temporal order. Various rhythms reach their peaks at different phases of the daily light–dark cycle, and show a specific waveform. Rhythmic changes in several physiological parameters monitored in human subjects provide a good example of this general principle (Fig. 9.1). Body temperature is elevated throughout the day. Plasma levels of growth hormone rise to a brief and sharp peak in the early night, while levels of cortisol display a broad peak during the late night and morning. The rate of urinary potassium

Molecular Biology of Circadian Rhythms. Edited by Amita Sehgal
ISBN 0-471-41824-2

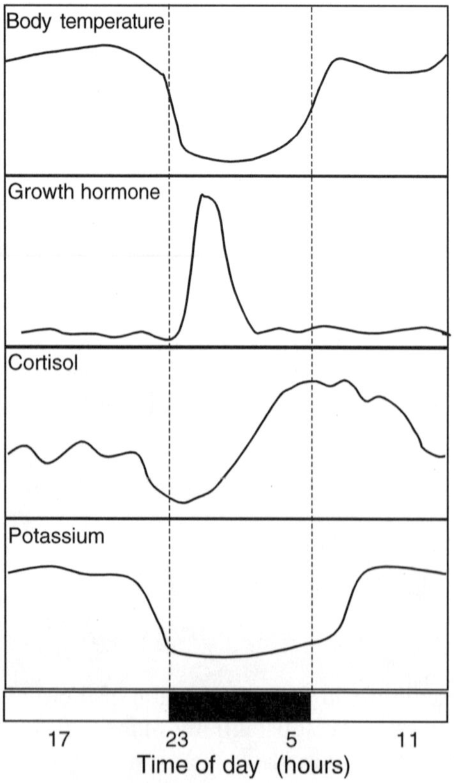

Figure 9.1. Approximate profiles of several rhythmic physiological parameters in a human subject monitored in light–dark cycles. Each rhythm has a characteristic phase relationship and waveform. [Modified from Moore-Ede et al. (1982).]

excretion is high during the day and low at night. Are different rhythms under the command of a single central oscillator, or are they governed by many separate ones? In 1960, Dr. Colin Pittendrigh predicted that "A multicellular system is literally a population of oscillators" based on very limited experimental data available at the time. Since then, much experimental evidence has accumulated in support of this statement.

Freerunning Period Variation

Different rhythms in one species may have divergent freerunning periods. Insects provide an excellent model system in which to explore circadian organization, because they possess a rich repertoire of overt rhythmic activities such as ecdysis, foraging, wandering, mating, and oviposition. How can one test whether one or more clocks drive the different rhythms? One way is to monitor several rhythms in the same species in constant darkness and temperature. In these conditions, devoid of zeitgebers, each rhythm should reveal its endogenous, freerunning period. Experiments of this kind performed in the fleshfly demonstrated that various rhythms had different freerunning periods. For example, the rhythm of larval wandering displayed a periodicity of about 20 hours, while the rhythm of pupal eclosion had a period close to 24 hours. It was concluded that functionally separable oscillators are likely to generate rhythms with different endogenous periods. However, it is conceivable that the same oscillator may change its endogenous periodicity in subsequent developmental stages.

Internal Desynchronization between Different Rhythms in an Organism

To test the relationships between different rhythms co-existing in the body, one can monitor their adjustments following a phase shift of a light–dark cycle. It may be expected that rhythms driven by the same oscillator would shift in a more coordinated fashion than would those driven by separate oscillators. When such experiments were performed in humans, some rhythms, such as rest-activity cycles reentrained within a few days, while other rhythms, such as fluctuations in body temperature or cation excretion, took longer to reentrain. Similar results were obtained in the squirrel monkey. After an 8-hour phase delay of the light–dark cycle the activity rhythm resynchronized in one day, the body temperature rhythm took 3 days, and the rhythm of urinary potassium excretion took 7 days. Different rates of reentrainment result in a transient internal desynchronization, during which coexisting rhythms display abnormal phase relations with each other.

Another interesting observation concerning the organization of circadian rhythms was made on humans living in isolation without any time cues. In conditions deprived of Zeitgebers, circadian rhythms in some subjects begun to freerun with independent periods, a phenomenon termed *spontaneous internal desynchronization*. For example, in one human subject, the rhythm of rest–activity and the rhythm of urinary calcium excretion spontaneously began to oscillate with a period of 32.6 hours, while the rhythm of body temperature continued to oscillate with a period of 24.7 hours. Similar phenomena have been observed in other mammals kept in constant conditions. Both transient and spontaneous internal de-synchronization suggests that separate oscillators may generate different rhythms in one organism.

Circadian Rhythms in Isolated Organs

A strong argument for the existence of multiple circadian oscillators in animals is the persistence of circadian rhythms in different organs isolated from the body and maintained in culture. Work on mollusks and insects suggests that oscillators in invertebrates reside in various tissues. Marine mollusks, *Aplysia* and *Bulla*, display rhythms of electrical activity in a group of neurons at the base of the eye. Those basal retinal neurons maintain robust rhythms when cultured in isolation from the rest of the body. In insects, some physiological rhythms are maintained in peripheral tissues that lack innervations. For example, the rhythm of sperm release from the testes to the vas deferens in moths persists in the testes–vas deferens complexes cultured in vitro.

Similar local and tissue-specific oscillators exist in cockroaches. A daily rhythm has been reported in the formation of the outer body cover, the cuticle, by underlying epidermal cells. When small pieces of cockroach epidermis were cultured in vitro, they maintained the rhythmicity of cuticle secretion.

The persistence of rhythmic activities in isolated tissues is not limited to invertebrates. A classic example of an oscillator that continues to operate in vitro is the pineal gland of many nonmammalian vertebrates. This gland is responsible for the rhythmic synthesis and release of melatonin (Chapter 10). It is well documented that the rhythmic production of melatonin persists in isolated pineal glands and even in dispersed pinealocytes of birds. Similarly, retinas of several vertebrates rhythmically produce melatonin, and these rhythms persist in cultured eyes. The major timing center of mammals, the suprachiasmatic nucleus (SCN) of the brain, displays strong rhythms in electrical activity, metabolic rate, and neuropeptide levels. Daily fluctuations in spontaneous neural activity continue in the SCN maintained in vitro and, impressively, even in a few cultured neurons derived from the SCN. Nonneural tissues of mammals also display physiological rhythms when isolated in vitro. More recently, molecular rhythms have been reported in such tissues (see below).

■ MOLECULAR EVIDENCE FOR MULTIOSCILLATORY SYSTEMS

Several clock genes have been identified and characterized in the fruitfly (Chapter 3) and mouse (Chapter 4), and their homologs have been cloned in other invertebrates and vertebrates. Studies of the spatial patterns of clock gene expression provide compelling support for the earlier physiological data that predicted the existence of multioscillatory circadian systems in animals. Rhythmic activities of clock genes and their products are found in a surprisingly broad range of tissues in all examined animals, from insects, through lower vertebrates to mammals.

Clock Genes in Drosophila

Various techniques used to study the expression of clock genes in the fruitfly consistently revealed wide distribution of *per* mRNA and protein in many tissues. Cycling PER expression was detected in certain areas of the CNS, including several subsets of brain neurons, glial cells, photoreceptors, and chemosensory hairs. The PER protein was also found in many peripheral tissues outside of the central nervous system, such as the digestive, excretory, and reproductive systems (Table 9.1). Tissues that express PER also express another clock protein, TIMELESS (TIM). Patterns of PER and TIM expression are similar in the cells of the CNS and peripheral organs and are consistent with their function as clock proteins (see Chapter 3). PER and TIM dimerize in the cytoplasm during the early night, and later accumulate in cell nuclei. In both the brain and peripheral tissues, TIM is degraded in response to the morning light, while PER remains in the nuclei through midday. A temporal profile of both proteins in the excretory organ of Drosophila, the Malpighian tubules, is shown in Figure 9.2.

To study the expression of *per* and *tim* genes in greater detail, several lines of transgenic flies were created which report the activity of clock genes. Particularly useful are flies that carry sequences coding for firefly luciferase fused to the promoter region of either *period* (*per*-luc) or *timeless* (*tim*-luc) genes. Tissues expressing clock genes emit measurable light (when luciferase acts on its substrate, luciferin) in proportion to the activity of the respective transgene. This allows for the observation of the activities of *per* and *tim* genes in real time. The sensitivity of this technique is so high that *per* or *tim* expression can be continually monitored in individual organs

TABLE 9.1. Expression of Clock Genes in Insects and Vertebrates

	Insects[a]			Vertebrates[b]		
		Cycling			Cycling	
Tissues and Organs	Present	In vivo	In vitro	Present	In vivo	In vitro
Brain neurons	+	+	?	+	+	+
Brain glial cells	+	+	?	?		
Eye	+	+	?	+	+	+
Chemosensory organs	+	+	+	+		
Skeletal muscles	–			+	+	+
Heart	?			+	+	+
Respiratory system	–			+	+	+
Digestive system	+	+	?	?		
Liver (fat body)[c]	+	+	?	+	+	+
Renal organs	+	+	+	+	+	+
Male reproductive system	+	+	+	+	?	?

[a] Data based mostly on Drosophila.
[b] Based on mammals and zebrafish.
[c] Fat body of insects is a functional equivalent of vertebrate liver.

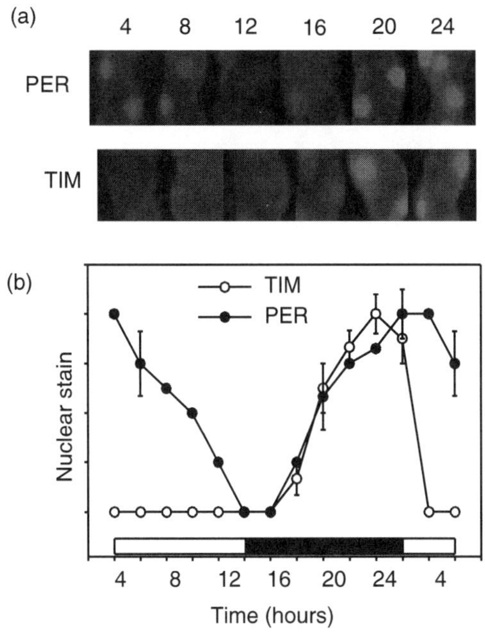

Figure 9.2. Expression of clock proteins PER and TIM in Malpighian (excretory) tubules of *D. melanogaster*: (a) fragments of tubules showing nuclear location of both proteins; (b) quantification of the intensity of PER and TIM signal [modified from Ivanczenko et al. (2001)].

over several days. An example of the record of *per* and *tim* activity in individually cultured Malpighian tubules and rectal tissues is shown in Figure 9.3. Results of such experiments confirmed that various fly organs rhythmically express clock genes and demonstrated that these rhythms are self-sustained.

Clock Genes in Vertebrates

Clock genes are rhythmic in many vertebrate organs. Several genes involved in the circadian mechanism are known in mammals, including *Clock* (*Clk*) and three orthologs of the *period* gene of Drosophila. Where in the body are those genes expressed? Many expected to find *Clk* and *per* mRNA in the SCN, which plays a central role in generating many mammalian rhythms. This indeed turned out to be the case, but, unexpectedly, high expression of clock genes was also detected in other parts of the brain, and in various peripheral tissues. Clock genes that show rhythmic activity in the SCN are also rhythmic in other tissues (Table 9.1). As in flies, transgenic technology helped to demonstrate that the rhythmic expression of clock genes persisted in isolated organs. Rhythmic bioluminescence was recorded in cultured livers, lungs, and skeletal muscles taken from transgenic rats expressing luciferase under the control of the *Per1* promoter. The circadian mechanism appears to exist not only in internal organs freshly explanted from the body but is also preserved in rat fibroblasts that existed for many years only as cell lines. These fibroblasts can initiate rhythmic expression of clock genes and some clock-controlled genes when high concentrations of serum are added to their culture medium.

Besides mammalian tissues, oscillations in the expression of clock genes were observed in the cultured organs of zebrafish, such as the heart and kidney. These examples of rhythmicity in phylogenetically distant organisms, from insects, through fish, to mammals, suggest that self-sustained peripheral oscillations may be a conserved feature in all complex animals. This phenomenon may be explained from an evolutionary perspective. Circadian mechanisms are very ancient; they have been identified in bacteria and unicellular eukaryotes. As these simple organisms evolved into tissues of higher animals, the descendent tissues apparently retained their rhythm-generating capacities.

■ ORGANIZATION OF MULTIOSCILLATORY SYSTEMS

The relative independence of different rhythms in a single organism and the persistence of clock functions in various

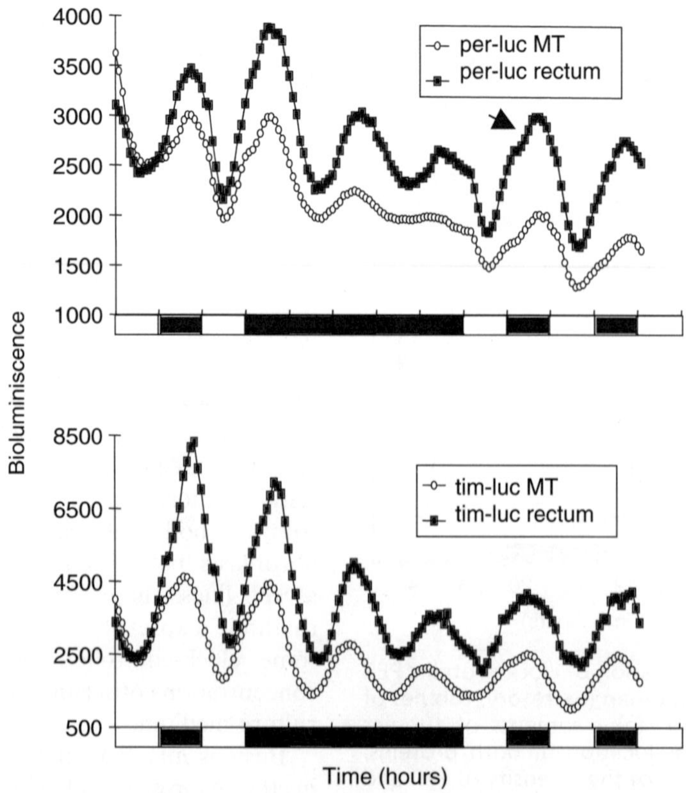

Figure 9.3. Real-time expression of *per*-luc and *tim*-luc reporter genes in individual body tissues. Bioluminescence emanating from individual Malpighian tubules (MT) or rectal tissues (rectum) was recorded hourly. The upper panel shows the average plot of *per*-luc expressing tissues, the lower panel shows the average plot of *tim*-luc expressing tissues. Black and white bars indicate 12-hour periods when the lights were on and off, respectively; shaded bars indicate dark periods, when the lights would be on in LD. An increase in the amplitude of cycling on reexposure of tissues to LD (arrow) indicates that the two peripheral oscillators are directly light-responsive. [Reprinted from *Current Biology* **10**, Giebultowicz et al., Transplanted Drosophila excretory tubules maintain circadian clock cycling out of phase with the host, pp 107–110, [copyright (2000) with permission from Elsevier Science.]

isolated organs demonstrated that animal circadian systems consist of multiple oscillators. How do these oscillators work together to ensure internal temporal order? There are sparse experimental data available from different taxonomic groups that show a broad spectrum of relationships between oscillators. At one end of this spectrum are insects, whose circadian systems include autonomous, independently entrained oscillators. At the other end are mammals with a hierarchical circadian system containing a master, or central, oscillator, which is entrainable by light and regulates the light-insensitive peripheral oscillator via neural or humoral signals. In the following paragraphs, we will review available data concerning the organization of the circadian system in different groups of animals.

Autonomous Oscillators in Insects

Among the many behavioral rhythms displayed by insects, the best studied are the rhythm of rest–activity and the rhythmic emergence of adults from pupal cases, a process known as *eclosion*. From early on, scientists were compelled to search for pacemaking centers controlling those rhythms. In a set of classic experiments performed on moths, it was demonstrated that the oscillator located in the central part of the brain controls the eclosion rhythm. Cross-transplantation of brains between two species of moths showing eclosion peaks at different times of the day, caused the recipient moths to eclose at the time characteristic for the donor species! Oscillators, which control rest–activity rhythms in cockroaches, crickets, and flies, are also located in the brain. The most extensive studies on the localization of the circadian oscillator controlling behavior have been performed in *Drosophila melanogaster*. The task was difficult because the fly brain contains many groups of neurons and glial cells that express clock genes. As described in Chapter 3 , several genetic and molecular approaches were used to locate the circadian pacemaker controlling behavioral rhythms in a cluster of ventrally located lateral neurons (LNv) at the border between the central brain and the optic lobes. Genetic removal of all LNvs renders flies behaviorally arrhythmic, demonstrating that these cells are essential for generating the rest–activity rhythm. LNvs are often referred to as the *central clock* in flies, because of their master role in the control of behavior and their location in the brain.

The clock in the LNvs is one of many components of the fly circadian system. Clock genes are widely expressed in *D. melanogaster* in a pattern consistent with the existence of other oscillators. Use of transgenic flies made it possible to demonstrate that such oscillators are indeed located in various organs and that they display a remarkable degree of autonomy. In one experiment, parts of per-*luc* flies such as the wings, legs, proboscis, and antennae were separately cultured in vitro for several days. Bioluminescence emitted from these appendages was rhythmic. Moreover, the phase of these rhythms shifted in response to a shift in the light–dark cycle applied in vitro. Further experiments attributed oscillatory properties to chemosensory organs, which are distributed on most fly appendages. In another set of experiments, various internal organs, such as the Malpighian tubules, rectum, and testes, taken from *per-luc* and *tim-luc* flies were cultured in vitro. Despite being separated from each other, different tissues displayed synchronous oscillations that persisted in constant darkness and increased in amplitude upon re-introduction of light–dark cycles (Fig. 9.3). Collectively, these experiments demonstrate that fly peripheral organs contain self-sustaining circadian oscillators, which can be directly entrained by light–dark cycles.

The fact that the fly circadian system is composed of the central oscillator in the brain, and an array of peripheral oscillators, prompted us to examine relations between those components of the timing system. Several lines of evidence suggest that oscillators in peripheral organs may be independent of the clock in LNvs or any other part of the brain. First, clock molecules cycle with similar phases in the brain and peripheral tissues. For example, rhythms of *per*, *tim*, and their respective proteins do not show any significant phase lag in renal tubules and rectal tissues relative to the brain clocks, when examined in vitro (Fig. 9.3) or in decapitated flies (Fig. 9.4a). The second argument against the dominant role of the brain comes from monitoring of the resetting of peripheral clocks in the Malpighian tubules. For this experiment, a group of flies was decapitated and subjected, along with intact flies, to a reversal of the LD cycle. The phase shift in the oscillation of the *per* gene in response to reversed LD occurred with a very similar time course in intact and decapitated flies

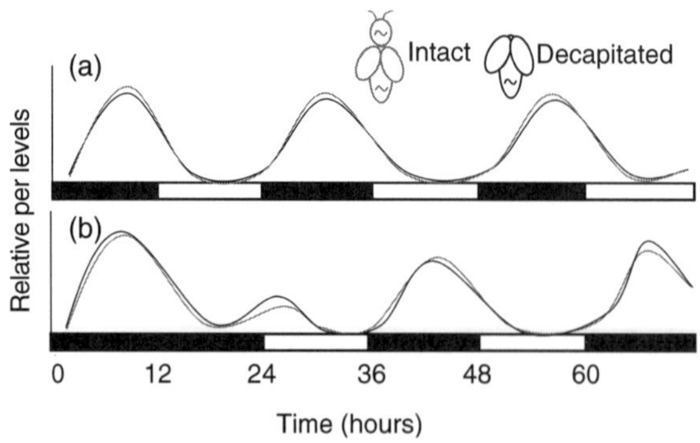

Figure 9.4. The temporal expression of a PER reporter in the Malpighian tubules of *D. melanogaster.* The onset of the first experimental dark period is termed 0 hour. Headless flies (black lines) were decapitated a few hours prior to time 0 hour and later treated as control intact flies (dotted lines). Malpighian tubules were assayed every 6 hours. (a) Flies maintained in LD conditions exhibit similar PER oscillations regardless of whether the head oscillators are present. (b) Intact and headless flies subjected to a phase shift of LD cycle, with a 12 hour extension of the first dark period, reset with a similar time course. [Adapted from Hege et al. (1997).]

(Fig. 9.4b). These data suggest that the central clock does not mediate resetting of the peripheral clock, but rather the Malpighian tubule clocks seem to be directly reentrained by environmental cycles. In a third experimental approach, tubules were transplanted into host flies entrained to an opposite light–dark cycle (relative to donor flies) and kept in constant darkness after the surgery. Under those conditions, the clock in the donor tubules cycled out of phase relative to host tubules, even though both sets of tubules shared the same hormonal milieu. Apparently, the rhythms of the transplanted organ were unaffected by the host fly. Thus, at least some peripheral oscillators in Drosophila seem to operate as independent units, achieving synchrony due to their sensitivity to a common Zeitgeber in the form of external light–dark signals (Fig. 9.6a).

Similar brain independence of peripheral oscillators is observed in other insects. It has been known for many decades that the circadian clock located in the brain controls behavioral rhythms in moths. When the rhythm of sperm release from testes to the vas deferens was described in moths, it was first assumed that the brain would also control it. Experiments, however, proved otherwise. The rhythm of sperm release persisted in brainless moths with a phase similar to that in intact moths. Furthermore, light applied to isolated testes–vas deferens complexes shifted the phase of sperm release rhythm. These experiments clearly demonstrated that the reproductive system of the male moth contains an autonomous oscillator that is directly photoresponsive and able to operate independently of the brain.

Despite the great autonomy of some peripheral clocks, specific insect oscillators may be subject to modulating signals from other oscillators, as was demonstrated in the hemipteran bug, *Rhodnius prolixus.* These insects display pronounced daily fluctuations in the levels of circulating molting hormone, ecdysone. Glands producing this hormone contain their own photosensitive circadian oscillator, which is capable of timing the release of ecdysone in vitro. In intact animals, however, this oscillator is entrained by the rhythmic release of tropic hormone that is controlled by another oscillator located in the brain. Similar relationships may exist between the brain and the endocrine glands in other insects. The experiments performed in *Rhodnius* show that some oscillators with direct sensitivity to light may be nevertheless modulated by internal messengers derived from other clocks in the body. Therefore, the conclusions about the independent status of a given oscillator should be made only after phase of the oscillations was determined both in vivo and in vitro.

Multioscillatory Circadian Organization in Lower Vertebrates

The results discussed above suggest that peripheral pacemakers in flies have a high degree of autonomy. The same situation seems to prevail in lower vertebrates as well. In the zebrafish, the expression of the *Clk* gene is rhythmic in the kidney, spleen, and heart, and the oscillations have similar phases in vivo and in vitro. Subsequent studies have shown that *Clk* rhythms in isolated zebrafish organs are entrainable directly by light–dark cycles. The resetting of the *Clk* mRNA rhythms in vitro suggests that the brain does not mediate this process. Thus the organization of the circadian system in the zebrafish appears to be similar to that of Drosophila in that many internal organs harbor self-sustained photoreceptive oscillators. The degree of autonomy these oscillators have in vivo and their possible coordination by any humoral messengers have not been investigated in the zebrafish.

Multioscillatory circadian organization also extends to other vertebrates. Experimental and surgical manipulations in a reptile, the green iguana, have shown that their timing system is composed of multiple circadian oscillators that reside in different tissues, and have specific and different roles. For example, the removal of the pineal gland abolishes the rhythm of body temperature but not the rhythm of rest–activity, or the rhythm of retinal sensitivity to light. Circadian organization has been also studied in several species of birds. The avian circadian system is composed of several interacting sites, each containing photosensitive circadian oscillators. These sites include the pineal gland, the suprachiasmatic nucleus of the brain, and in some species, the eyes. The pineal gland seems to be the locus of a major circadian pacemaker, which may control the period and the phase of other circadian oscillators. In the absence of this gland, some birds, such as sparrows, are unable to maintain a coherent circadian organization. However, significant variations occur among birds in the relative importance of this organ. For example, removal of the pineal gland does not affect body temperature or activity rhythms in the Japanese quail.

Hierarchical Organization of Circadian System in Mammals

Mammals provide compelling evidence for a hierarchically organized circadian system. Virtually all mammals display daily rest–activity cycles. These rhythms have been studied particularly closely in rodents such as the hamster and rat, where long-term behavioral records can be easily obtained. The realization that these animals are able to maintain an activity cycle of about 24 hours without any environmental cues led to a search for the location of the "master clock." Selective destruction of a part of the hypothalamus, the suprachiasmatic nucleus (SCN), resulted in the abolishment of all behavioral and endocrine rhythms (see also Chapter 4). Removal of other parts of the brain, endocrine glands, or peripheral tissues did not cause such global arrhythmia as the removal of the SCN. The central status of the SCN was confirmed using mutant hamsters with a significantly shorter circadian period. When the SCNs of such mutants were transplanted into the hypothalamus of the SCN-lesioned wild-type hamsters, the host animals developed rhythms with a shorter period characteristic of the SCN donors.

How does the SCN control multiple rhythms in the body? Some rhythms, such as behavioral activity, can be restored by a transplanted SCN, which is experimentally prevented from establishing neural connections with the host nervous system. This suggests that the SCN pacemaker produces humoral outputs affecting target tissues. Other rhythms, such as cortisol or melatonin secretion, are not restored by transplants and may require formation of precise connections between the SCN graft and the target neurons. Thus, the SCN may transmit clock signals to various tissues by both humoral and neural pathways. Taken together, these experiments left little doubt that the mammalian circadian system is hierarchically organized, and that the SCN is at the top of hierarchy.

While the SCN is the central pacemaker, the mammalian timing system appears to also have other components: (1) some physiological rhythms persist in mammals with lesioned SCN, and even in isolated peripheral organs in vitro; and (2) many organs outside of CNS rhythmically express clock genes (see previous section). Thus, peripheral oscillators exist in mammals. The status of those oscillators relative to the SCN was studied in a series of interesting experiments. To monitor oscillations in peripheral tissues, researchers used transgenic rats carrying the promoter of the *Per1* gene linked to a luciferase reporter. Cultured SCNs taken from transgenic rats showed robust oscillations, which persisted for several months. Other organs such as liver, lung, and skeletal muscles also expressed circadian rhythms in vitro, but those rhythms tend to dampen sooner than in the SCN (Fig. 9.5). The phase of the oscillations in peripheral organs was delayed by several hours relative to the SCN. Moreover, in response to advances or delays in the environmental light cycle, the circadian rhythm in the SCN shifted more rapidly than the rhythms in peripheral tissues. These studies confirm the special status of the SCN oscillator, and suggest that the light-entrainable SCN may coordinate phases of peripheral oscillators, which are not entrainable by light input, as depicted in Figure 9.6b.

While the SCN acts as a master clock in conveying light–dark information to the rest of the body, there may be other entraining factors that affect rhythmic behavior irrespective of the SCN. It has been known for a number of years that cycles of food availability exert powerful entraining effects on the rhythm of locomotor activity in both intact and SCN-lesioned rodents. One study compared the effects of restricted feeding and light–dark cycles on the rhythms of clock genes in the SCN and peripheral oscillators. Restricted feeding rapidly entrained the liver, shifting its rhythm by 10 hours within 2 days, while the oscillations in the SCN remained phase-locked to the light-dark cycle. It is not yet known which kind of signals associated with feeding affect the phase of the liver oscillator. Nevertheless, these experiments demonstrate that individual peripheral clocks may be sensitive to timing signals that are relevant to physiological functions of a given organ (Fig. 9.6b). Thus, the hierarchy of the mammalian circadian system is not rigid but flexible; direct sensitivity of peripheral oscillators to external resetting factors may aid in the adaptation of the organism to its surroundings.

Interspecies Variation in Multioscillatory Communication

The degree of communication between different oscillators within one circadian system depends on the phylogenetic position of the organism. Evolution seems to have proceeded from autonomous oscilla-

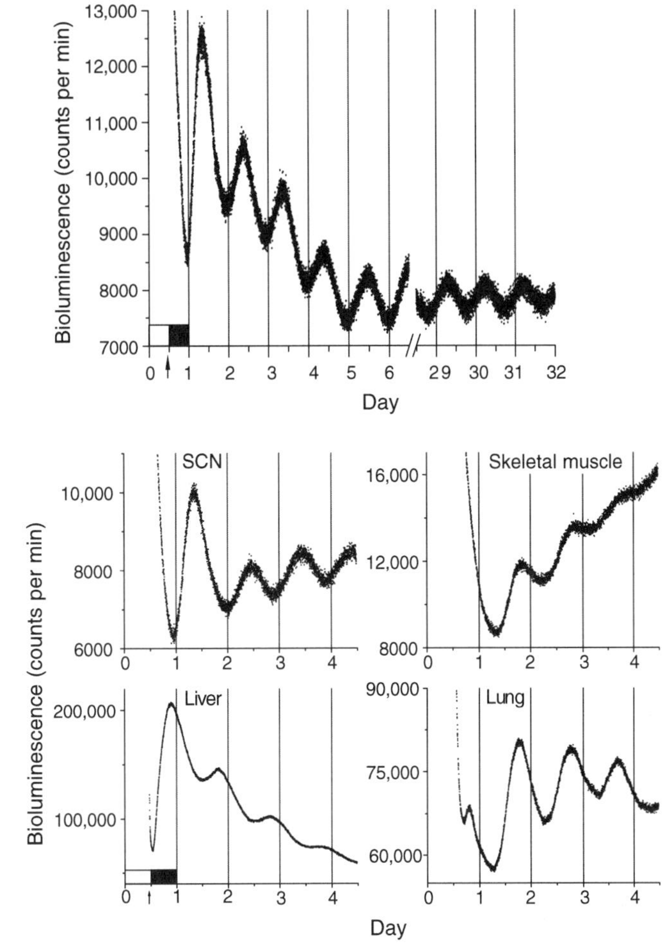

Figure 9.5. The circadian rhythm of bioluminescence from tissues of *Per1*-luc transgenic rats. In the SCN the near-24-hour rhythm was observed, which peaked in the middle of the subjective day and persisted for more than 32 days in vitro (upper panel). Circadian rhythms were expressed in vitro in different tissues from the same animal (lower panel). Note that oscillations in peripheral tissues peaked several hours later than in the SCN. In contrast to SCN, rhythms in peripheral tissues dampened after only a few cycles in vitro. [Reprinted with permission from Yamazaki et al. (2000). Copyright (2000) American Association for the Advancement of Science.]

tors that are independent of each other, and synchronized by external light–dark cycles, to a less autonomous and more coupled system of oscillators. An extreme autonomy has been documented in plants. The rhythmic expression of a single gene could be set at three phases in different anatomical locations of a single *Arabidopsis* plant, by applying different light–dark treatments to restricted tissue areas. The initial phases were stably maintained after the entraining treatments ended, indicating that circadian oscillators in intact plants are autonomous and do not communicate with each other to coordinate the phases of their oscillations.

A similar lack of communication between oscillating organs has been docu-

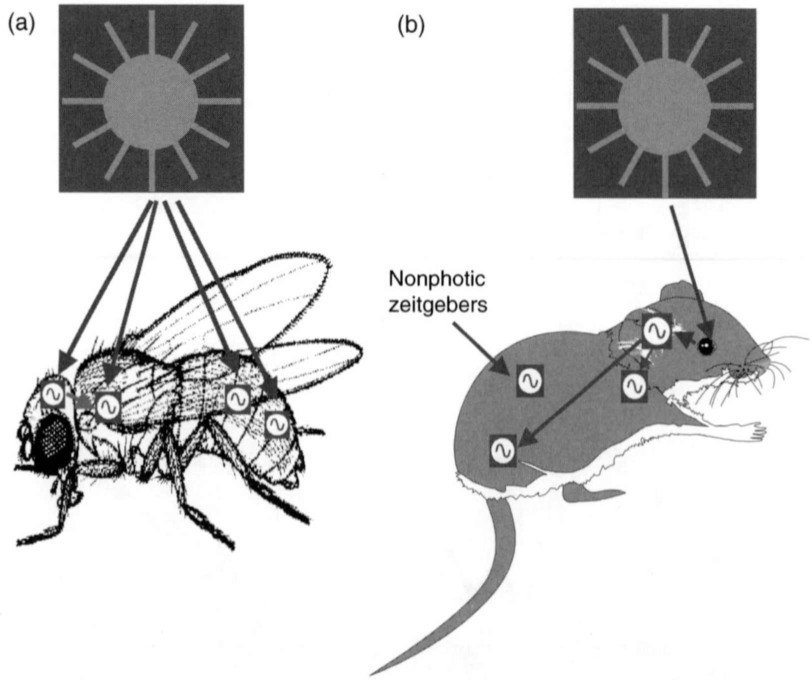

Figure 9.6. Relationship between multiple oscillators in circadian systems. (a) Multiple oscillators are independently responsive to light–dark cycles; however, the output of the brain oscillator may modulate the phase of another oscillator. Such relationships are observed in insects and may also occur between different oscillators in lower vertebrates. (b) In mammals, the SCN oscillator assumes the central role. This oscillator receives light input and coordinates the phases of light-insensitive peripheral oscillators via its rhythmic neuronal and humoral outputs. Peripheral oscillators may be independently entrained by external factors other than light.

mented in insects, and may also occur in lower vertebrates. The circadian system in these animals appears to function as decentralized collection of oscillators. The whole system achieves coordination by the sensitivity of its components to common light–dark cycles. However, insects have a well-developed neuroendocrine system, and at least some insect hormones show daily fluctuations in their concentrations. Rhythmic humoral factors may play a role in the internal coupling of some of the peripheral oscillators. However, they do not seem to affect the phase of clock gene cycling in the Malpighian tubules of Drosophila (see previous section).

Humoral factors may be involved in the communication between different oscillators in vertebrate circadian systems (see also Chapter 10). Blood levels of many vertebrate hormones display circadian rhythms (see Fig. 9.1). While these hormonal fluctuations must have profound effects on the physiology of the target tissues, it is not clear to what degree they affect the phases of peripheral oscillators. A good candidate for a humoral circadian messenger in vertebrates, especially in birds, is melatonin. This hormone, which is rhythmically produced by pineal glands and retinas, shows robust circadian fluctuations in the blood of all examined vertebrates.

Exogenous administration of melatonin can cause either arrhythmicity or changes in the period of freerunning rhythms. Conversely, periodic infusion of melatonin restores rhythmicity to birds with a lesioned pineal gland. Many regions of the brain, as well as cells of peripheral organs, have melatonin receptors, and therefore, could be affected by this hormone's changing levels. The mechanism of melatonin action as a transducer of biological time is not yet well understood.

The information channels within the multioscillatory system seem to be most complex in mammals. What sets the mammalian circadian system apart is that peripheral oscillators do not appear to be entrained by light–dark cycles. Instead they seem to rely on humoral and neural timing signals distributed by the light-entrainable master oscillator in the SCN. Experiments comparing immortalized cells derived from the SCN and a fibroblast line demonstrated the importance of the SCN-specific outputs for circadian coordination (Fig. 9.7). SCN cells endogenously generate circadian rhythms in *Per* gene expression and in metabolic activity, and impose those rhythms on cocultured fibroblasts via a diffusible signal. In contrast, cultured fibroblasts, when induced by a serum shock, generate a circadian rhythm in *Per* gene expression but not in metabolic activity. Serum-shocked fibroblasts could not confer rhythmic functions on other fibroblasts.

Although the special pacemaking properties of the SCN are very important in circadian coordination, other channels of communication also exist in the mammalian circadian systems. Peripheral oscillators were shown to be entrained by nonphotic external factors, which may cause their uncoupling from the master clock. Internally, peripheral oscillators may communicate with each other via their rhythmic outputs. For example, corticosteroids, which are rhythmically produced by the adrenal gland, can transiently reset the phase of oscillations in other organs.

In addition to hormonal signals, oscillators in specific tissues may be regulated by rhythmically fluctuating metabolites produced within or outside these tissues. This idea gained support from the demonstrated molecular link between circadian oscillations and energy homeostasis. It appears that cellular redox states are under circadian control and, via a feedback mechanism, may affect the activities of the molecular components of the circadian clock. It is possible that other channels of communication within the circadian system will be revealed as we learn more about functions and outputs of peripheral oscillators in animals.

Variation in Molecular Clock Components in Tissues of the Same Organism

While the core clock mechanism is shared in cells forming various tissues, some clock components and light-entrainment pathways appear to have tissue-specific roles. In *Drosophila melanogaster*, disruption of the night by even short light exposures results in degradation of the clock protein TIMELESS (TIM) leading to shifts in the fly molecular and behavioral rhythms (Chapter 3). The blue-light photoreceptor cryptochrome (CRY) is involved in TIM-mediated entrainment of the oscillators in the LNs and in the Malpighian tubules. Experiments on flies with a mutated *cryptochrome* gene demonstrated that freerunning clock oscillations can continue without CRY in LNs but not in Malpighian tubules or antennae. Thus, besides its role as a photoreceptor, cryptochrome is also an indispensable component of the endogenous clock mechanism in the fly excretory system, but not in the brain. The lack of CRY may be compensated for by as yet unrevealed LN-specific clock components, or by input from other structures in the central nervous system. LNs are part of a complicated neural network, undoubtedly interacting with other cells via chemical and electrical signaling. In contrast, Malpighian tubules consist of noninnervated epithelium in which clock functions do not seem to be affected by the fly internal milieu. Thus, there may be more redundant mechanisms keeping freerunning clocks in gear in the brain than in the peripheral organs.

The molecular components of the clock mechanism may also differ in various oscillators of mammals. In the central SCN pacemaker, two transcription factors, Clock and BMAL1, form heterodimers essential for function of the oscillating feedback loop. It has been found that another transcription factor constitutes a component of oscillators present in the forebrain and in the smooth muscle of the vasculature. In these tissues MOP4 (NPAS2), a transcription factor highly related to Clock, can replace the latter in forming heterodimers with BMAL1. Despite the commonality of the molecular circadian mechanism, various oscillators play different roles in the organism. This seems to be linked to the nature of the output rhythms produced by oscillating cells. For example, a cluster of neurons in the fly brain or in the mammalian SCN may affect many target tissues, via neural and humoral outputs, producing a response at the level of the whole organism. On the other hand, an oscillator in the excretory epithelial cells will likely exert only local effects, controlling levels of

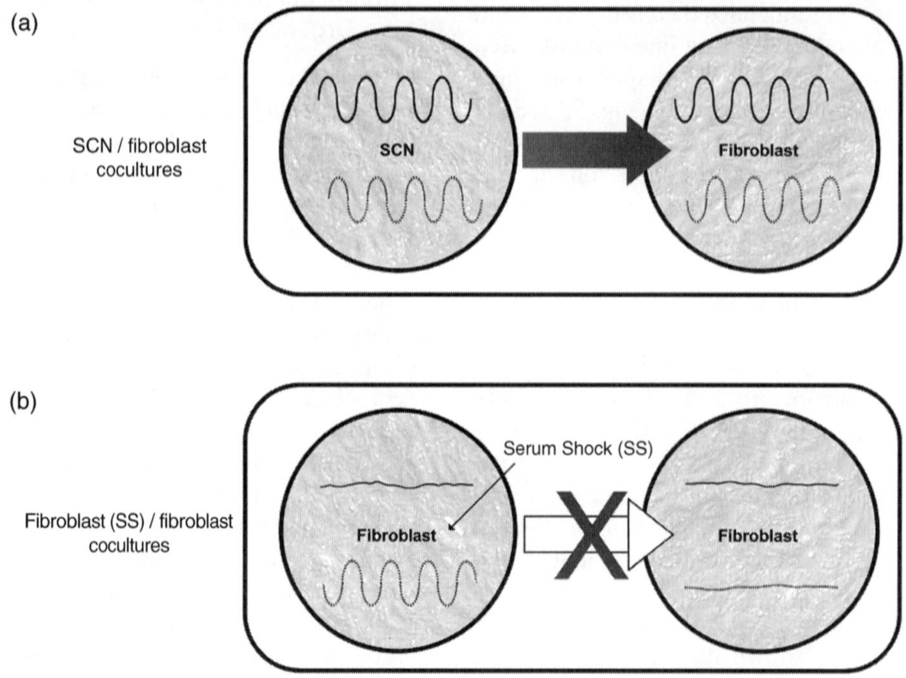

Figure 9.7. Experiment demonstrating that SCN cells are distinguished by special pacemaking properties. (a) Immortalized cells derived from the rat SCN (SCN2.2 cell line) generate metabolic (solid line) and molecular (dashed line) rhythms and are able to confer these rhythms onto cocultured fibroblasts (NIH/3T3 cell). (b) Serum-shocked fibroblasts generate only molecular oscillations and cannot drive rhythmicity in cocultures of untreated fibroblasts. (Diagram courtesy of David J. Earnest.)

enzymes, ion channels, and other components in those cells.

ROLES OF PERIPHERAL CLOCKS IN PHYSIOLOGY

Rhythmic activities of clock genes and their products are found in a surprisingly broad range of organs. Clock molecules cycle in many loci within the CNS, in tissues involved in food intake, metabolism, excretion, and reproduction. This strongly suggests that peripheral circadian oscillators may be involved in coordinating many physiological processes in a tissue-autonomous fashion. However, many peripheral oscillators were identified on the basis of cyclic expression of clock genes, and their relevance to the organism's physiology is not yet clear.

There are a few cases in insects where the physiological role of peripheral oscillators is relatively well understood. The importance of the circadian coordination of physiological processes is manifested dramatically in moths, in which constant light (a known disrupter of insect rhythms) leads to male sterility. This intriguing discovery was followed by the realization that the testes–vas deferens complex displays many coordinated rhythms associated with a daily cycle of sperm release and maturation. Clones of differentiated spermatozoa (sperm bundles) are released from the testis to the vas deferens by penetrating the epithelial barrier separating the two organs during the circadian gate at the end of the day. After nighttime retention in the vas deferens lumen, sperm is moved from this compartment because of the morning increase in the intensity of contraction of the vas deferens muscles. The peak of sperm accumulation in the vas deferens lumen is correlated with periodic acidification of this compartment (Fig. 9.8). Acidification appears to be caused by a rhythm in the levels of the proton pump, vacuolar H^+ATPase. Rhythms associated with sperm release are driven by the *per*-based circadian mechanism. *per* mRNA and PER protein are rhythmically expressed in all epithelial cells of the moth vas deferens. Thus, this circadian system appears to provide a temporal framework for coordinated maturation of sperm.

Another example of a peripheral oscillator with a known function is found in Drosophila. A defined physiological output has been assigned to oscillators located in the chemosensory hairs on the fly's antennae. These organs display a rhythm in electrophysiological responses to two different classes of olfactory stimuli. Olfactory rhythms are driven by clock genes expressed locally in the antenna, and they persist when the fly's central clock in the LNvs is genetically removed.

The links between clocks and their physiological outputs include several intermediate steps leading from clock genes via other transcription factors to the effector genes that are involved in cellular physiology. Some steps are known in both central and peripheral oscillators; a handful of rhythmic transcription factors and effector genes have been identified, but in no case do we understand the whole clock-to-physiology cascade. Given this paucity of information it is not clear whether circadian clocks control diverse cellular processes in different tissues, or whether many cell types share specific rhythmic aspects of cellular physiology. Convergent clock-controlled output pathways seem to occur in neurons and epithelial cells. For example, one of the mammalian transcription factors, DBP, shows a circadian rhythm in both central (SCN) and peripheral (liver) oscillators. In *D. melanogaster*, clock-controlled oscillatory expression of the gene *takeout*, which is implicated in the control of feeding, occurs in the brain but also in segments of the alimentary tract. We are probably seeing only the tip of the iceberg when it comes to clock-controlled genes. Microchip

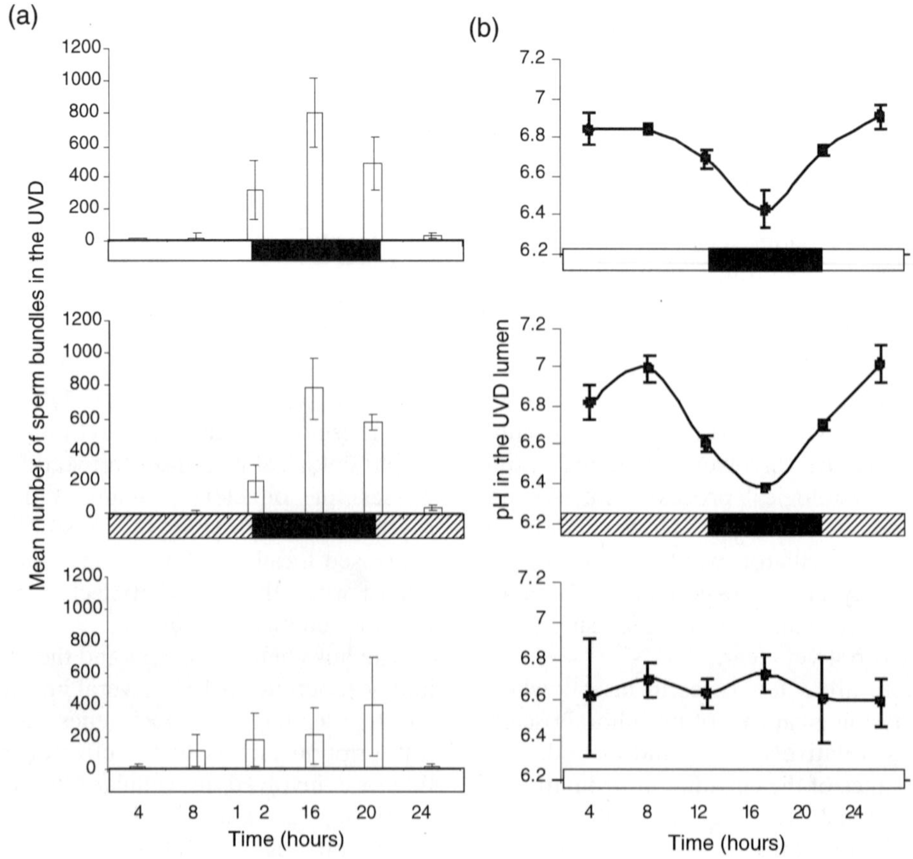

Figure 9.8. Pattern of sperm release (a) and pH changes (b) in the vas deferens (UVD) of *S. littoralis.* In LD and DD, sperm bundles accumulate in the UVD at night, and this rhythm correlates with maximum acidification of the UVD lumen in the middle of the dark phase. The rhythms in both sperm release and pH are abolished in constant light. Horizontal bars represent day (white), night (black), and former day (hatched) portions of the photoregime. (Composite figure from the author's result.)

technology is revealing hundreds of rhythmically expressed genes in Arabidopsis, Drosophila, and mammals. Further studies of the clock-controlled genes should help reveal how the timing system ensure optimization of everyday functioning of the body, as well as adaptability to external environment.

FURTHER READING

Allen, G, Rappe, J, Earnest, DJ, Cassone, VM (2001): Oscillating on borrowed time: Diffusible signals from immortalized suprachiasmatic nucleus cells regulate circadian rhythmicity in cultured fibroblasts. *J Neurosci* **21**: 7937–7943.

Buijs, RM, Kalsbeek, A (2001): Hypothalamic integration of central and peripheral clocks. *Nature Rev Neurosci* **2**: 521–526.

Giebultowicz, JM (2001): Peripheral clocks and their role in circadian timing: Insights from insects. *Philos Trans Roy Soc Lond B Biol Sci* **356**: 1791–1799.

Glossop, NR, Hardin, PE (2002): Central and peripheral circadian oscillator mechanisms in flies and mammals. *J Cell Sci* **115**: 3369–3377.

Hege, DM, Stanewsky, R, Hall, JC, Giebultowicz, JM (1997): Rhythmic expression of a PER-reporter in the Malpighian tubules of decapitated Drosophila: Evidence for a brain-independent circadian clock. *J Biol Rhythms* **12**: 300–308.

Ivanchenko, M, Stanewsky, R, Giebultowicz, JM (2001): Circadian photoreception in Drosophila: Functions of cryptochrome in peripheral and central clocks. *J Biol Rhythms* **16**: 205–215.

Krishnan, B, Levine, JD, Lynch, MK, Dowse, HB, Funes, P, Hall, JC, Hardin, PE, Dryer, SE (2001): A new role for cryptochrome in a Drosophila circadian oscillator. *Nature* **411**: 313–317.

Moore-Ede, MC, Sulzman, FM, Fuller, CA (1982): *The Clocks that Time Us*. Cambridge, MA: Harvard University Press.

Pittendrigh, CS (1960): Circadian rhythms and circadian organization of the living systems. *Cold Spr Harb Symp Quant Biol* **25**: 159–182.

Schibler, U, Sassone-Corsi, P (2002): A web of circadian pacemakers. *Cell* **111**: 919–922.

Tosini, G, Menaker, M (1998): Multioscillatory circadian organization in a vertebrate, *Iguana iguana*. *J Neurosci* **18**: 1106–1114.

Underwood, HA, Wassmer, GT, and Page, TL (1997): Daily and seasonal rhythms. In *Handbook of Physiology*, Dantzler. WH, ed, Oxford University Press, New York, pp 1653–1763.

Whitmore, D, Foulkes, NS, Sassone-Corsi, P (2000): Light acts directly on organs and cells in culture to set the vertebrate circadian clock. *Nature* **404**: 87–91.

Yamazaki, S, Numano, R, Abe, M, Hida, A, Takahashi, R, Ueda, M, Block, GD, Sakaki, Y, Menaker, M, Tei, H (2000): Resetting central and peripheral circadian oscillators in transgenic rats. *Science* **288**: 682–685.

10

HORMONAL RHYTHMS

Peter McNamara

■ INTRODUCTION

The ubiquity of circadian rhythms in all aspects of eukaryotic life emphasizes their evolutionary importance for the survival of the species. The functional significance of circadian rhythms may provide synchronization with the pronounced periodic fluctuations of the external environment and may also organize the internal milieu so that there is coordination and synchronization of internal processes. Many behaviors such as feeding, drinking, sleep, exploration, and reproductive activity that change on a daily basis are correlated with daily changes in the physical environment (e.g., temperature, illumination, humidity) as well as changes in the biological environment (e.g., food availability, presence of predators, parasites, competitors, reproductive mates). Associated with these behavioral changes are changes in perception, sensation, learning, and performance that occur on a daily basis in many animal species, including humans. Underlying the behavioral adaptations to the daily changes in the external environment are a multitude of metabolic, hormonal, and biochemical rhythms. For example, changes in the digestive system occur before the behavioral changes that lead to food intake, and the rise in body temperature occurs before an animal wakes up in anticipation of the increased metabolic demands. This intricate temporal organization provides the endocrine system with remarkable flexibility. Not only can specific physiological processes be turned on and off depending on the presence or absence of a particular hormone, but also the precise pattern of hormonal release may provide specific signaling information.

There is virtually neither tissue nor function within the human organism that does not manifest regular changes from day to night. Apart from rhythms in the endocrine system, which is the focus of this chapter,

Molecular Biology of Circadian Rhythms. Edited by Amita Sehgal
ISBN 0-471-41824-2

circadian variations outside the endocrine system include innumerable physiological variables, such as body temperature, heart rate, blood pressure, urinary volume, propensity for rapid eye movement (REM) sleep, intraocular pressure, and taste threshold for salt; a wide variety of blood constituents, such as white blood cells, amino acids, and phosphorus; as well as behavioral parameters such as time of birth, time of death, mood, reaction time, performance on learning tasks, computation skills, pattern recognition, and relative coordination. There are also rhythms in responsive to various challenges such as drugs and stress. Circadian rhythmicity is maintained when subjects are sleep-deprived, when they are starved, and when they receive equal amounts of food at short intervals over the day. The timing of single meals, however, can have effects on the pattern of at least some variables, including hormones, and the sleep–wake cycle can have phase-setting as well as phase-masking effects on many rhythms, especially those of the endocrine system.

Many diurnal hormonal patterns are dependent, to some degree, on the circadian clock. The relative contributions of circadian rhythmicity versus homeostatic control in the temporal organization of hormonal release differ from one endocrine axis to another. For most pituitary hormones, the 24-hour profiles reflect the superposition of circadian signals on an ultradian, or pulsatile, release and result from interaction of the circadian clock with sleep–wake homeostasis (ultradian rhythm is used to designate rhythmicities with periods ranging from fraction of hours to several hours). Several rhythmic and non-rhythmic facets, such as periodic food intake, postural changes and levels of physical activity may also exert modulatory effects on diurnal hormonal secretary patterns.

However, for practical purposes, there is little reason to make a distinction between diurnal and circadian rhythms, since almost all diurnal rhythms are expressed under constant environmental conditions in the laboratory. In this chapter the use of the term *circadian rhythm* is extended to mean all diurnal variations recurring regularly at a time interval of approximately 24-hour, and we will concentrate on hormonal rhythms in humans with in-depth descriptions of some of the more well-characterized hormonal axes.

NIGHTTIME MELATONIN PRODUCTION: A KEY HORMONAL OUTPUT FROM THE CLOCK

In vertebrates melatonin is synthesized only by the retina and pineal gland, which is a small structure located in the center of the skull between the two cerebral hemispheres. In birds, reptiles, amphibia, and fish the pineal gland is directly light-sensitive and possesses an independent circadian clock. In contrast, in mammals, pinealocytes neither are light-sensitive nor possess a clock, which is instead located in the SCN. The nocturnal rise starts early in the evening, between 8 and 10 pm, and the maximum occurs around the middle of the sleep period. The melatonin rhythm and the rhythm of body temperature undergo equivalent phase shifts following a regimen of scheduled exposures to bright light and darkness, supporting the concept that both rhythms are controlled by the central circadian pacemaker. Exposure to light of sufficient intensity results in an acute dose-dependent inhibition of nocturnal melatonin secretion, with a rapid return to high nighttime levels when darkness resumes. The melatonin rhythm appears to be largely insensitive to stimuli other than light, such as sleep and food intake. Among the primary roles of melatonin in mammals are the regulation of seasonal changes in reproductive activity in response to changes in daylength and in retinal regulation of photopigment disk shedding, phagocytosis, and the inhibition of retinal dopamine where melatonin is also synthesized.

Synthesis of Melatonin

The synthesis of melatonin begins with the *N*-acetylation of serotonin followed by the addition of a methyl group at the 5-hydroxy position via the enzyme hydroxyindole-*O*-methyltransferase (HIOMT) (Fig. 10.1). Melatonin mediates its biological effects by binding to high-affinity and low-affinity receptors ML1 and ML2 belonging to the GPCR superfamily. The pars tuberalis (PT) of the pituitary shows the highest density of melatonin receptors and apart from the SCN is the only structure to be found consistently labeled in all mammalian studies so far. Thus, these are considered the two major sites of melatonin action. In the pineal gland there is a strong circadian rhythm of melatonin synthesis. Whereas serotonin levels are much lower at night than during the day, melatonin concentrations display a reverse rhythm with highest concentration at night associated with elevated circulating levels of melatonin. The link between these two reciprocal rhythms is the rate limiting enzyme for melatonin synthesis, serotonin *N*-acetyltransferase (AANAT) (Fig. 10.1). This displays a diurnal rhythm of activity with levels at the nighttime peaking up to 100 times higher than in the daytime. The AANAT and melatonin rhythms derive from activation at night of the pineal's sympathetic innervation in mammals. Norepinephrine binds to β-adrenoreceptors and thus stimulates adenylate cyclase activity. The resulting increase in cAMP levels has been shown to stimulate AANAT transcription and translation and also maintains the enzyme in an

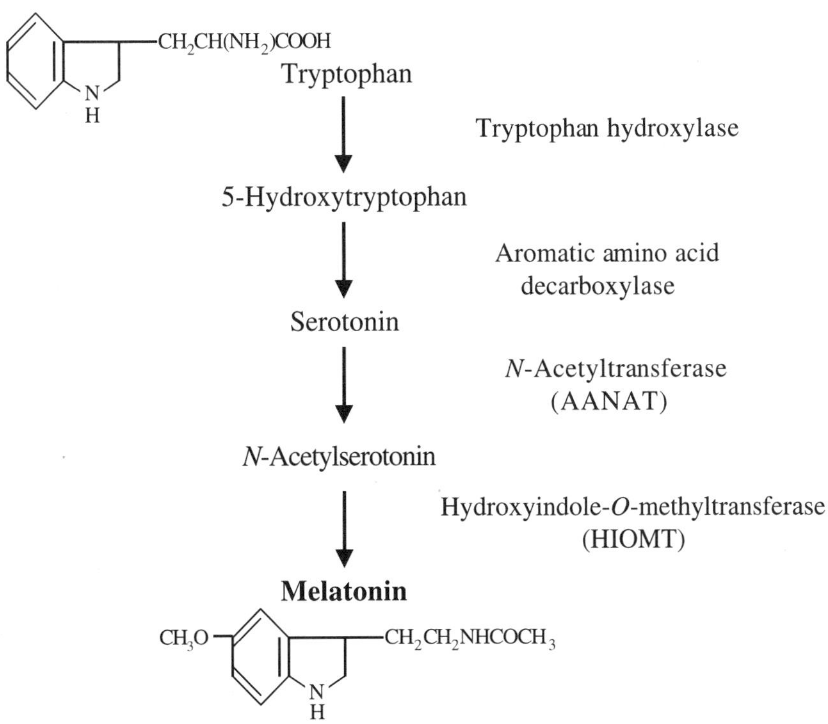

Figure 10.1. The melatonin synthesis pathway.

active form. Alpha-1 adrenergic receptors also participate in AANAT stimulation, apparently by activating the phosphoinositide (PI) cycle and protein kinase A and C (PKA and PKC), which potentiates β-receptor-induced cAMP production.

Targets of cAMP include a group of transcription factors that modulate the expression of cAMP responsive genes. These bZip factors constitute a family of both activators and repressors that bind as homo- and heterodimers to cAMP-responsive elements (CREs) and include CREB, CREM, and the CREM product ICER. Nighttime adrenergic signals, originating from the SCN, activate PKA and thus phosphorylate CREB. Phosphorylated CREB activates the P2 promoter of the CREM gene and thus induces the expression of ICER. ICER downregulates its own expression constituting the CREM feedback loop. Like AANAT, ICER mRNA expression displays diurnal rhythmicity in the pineal gland. The peak of ICER mRNA occurs during the second part of the night, just preceding the decline of melatonin synthesis. Via binding to the CRE in the AANAT promoter and exerting a dampening effect, ICER modulates the rate and magnitude of melatonin induction in response to adrenergic signals. The balance between the proportion of phosphorylated CREB and ICER protein levels determines the transcriptional activity of the AANAT promoter. Thus the promoter cycles between activated and repressed states as a function of time. In this way, AANAT mRNAs oscillates between high nighttime and low basal daytime levels and determines the characteristic day–night oscillation of AANAT activity. This ensures rhythmic melatonin synthesis (Fig. 10.2).

Functional Significance of the Melatonin Rhythm

This has been well established for a wide variety of seasonal vertebrates. Environmental lighting, acting through the eyes in mammals and in part directly on the pineal

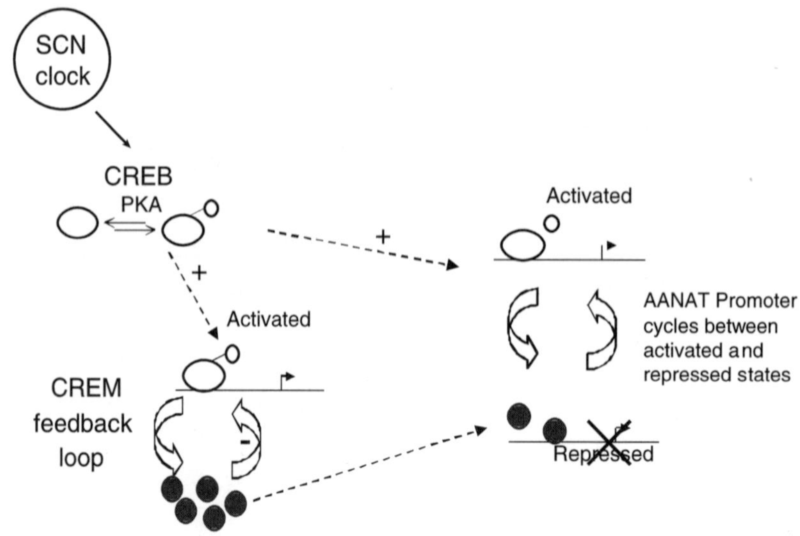

Figure 10.2. The role of the CREM feedback loop in transducing a rhythmic clock-directed signal into rhythmic hormone synthesis. Cartoon representation of the mechanism by which melatonin rhythms are induced. Activated PKA phosphorylates CREB, which in turn activates CREM and induces expression of ICER. ICER represses its own expression. The level of AANAT mRNA is under positive and negative control by CREB and ICER, respectively.

in some birds, lizards, and fish, has a profound effect on the rhythm of melatonin synthesis. The duration of melatonin secretion is positively correlated with the length of the night period. It is through these changes in the duration of melatonin synthesis that the brain is able to integrate photoperiodic information. This relationship explains the present use of melatonin in farming to control seasonal function (e.g., growth, reproduction, milk production). The exact mechanism of action is unclear. The duration of nocturnal melatonin production is the key signal, but the existence, within this signal, of a melatonin-driven circadian rhythm of sensitivity to melatonin has been proposed to explain the photoperiodic response. The melatonin receptors involved most probably are of the mt1 subtype as the gene of the other subtype in mammals, mt2, is nonfunctional in two highly photoperiodic species, the Siberian and Syrian hamsters. The target sites mediating the control by melatonin of photoperiod-dependent seasonal functions and, in particular, the annual sexual cycle have not yet been clearly determined, although the par tuberalis is suspected of being the target for melatonin seasonal effects. There is growing evidence to indicate that while the rhythm of melatonin secretion is driven by the circadian pacemaker, this rhythm also feeds back on the clock. Indeed, to a minor extent exogenous administration of melatonin has been shown to resynchronize certain overt rhythms, and the sleep–wake cycle in a variety of conditions, including jet lag, and freerunning rhythms in blind subjects.

■ CONTROL OF HORMONAL RHYTHMS BY THE HYPOTHALAMUS

The hypothalamus is located in the middle of the base of the brain, and encapsulates the ventral portion of the third ventricle. The hypothalamus may be divided into four regions (*anterior*, *dorsalis*, *intermedia*, and *posterior*) or into three longitudinal zones (*periventricular zone*, *medial zone*, and *lateral zone*). The hypothalamic nuclei constitute that part of the corticodiencephalic mechanism that activates, controls, and integrates the peripheral autonomic mechanisms, endocrine activity, and many somatic functions, such as a general regulation of water balance, body temperature, sleep, and food intake, and the development of secondary sex characteristics. The hypothalamus secretes vasopressin and oxytocin, which are stored in the pituitary, as well as many releasing factors (hypophysiotropic hormones), including thyrotropin releasing hormone (TRHs), gonadotropin releasing hormone (GnRH), growth hormone releasing hormone (GHRH), Ghrelin, corticotropin releasing hormone (CRH), somatostatin, and dopamine. All these agents are released into the blood and travel immediately to the anterior lobe of the pituitary, where they exert their effects (Fig. 10.3). All of them are released in an ultradian rhythm in pulsatile spurts. In fact, replacement hormone therapy with these hormones does not work unless the replacements are also given in a similar fashion.

Pituitary Gland

The pituitary gland is often portrayed as the "master gland" of the body. Such praise is justified in the sense that the anterior and posterior pituitary secrete a battery of hormones that collectively influence all cells and affect virtually all physiologic processes. The pituitary gland, also known as the *hypophysis*, is a roundish organ that lies immediately beneath the hypothalamus, resting in a depression of the base of the skull called the *sella turcica* ("Turkish saddle"). In an adult human or sheep, the pituitary is roughly the size and shape of a large garbanzo bean. Careful examination of the pituitary gland reveals that it is composed of two distinctive parts: (1) the

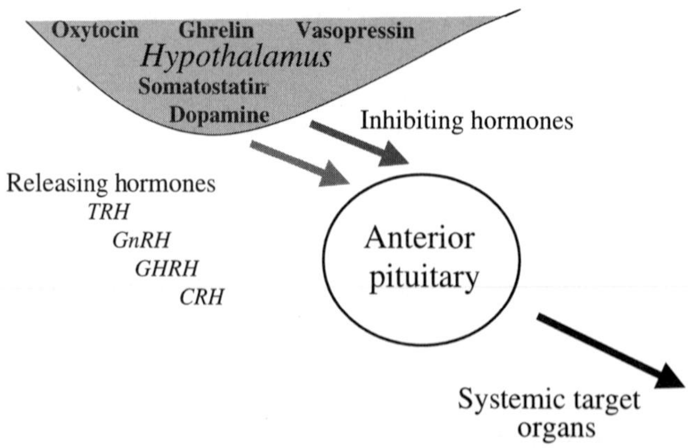

Figure 10.3. Control of hormonal rhythms by the hypothalamus. Specific interactions between the hypothalamus, pituitary, and target endocrine glands form a series of closed-loop regulatory units that are the core of the endocrine system. The hypothalamic hormones involved in this type of regulation are small, labile peptides that are delivered to the anterior pituitary through a special vascular portal system and stimulate the release of anterior pituitary hormones, which are delivered to target glands by the systemic circulation.

anterior pituitary (adenohypophysis), which is composed predominantly of cells that secrete protein hormones; and the *posterior pituitary* (neurohypophysis), which is not really an organ but an extension of the hypothalamus, composed largely of the axons of hypothalamic neurons that extend downward as a large bundle behind the anterior pituitary. It also forms the so-called pituitary stalk, which appears to suspend the anterior gland from the hypothalamus.

The pituitary gland may be king, but the power behind the throne is clearly the hypothalamus. Some of the neurons within the hypothalamus—neurosecretory neurons—secrete hormones that strictly control secretion of hormones from the anterior pituitary. The hypothalamic hormones are referred to as *releasing hormones* and *inhibiting hormones*, reflecting their influence on anterior pituitary hormones. Specific hypothalamic hormones bind to receptors on specific anterior pituitary cells, modulating the release of the hormone they produce. The anterior pituitary hormones enter the systemic circulation and bind to their receptors on other target organs. Clearly, robust control systems must be in place to prevent over- or undersecretion of hypothalamic and anterior pituitary hormones. A prominent mechanism for control of the releasing and inhibiting hormones is negative feedback.

A key to understanding the endocrine relationship between hypothalamus and the anterior pituitary is to appreciate the vascular connections between these organs. Secretion of hormones from the anterior pituitary is under strict control by hypothalamic hormones. These hypothalamic hormones reach the anterior pituitary through the following route. A branch of the hypophyseal artery extends into a capillary bed in the lower hypothalamus, and hypothalamic hormones destined for the anterior pituitary are secreted into that capillary blood. Blood from those capillaries drains into hypothalamic–hypophyseal portal veins. *Portal veins* are defined as veins between two capillary beds; the hypothalamic–hypophyseal portal veins branch again into another series of capillaries within the anterior pituitary. Capillaries within the anterior pituitary, which carry hormones secreted by that gland, coalesce into veins that drain into the systemic venous blood. Those veins also collect capillary blood from the posterior pituitary gland. The utility of this unconventional vascular system is that minute quantities of hypothalamic hormones are carried in a concentrated form directly to their target cells in the anterior pituitary, and are not diluted out in the systemic circulation.

Adrenal Glands

The adrenal glands secrete epinephrine, norepinephrine, and dozens of different steroid molecules. The two adrenal glands, lie very near the kidneys. In mammals each adrenal is actually a double gland, composed of an inner corelike medulla and an outer barklike cortex. The medulla is responsible for secreting epinephrine and norepinephrine, and the cortex synthesizes different steroid molecules, but only a few of these have biological activity. These sort into three classes of hormones: glucocorticoids, mineralocorticoids, and androgens. These hormones initiate their action by activating members of the nuclear receptor superfamily. The glucocorticoids are 21-carbon steroids with many actions, the most important of which is to promote gluconeogenesis in addition to being an essential component of adaptation to severe stress. Cortisol is the predominent glucocorticoid in humans. Corticosterone is less abundant in humans but is the dominent glucocorticoid in rodents. The adult cortex has three distinct layers or zones. The subcapsular area is called the zona glomerulosa and is associated with the production of mineralocorticoids. Next is the zona fasciculata, which, with the zona reticularis, produces glucocorticoids and androgens. The adrenal steroid hormones are synthesized from cholesterol that is derived mostly from the plasma, but a small portion is synthesized in situ from acetyl-CoA via mevalonate and squalene.

Pituitary–Adrenal Secretion

Among all endocrine circadian rhythms known to occur in normal humans, the 24-hour periodicity of pituitary–adrenal secretion has been studied most extensively. It is endogenous in nature, largely unaffected by short-term manipulations of the sleep–wake cycle, and it persists during complete fast or continuous feeding. To cause a rapid shift in the rhythm, crucially timed exposure to bright light is necessary. Stimulation of the adrenal by adrenocorticotropic hormone (ACTH) (Fig. 10.4) or cAMP, activates an esterase that converts stored esterified cholesterol to pregnenolone and starts the series of reactions that occur in either the mitochondria or endoplasmic reticulum. There is little, if any, storage of steroid hormones within the adrenal cell, since these hormones are released into the plasma when they are made.

Cortisol release occurs with a periodicity that is regulated by the diurnal rhythm of

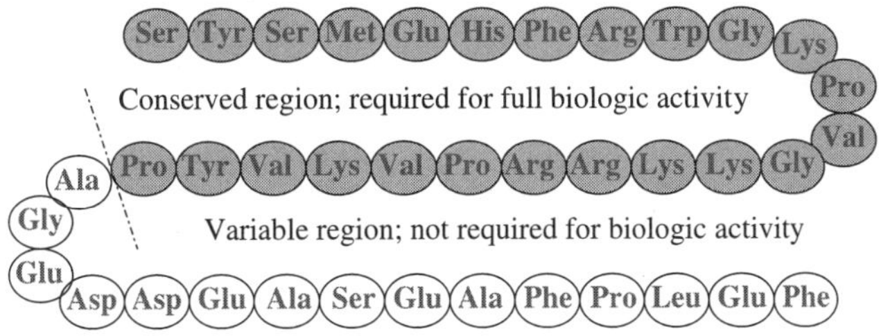

Figure 10.4. Structure of human ACTH.

ACTH release. ACTH has little control over secretion of aldosterone, the other major steroid hormone from the adrenal cortex. Cortisol circulates in plasma in protein-bound and free forms. The main plasma binding protein is an α-globulin called *transcortin*. ACTH is secreted from the anterior pituitary in response to corticotropin releasing hormone (CRH) from the hypothalamus. CRH is a peptide of 41 amino acids and is secreted in response to many types of stress, which makes sense in view of the "stress management" functions of glucocorticoids. Corticotropin releasing hormone itself is inhibited by glucocorticoids, making it part of a classical negative feedback loop. Within the pituitary gland, ACTH is produced in a process that also generates several other hormones. A large precursor protein named *proopiomelanocortin* (POMC) is synthesized and proteolytically chopped into several fragments as depicted (Fig. 10.5). Not all of the cleavages occur in all species and some occur only in the intermediate lobe of the pituitary.

Figure 10.6 shows 24-hour profiles of ACTH and cortisol typical of young adults. Changes in plasma cortisol occur in parallel with those in ACTH. Their pattern shows an early morning maximum, declining levels throughout daytime, a quiescent period of minimal secretory activity centered around midnight, and an abrupt elevation during late sleep. This circadian profile is produced by modulation of the height of successive secretory pulses. The 24-hour rhythm of adrenal secretion is primarily dependent on the circadian pattern of ACTH release. The rhythm in ACTH release results, in turn, from periodic changes in the level of stimulation by CRH. A circadian variation parallel to that of cortisol has been demonstrated for the plasma levels of adrenal steroids (Fig. 10.6). Pulses of plasma concentration of adrenal steroids occur in remarkable synchrony with bursts of cortisol secretion. A distinct rhythm of serum cortisol levels emerges at approximately 6 months of age. Once this periodicity is established, it persists throughout adulthood. In humans the 24-hour profile of cortisol appears to be similar for both sexes. In females, oral contraceptive therapy results in a large increase of the mean cortisol level and the amplitude of the rhythm, resulting from estrogen-induced elevation of binding capacity of the cortisol binding protein: transcortin. Disease states in which alterations of the cortisol rhythm have been observed include

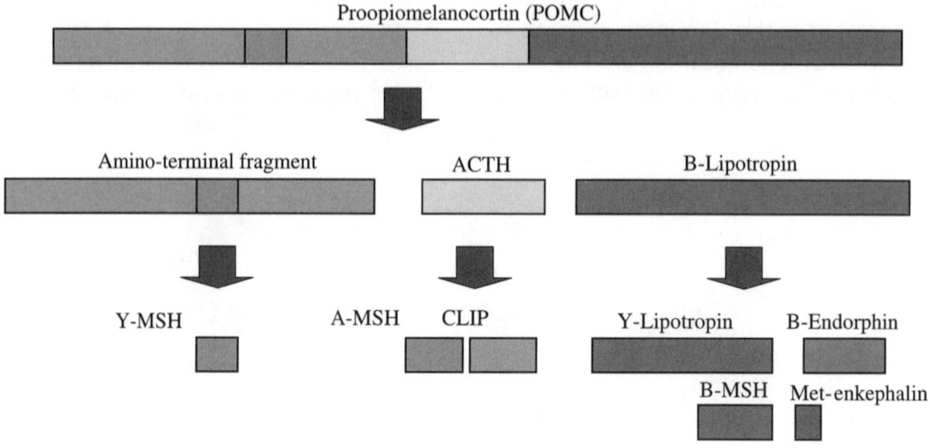

Figure 10.5. POMC is synthesized and proteolytically chopped into several fragments.

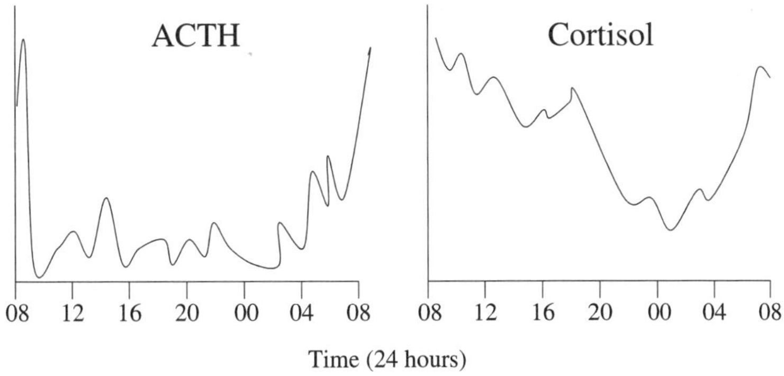

Figure 10.6. Mean 24-hour profiles of plasma ACTH and cortisol.

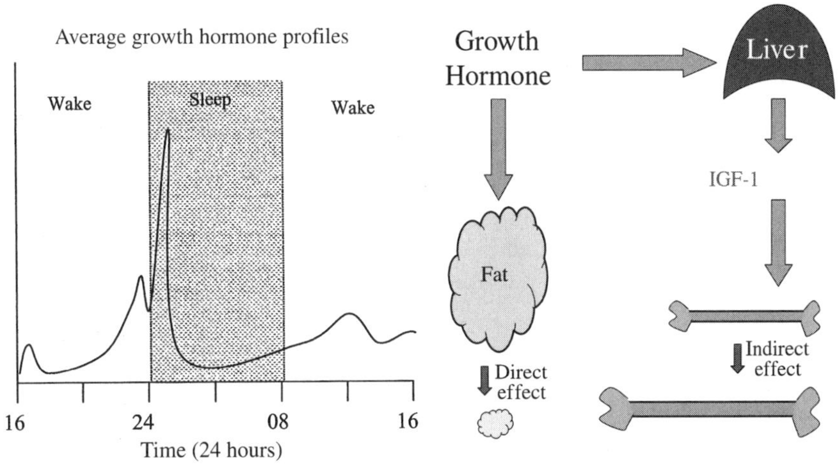

Figure 10.7. A 24-hour profile of plasma growth hormone (GH) and cartoon of GH effects. In normal adult subjects, the 24-hour profile of GH levels consists of stable low levels abruptly interrupted by bursts of secretion. The most reproducible pulse occurs shortly after sleep onset, in association with the first phase of SW sleep.

primarily (1) disorders involving abnormalities in binding and/or metabolism of cortisol, (2) the various forms of Cushing's syndrome, and (3) affective illness (i.e., depression).

Growth Hormone Release

Growth hormone release is affected by sleep and circadian influences. GH is a protein hormone of about 190 amino acids that is synthesized and secreted by cells called *somatotrophs* in the anterior pituitary. In normal adult subjects, the 24-hour profile of plasma GH levels consists of stable low levels abruptly interrupted by bursts of secretion (Fig. 10.7). The most reproducible pulse occurs shortly after sleep onset, in association with the first phase of slow-wave (SW) sleep. Other pulses may occur in later sleep and during wakefulness, in the absence of any identifiable stimulus. In men, the sleep-onset GH pulse is generally the largest and often the only pulse observed over the 24-hour span. In women, daytime GH pulses are more frequent, and the sleep-associated pulse, while still present in most cases, does not generally account for the majority of the 24-hour GH release. Circulating estradiol levels play an important role in determining overall levels of spontaneous GH secretion. Sleep onset will elicit a pulse of GH secretion regardless of whether sleep is advanced, delayed, interrupted, or fragmented. While SW sleep is clearly a major determinant of the 24-hour profile of GH secretion in humans, there is also evidence of a circadian modulation. Observation of the nocturnal profile of the GH pulse reveals that it occurs within 1 hour of the usual bedtime even in normal subjects subjected to sleep delay. The total amount and the temporal distribution of GH release is strongly dependent on age. Spontaneous GH secretion is detectable in term infants, who appear to have a high level of tonic secretion. As the infant matures, GH pulse frequency and pulse amplitude decrease, and tonic secretion diminishes. A pulsatile pattern of GH release, with increased pulse amplitude during sleep, is present in prepubertal boys and girls. During puberty, the amplitude of the pulses, but not the frequency, is increased, particularly at night. Maximal overall GH concentrations are reached in early puberty in girls and in late puberty in boys. Age-related decreases in GH secretion have been well documented in both men and women. This decline in overall GH secretion appears to be achieved by a decrease in amplitude rather than frequency of GH pulses.

Growth Hormone Activity

GH activity has two distinct types of effects. Direct effects are the result of GH binding its receptor on target cells. Fat cells (adipocytes), for example, have GH receptors, and GH stimulates them to break down triglyceride and suppresses their ability to take up and accumulate circulating lipids (Fig. 10.7). GH is one of a battery of hormones that serves to maintain blood glucose within a normal range. GH is often said to have antiinsulin activity, because it suppresses the abilities of insulin to stimulate uptake of glucose in peripheral tissues and enhance glucose synthesis in the liver. Somewhat paradoxically, administration of GH stimulates insulin secretion, leading to hyperinsulinemia. GH has important effects on protein, lipid, and carbohydrate metabolism. In general, GH stimulates protein anabolism in many tissues. This effect reflects increased amino acid uptake, increased protein synthesis and decreased oxidation of proteins. Indirect effects are mediated primarily by insulinlike growth factor 1 (IGF1), a hormone that is secreted from the liver and other tissues in response to GH (Fig. 10.7). A majority of the growth promoting effects of GH are actually due to IGF1 acting on its target cells. Growth is a very complex process, and requires the coordinated action of several hormones. The major role of GH in stimulating body growth is to stimulate the liver and other tissues to secrete IGF1. IGF1 stimulates proliferation of chondrocytes (cartilage cells), resulting in bone growth, and also appears to be the key player in muscle growth (Kato et al. 2001).

GHRH and Somatostatin

The primary controllers of GH secretion are two hypothalamic hormones: growth hormone releasing hormone (GHRH) and somatostatin. Production of growth hormone is modulated by many factors, including stress, exercise, nutrition, sleep, and growth hormone itself. GHRH is a mixture of two peptides, one containing 40 amino acids; the other, 44. As the term indicates, GHRH stimulates cells in the anterior lobe of the pituitary to secrete growth hormone (GH). Somatostatin is produced by several tissues in the body, including the hypothalamus and is a mixture of two peptides, one of 14 amino acids, the other of 28.

It acts on the anterior lobe of the pituitary to inhibit the release of growth hormone (GH) and of thyroid stimulating hormone (TSH) in response to GHRH and to other stimulatory factors such as low blood glucose concentration. Growth hormone secretion is also part of a negative-feedback loop involving IGF1. High blood levels of IGF1 lead to decreased secretion of growth hormone not only by directly suppressing the lactotroph but also by stimulating release of somatostatin from the hypothalamus. Growth hormone also feeds back to inhibit GHRH secretion and probably has a direct (autocrine) inhibitory effect on secretion from the lactotroph.

Abnormalities in the 24-hour profile of plasma GH have been reported in a variety of metabolic, endocrine, neurological, and psychiatric conditions. There is an inverse relationship between adiposity and GH release, which results in a marked suppression of GH levels (reduction in both pulse frequency and GH half-life) throughout the 24-hour span in obese subjects. States of both growth hormone deficiency and excess provide very visible testaments to the role of this hormone in normal physiology. Such disorders can reflect lesions in the hypothalamus, the pituitary or in target cells. A deficiency state can result not only from a deficiency in production of the hormone but also in the target cell's response to the hormone. Clinically, deficiency in growth hormone or its receptor is manifested as growth retardation or dwarfism. The severity depends on the age of onset of the disorder and can result from either heritable or acquired disease. The effect of excessive secretion of growth hormone is also very dependent on the age of onset and is seen as two distinctive disorders:

1. *Giantism*, which is the result of excessive growth hormone secretion that begins in young children or adolescents. It is a very rare disorder, usually resulting from a tumor of somatotropes.
2. *Acromegaly*, which results from excessive secretion of growth hormone in adults with a pulsatile pattern superimposed over elevated basal levels. The onset of this disorder is typically insideous. Clinically, an overgrowth of bone and connective tissue leads to a change in appearance that might be described as having "coarse features."

The excessive growth hormone and IGF1 also lead to metabolic derangements, including glucose intolerance. Diurnal and nocturnal episodes of GH secretion are more frequent and of higher amplitude in adult subjects with hyperthyroidism, who have an overall daily GH production rate fourfold above normal.

Prolactin

Prolactin, a single-chain protein hormone closely related to GH, is secreted by the lactotrophs in the anterior pituitary. In addition to the lactotrophs, prolactin is also synthesized and secreted by a broad range of other cells in the body, most prominently various immune cells, the brain and the decidua of the pregnant uterus. The conventional view of prolactin is that its major target organ is the mammary gland, and stimulating mammary gland development and milk production pretty well defines its functions. Such a picture is true as far as it goes, but it fails to convey an accurate depiction of this multifunctional hormone. It is difficult to point to a tissue that does not express prolactin receptors, and although the anterior pituitary is the major source of prolactin, the hormone is synthesized and secreted in many other tissues.

Under normal conditions, the 24-hour profile of prolactin follows a bimodal pattern, with minimal concentrations around noon; an afternoon phase of slightly augmented secretion; and a major nocturnal elevation starting shortly after sleep onset and culminating around midsleep. In men there is an average increase of more than 250% above the minimum noon level. Pulses occur during daytime as well as during the night. Their number appears to be a reproducible individual characteristic, ranging between 7 and 22 per 24-hour span.

Studies of prolactin levels during daytime naps or after shifts of the normal sleep period have consistently demonstrated increased prolactin secretion associated with sleep onset. Pharmacological studies have tended to implicate dopaminergic as well as serotoninergic mechanisms in this sleep-related elevation of secretion. While early reports described the 24-hour rhythm of plasma prolactin as "entirely dependent on sleep," the existence of a circadian component in the secretary pattern of prolactin is now well recognized. Abrupt shifts of the sleep–wake cycle during real or stimulated jet lag indeed reveal the existence of a sleep-independent secretory rise. Under normal conditions, this circadian rise is synchronized with the early part of sleep, and both circadian and sleep effects are superimposed. Current data indicate that the day–night difference in prolactin levels is present in newborns and persists into the ninth decade. During pregnancy, serum prolactin levels rise, but the 24-hour pattern of secretion is maintained, albeit at a higher level. The nocturnal rise is also maintained throughout the postpartum period.

Mammary Gland Development, Milk Production, and Reproduction

Overall, several hundred different actions have been reported for prolactin in various species. Some of its major effects are summarized here. Prolactin induces lobuloalveolar growth of the mammary gland and stimulates lactogenesis or milk production after giving birth. Prolactin, along with cortisol and insulin, stimulates transcription of the genes that encode milk proteins. The critical role of prolactin in lactation has been confirmed in mice with targeted deletions in the prolactin gene. Female mice that are heterozygous for the deleted prolactin gene (and produce roughly half the normal amount of prolactin) show failure to lactate after their first pregnancy. Prolactin also appears important in several nonlactational aspects of reproduction. In some species (rodents, dogs, skunks), prolactin is necessary for maintenance of corpora lutea (ovarian structures that secrete progesterone, the "hormone of pregnancy"). Mice that are homozygous for an inactivated prolactin gene and thus incapable of secreting prolactin are infertile because of defects in ovulation, fertilization, preimplantation development, and implantation.

Finally, prolactin appears to have stimulatory effects in some species on reproductive or maternal behaviors such as nest building and retrieval of scattered young. The prolactin receptor is widely expressed by immune cells, and some types of lymphocytes synthesize and secrete prolactin. These observations suggest that prolactin may act as an autocrine or paracrine modulator of immune activity. Interestingly, mice with homozygous deletions of the prolactin gene fail to show significant abnormalities in immune responses. A considerable amount of research is in progress to delineate the role of prolactin in normal and pathologic immune responses. At the current time (2003), however, the significance of these potential functions remains poorly understood.

Excessive secretion of prolactin—hyperprolactinemia—is a relatively common disorder in humans. This condition has numerous causes, including prolactin secreting tumors and therapy with certain drugs. Common manifestations of hyperprolactinemia in women include amenorrhea (lack of menstrural cycles) and galactorrhea (excessive or spontaneous secretion of milk). Men with hyperprolactinemia typically show hypogonadism, with decreased sex drive, decreased sperm production, and impotence.

Luteinizing and Follicle Stimulating Hormones

Luteinizing hormone (LH) and follicle-stimulating hormone (FSH)—the gonadotropins—although not necessary for life, are essential for reproduction. These two hormones are secreted from cells in the anterior pituitary called *gonadotrophs*. Most gonadotrophs secrete only LH or FSH, but some appear to secrete both hormones. They are large glycoproteins composed of alpha and beta subunits. The alpha subunit is identical in both of these anterior pituitary hormones, while the beta subunit is unique and endows each hormone with the ability to bind its own receptor. Physiologic effects of the gonadotrophins are known only in the ovaries and testes. Together, they regulate many aspects of gonadal function in both males and females (Fig. 10.8).

In both sexes, LH stimulates secretion of sex steroids from the gonads. In the testes, LH binds to receptors on Leydig cells, stimulating synthesis and secretion of testosterone. Theca cells in the ovary respond to LH stimulation by secretion of estrogens. In females, ovulation of mature follicles in the ovary is induced by a large burst of LH secretion known as the *preovulatory LH surge*. Residual cells within ovulated follicles proliferate to form corpora lutea, which secrete the steroid hormones progesterone and estradiol. Progesterone is necessary for maintenance of pregnancy, and, in most mammals, LH is required for continued development and function of corpora lutea. The name luteinizing hormone derives from this effect of inducing luteinization of ovarian follicles.

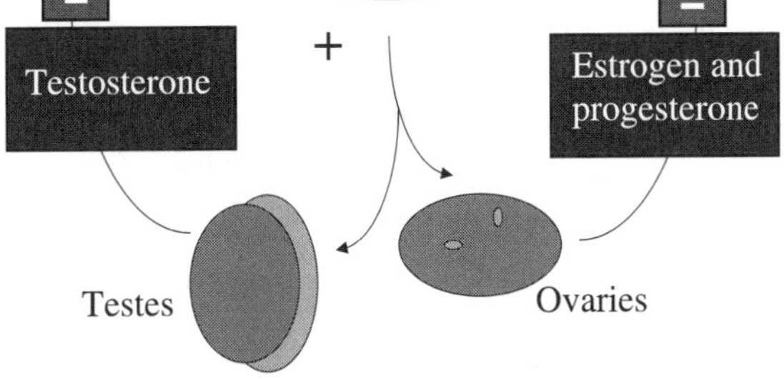

Figure 10.8. Physiologic effects of the gonadotrophins are known only in the ovaries and testes. These hormones are responsible for gametogenesis and steroidogenesis in the gonads and are all synthesized as prohormones, which are subject to posttranslational processing within the cell to yield glycosylated proteins. Testosterone, estrogen, and progesterone provide feedback control at the hypothalamus and pituitary through inhibition of GnRH, LH, and FSH release and/or production.

As its name implies, FSH stimulates the maturation of ovarian follicles. Administration of FSH to humans and animals induces "superovulation," or development of more than the usual number of mature follicles and hence an increased number of mature gametes. FSH is also critical for sperm production. It supports the function of Sertoli cells, which in turn support many aspects of sperm cell maturation.

The Gonadotropic Axis

Rhythms in the gonadotropic axis cover a wide range of frequencies, from ultrafast oscillations of LH levels recurring at intervals of a few minutes, to episodic release in the hourly range, to diurnal rhythmicity, and finally, to monthly and seasonal cycles. These various rhythms interact to provide a coordinated temporal program governing the development of the reproductive axis and its operation at every stage of maturation. Both LH and FSH are secreted in a pulsatile pattern, and an augmentation of this pulsatile activity is associated with sleep onset in a majority of boys and girls. In pubertal children the magnitude of the nocturnal pulses of LH and FSH is consistently increased during sleep. As the pubescent child enters adulthood, the daytime pulse amplitude increases as well, eliminating or diminishing the diurnal rhythm. When the sleep–wake cycle is reversed in pubertal boys, LH augmentation occurs during the daytime sleep period, although there is also an elevation in LH concentrations at the time when sleep occurred under basal conditions, indicating that although the rhythm is sleep-dependent, there is an underlying inherent circadian component. In pubertal boys, the nocturnal rise in testosterone coincides with the elevation of gonadotropins (Mitamura et al. 1999).

Patterns of LH release in adult men exhibit large interindividual variability. The diurnal rhythm is dampened and may become undetectable, although a number of studies point to modest elevation of nocturnal LH and FSH. A marked diurnal rhythm in circulating testosterone levels in young adults, with minimal levels in the late evening and maximal levels in the early morning, has been well demonstrated. With a 15-minute sampling interval, 17–18 testosterone pulses per 24-hour span can be detected. Since the 24-hour pattern of peripheral LH concentration is relatively inconsistent and of modest magnitude, the robust circadian rhythm of plasma testosterone may be partially controlled by other factors such as sleep-associated variations in testicular blood flow, diurnal changes in Leydig cell response to LH, and/or circadian fluctations in other hormones.

In adult women, the 24-hour variation in plasma LH is markedly modulated by the menstrual cycle. In the early follicular phase, LH pulses are large and infrequent, and a slowing of the frequency of secretory pulses occurs during sleep. In the midfollicular phase, pulse amplitude is decreased, pulse frequency is increased, and the frequency modulation of LH pulsatility by sleep is less apparent. Pulse amplitude increases again by the late follicular phase. No modulation by sleep is apparent until the early luteal phase, when nocturnal slowing of pulsatility is again evident. During the luteal–follicular transition, there is four- to fivefold increase in LH pulse frequency, which accompanies the selective FSH rise necessary for normal folliculogenesis. An interaction between the menstrual cycle and circadian rhythmicity is involved in the timing of the preovulatory LH surge. In normal women, onset of the LH surge occurs most often in late sleep or early morning. Moreover, a seasonal variation of this circadian timing has been reported, consisting of the appearance of a biphasic pattern of occurrence in spring only. Toward menopause, gonadotropin levels are elevated but show no consistent circadian pattern.

Gonadotropin Releasing Hormone

The principal regulator of LH and FSH secretion is gonadotropin releasing hormone or GnRH (also known as *LH releasing hormone*). GnRH is a ten amino acid peptide that is synthesized and secreted from hypothalamic neurons and binds to receptors on gonadotrophs (Fig. 10.8). GnRH stimulates secretion of LH, which in turn stimulates gonadal secretion of the sex steroids testosterone, estrogen, and progesterone. In a classical negative-feedback loop, sex steroids inhibit secretion of GnRH and also appear to have direct negative effects on gonadotrophs. This regulatory loop leads to a pulsatile secretion of LH and, to a much lesser extent, FSH. A hyposecretion of GnRH may result from intense physical training and anorexia nervosa.

Numerous hormones influence GnRH secretion, and positive and negative control over GnRH and gonadotropin secretion is actually considerably more complex than depicted in Figure 10.8. For example, the gonads secrete at least two additional hormones—inhibin and activin—which selectively inhibit and activate FSH secretion from the pituitary. Diminished secretion of LH or FSH can result in failure of gonadal function (hypogonadism). This condition is typically manifest in males as failure in production of normal numbers of sperm. In females, cessation of reproductive cycles is commonly observed. Elevated blood levels of gonadotropins usually reflect lack of steroid negative feedback. Removal of the gonads from either males or females, as is commonly done to animals, leads to persistent elevation in LH and FSH. In humans, excessive secretion of FSH and/or LH is most commonly the result of gonadal failure or pituitary tumors. In general, elevated levels of gonadotropins per se have no biological effect.

Regulation of Thyrotropin Hormones

Increased Oxidative Metabolism

The most characteristic effect of thyrotropin hormones (TH) is stimulation of oxidative metabolism of most tissues of the body. The secretion of thyroid hormones (T3, T4 and calcitonin) is controlled by thyrotropin stimulating hormone (TSH) and thyrotropin releasing hormone (TRH). TSH, also known as *thyroid secreting hormone* is secreted from cells in the anterior pituitary called the *thyrotrophs*, finds its receptors on epithelial cells in the thyroid gland, and stimulates that gland to synthesize and release thyrotropic (thyroid) hormones (Fig. 10.9). TSH is a glycoprotein hormone composed of two subunits, which are noncovalently bound to one another. The alpha subunit of TSH is also present in two other pituitary glycoprotein hormones, follicle stimulating hormone and luteinizing hormone, and in primates, in the placental hormone chorionic gonadotropin. Each of these hormones

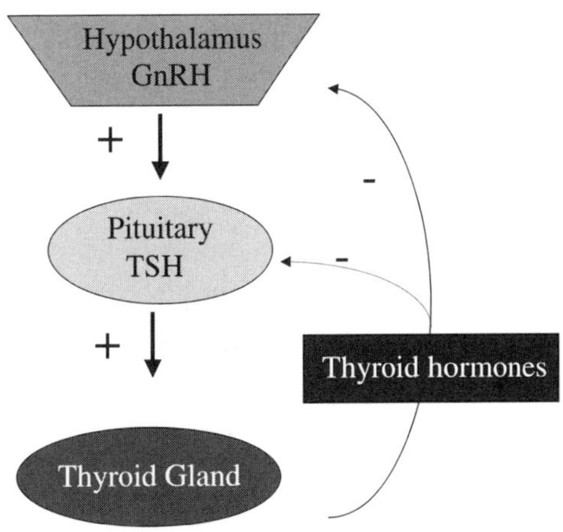

Figure 10.9. The thyrotropin hormone axis. Feedback mechanisms that function within this axis are indicated.

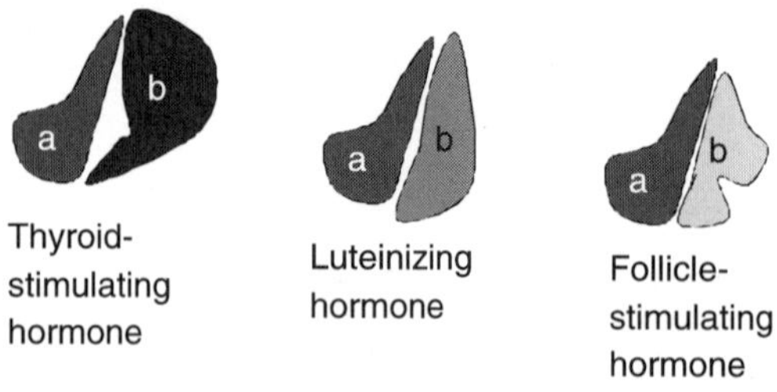

Figure 10.10. Common structure of pituitary glycoprotein hormones TSH, LH, and FSH.

also has a unique beta subunit, which provides receptor specificity (Fig. 10.10).

The most important controller of TSH secretion is TRH. TRH is secreted by hypothalamic neurons into hypothalamic–hypophyseal portal blood, finds its receptors on thyrotrophs in the anterior pituitary, and stimulates secretion of TSH and prolactin. One interesting aspect of TRH is that it is only three amino acids long. Its basic sequence is glutamic acid–histidine–proline, although both ends of the peptide are modified. Secretion of TRH and hence TSH is inhibited by high blood levels of thyroid hormones in a classical negative-feedback loop. Most vertebrates have two thyroid glands, located in the neck; in humans the two have fused to form a single gland. Thyroxin (T4) is an amino acid containing four atoms of iodine. Triiodothyronine (T3) contains only three atoms of iodine but is 3–5 times more active than thyroxin, but is secreted in smaller amounts. Calcitonin (another thyroid hormone) is functionally unrelated to T3, is also secreted by the thyroid. Its chief role is to prevent excessive concentration of calcium in the blood.

Robust Circadian Rhythms

All aspects of thyrotropic (thyroid) hormone signaling display robust circadian rhythms. In normal adult men and women, TSH levels are low throughout the daytime and begin to increase in the late afternoon or early evening. Maximal levels occur shortly before sleep. During sleep, TSH levels generally decline slowly. A sharp decrease occurs in the early morning hours. Studies involving sleep deprivation and shifts of the sleep–wake cycle have consistently indicated that an inhibitory influence is exerted on TSH during sleep. The timing of the evening rise seems to be controlled by circadian rhythmicity. The temporal pattern of TSH secretion seems to reflect both tonic and pulsatile release, with both the frequency and amplitude of the pulses increasing during the nighttime. Pulses of TSH secretion persist under both somatostatin and dopamine treatment, suggesting that the control of pulsatility is largely TRH-dependent. When the depth of sleep is increased by prior sleep deprivation, the nocturnal TSH rise is markedly decreased, suggesting that slow-wave sleep is probably the primary determinant of the sleep-associated fall.

Studies measuring free T3 or the T3/T4 ratio have shown that changes in thyroid hormone secretion occur in parallel with those of plasma TSH, but the amplitude of this rhythm is very modest. However, under conditions of sleep deprivation, the increased amplitude of the TSH rhythm results in an increased amplitude of the T3 rhythm. The diurnal rhythmicity of plasma TSH levels is not apparent until the second month of life but then seems to remain stable throughout childhood, without notable changes at the time of puberty. The nocturnal TSH surge is diminished or absent in hyperthyroidism, central hypothyroidism, and conditions of hypercortisolism. In poorly controlled diabetic states the surge also disappears.

The thyroid gland releases inactive thyroxine (T4) into blood circulation, which is subsequently dionized to the active T3 form by the liver enzyme deiodinase 1. Circadianly regulated synthesis and release of T4 from the thyroid has long been considered the underlying mechanism in maintaining a daily rhythm in serum T3. In addition, we have observed a rhythm in deiodinase 1 that further explains the observed rhythm in serum T3. In addition, studies and our own observations have demonstrated that the thyroid hormone receptor, a member of the nuclear receptor superfamily and the intracellular receptor for T3 and T4, undergoes marked circadian variation at the gene expression level.

Vasopressin

Vasopressin, which functions to conserve body water by reducing the output of urine, is a nine-amino-acid peptide secreted from the posterior pituitary (Fig. 10.11). Within hypothalamic neurons, the hormone is packaged in secretory vesicles with a carrier protein called *neurophysin*, and both are released on hormone secretion. Vasopressin binds to receptors in the distal or collecting tubules of the kidney and promotes reabsorbtion of water back into the circulation. In its absence, the kidney tubules are virtually impermeable to water, and it flows out as urine. In many species, high concentrations of vasopressin cause widespread constriction of arterioles, which leads to increased arterial pressure. It was for this effect that the term *vasopressin* was coined. In healthy humans, vasopressin has minimal pressor effects. The most important variable regulating vasopressin secretion is plasma osmolarity, or the concentration of solutes in blood. Osmolarity is sensed in the hypothalamus by neurons known as *osmoreceptors*, and those neurons, in turn, simulate secretion from the neurons that produce vasopressin. Secretion of vasopressin is also simulated by decreases in blood pressure and volume, conditions sensed by stretch receptors in the heart and large arteries. Changes in blood pressure and volume are not nearly as sensitive a stimulator as increased osmolarity, but are nonetheless potent in severe conditions. For example, loss of 15 or 20% of blood volume by hemorrhage results in massive secretion of antidiuretic hormone. Another potent stimulus of antidiuretic hormone is nausea and vomiting, both of which are controlled by regions in the brain with links to the hypothalamus. The most common vasopressin-related disease in humans and animals is diabetes insipidus. This condition can arise from either of two situations: (1) *hypothalamic* ("central") *diabetes insipidus*, resulting from a deficiency in secretion of vasopressin from the posterior pituitary, caused by head trauma, infections or tumors involving the hypothalamus, and other disorders; and (2)

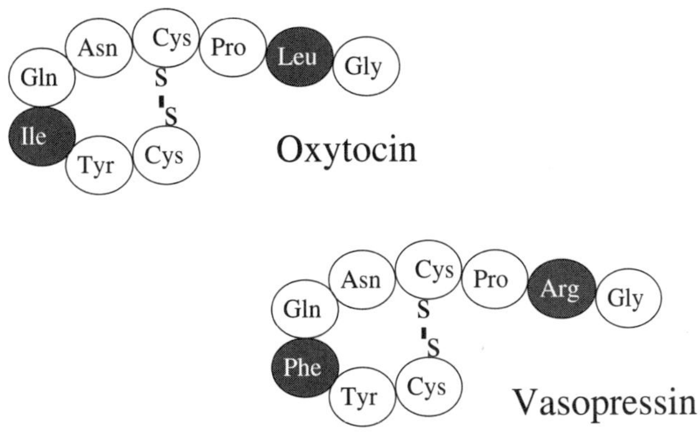

Figure 10.11. Amino acid sequence of oxytocin and vasopressin.

nephrogenic diabetes insipidus, which occurs when the kidney is unable to respond to antidiuretic hormone. Most commonly, this results from some type of renal disease, but mutations in the arginine vasopressin (AVP) receptor gene or in the gene encoding aquaporin-2 have also been demonstrated in affected humans.

Oxytocin

Oxytocin, a nine-amino-acid peptide synthesized in hypothalamic neurons, has profound physiological effects. Oxytocin is transported down axons of the posterior pituitary for secretion into blood. It is also secreted within the brain and from a few other tissues, including the ovaries and testes. Oxytocin differs from vasopressin in two of the nine amino acids (Fig. 10.11). In years past, oxytocin had the reputation of being an "uncomplicated" hormone, with only a few well-defined activities related to birth and lactation. As has been the case with so many hormones, further research has demonstrated many subtle but profound influences of this little peptide. Nevertheless, it has been best studied in females, where it clearly mediates three major effects: stimulation of milk ejection, stimulation of uterine smooth-muscle contraction at birth, and establishment of maternal behavior. While there is no doubt that oxytocin stimulates all the effects described above, doubt has been cast on its necessity in parturition and maternal behavior. Mice that are unable to secrete oxytocin because of targeted disruptions of the oxytocin gene will mate, deliver their pups without apparent difficulty, and display normal maternal behavior. However, they do show deficits in milk ejection and have subtle derangements in social behavior. It may be best to view oxytocin as a major facilitator of parturition and maternal behavior rather than a necessary component of these processes. Both sexes secrete oxytocin—what about its role in males? Males synthesize oxytocin in the same regions of the hypothalamus as in females, and also within the testes and perhaps other reproductive tissues. Pulses of oxytocin can be detected during ejaculation. Current evidence suggests that oxytocin is involved in facilitating sperm transport within the male reproductive system and perhaps also in the female, because of its presence in seminal fluid. It may also have effects on some aspects of male sexual behavior.

Oxytocin and Vasopressin Secretions during Sleep

The neurohypophysial hormones oxytocin and vasopressin show daily rhythms of secretion with elevated hormone release during the hours of sleep in both rats and humans. The changes in plasma hormone concentrations in the rat are reflected by changes in the posterior pituitary and hypothalamic hormone content. The amount in the pituitary falls as the plasma concentration increases, while the hypothalamic content is in phase with the plasma concentration. This pattern can be modulated by ovarian steroids and alters with age. The pattern appears to be due in part to the nocturnal increases in melatonin synthesis, which stimulate hormone release in humans but is inhibitory in rats. In women, plasma vasopressin concentrations and the nocturnal peak are highest in the follicular phase of the menstrual cycle and attenuated in women on oral contraceptives.

■ OTHER HORMONAL OR HORMONE-DEPENDENT RHYTHMS

The Parathyroid Hormone

The parathyroid hormone (PTH) helps to regulate the calcium-phosphate balance between the blood and the other tissues. The parathyroid glands in humans are small pealike organs, usually four in number, located on the surface of the thyroid. PTH increases the concentration of calcium in the blood (thus functioning as a calcitonin antagonist) and decreases the concentration of phosphate, by acting on at least two organs—the kidneys and the intestines—and it stimulates the release of calcium into the blood from bone. The parathyroid gland quadruplet was among the most recent endocrine organs for which a pulsatile secretion mode was demonstrated in both animals and humans. It is now clear that the plasma concentrations of PTH fluctuate episodically. The pulses are of small amplitude and have rapid decay rates, explaining why they escaped some early studies. PTH, which normally increases during sleep, also has an endogenous circadian component that persists in the absence of sleep. The primary peak of the endogenous component occurs at approximately 8 pm, with a minimum around 10 am; the amplitude and shape of the rhythm are different under CR conditions than under ambulatory conditions.

The PTH receptor is coupled to both adenylate cyclase and phospholipase C, thus stimulating both in parallel. PTH is an 84-amino-acid peptide, and full biological activity resides in the amino terminal third of the molecule. In humans PTH is apparently secreted by two separate mechanisms. Under normocalcemic conditions about 30% of total PTH secretion is released in an episodic or pulsatile fashion with a mean frequency of 6–7 pulses per hour and a burst half-duration of approximately 2.5 minutes. The average maximal secretion rate during spontaneous bursts is approximately 25% of the prevailing mean plasma level. The remaining 70% of total PTH secretion is attributable to tonic (time-invariant) hormone release. At present the physiological mechanisms underlying tonic and pulsatile PTH secretion are largely speculative. PTH is stored in cytoplasmic granules before release. The most important physiological regulator of instantaneous PTH secretion is ionized Ca^{2+}. A specific cell membrane receptor for Ca^{2+} has been identified on parathyroid cells. Activation of this receptor by acute hypercalcemia instantaneously suppresses PTH secretion, whereas hypocalcemia results in an immediate increase of PTH release. Besides ionized calcium, active vitamin D (calcitrol, 1a,25-dihyroxy-vitamin D_3) is an important regulator of PTH secretion. Calcitrol reduces PTH availability by suppressing PTH transcription. These effects are mediated both directly via a specific vitamin D receptor and indirectly by stimulation of intestinal Ca^{2+} resorption.

Rhythms in Glucose Regulation

Robust variations in glucose regulation across the 24-hour cycle are now well demonstrated in both normal conditions and states of impaired glucose tolerance. A large number of studies have documented reproducible changes in daytime and/or nighttime glucose utilization (both peripheral and central), insulin secretion, and insulin sensitivity in healthy subjects.

Plasma glucose responses to oral glucose, intravenous glucose, or meals are markedly higher in the evening than in the morning. Diminished insulin sensitivity and decreased insulin secretion in relation to elevated glucose levels are both involved in causing reduced glucose tolerance later in the day. It has been shown that glucose tolerance further deteriorates as the evening progresses, reaching a minimum around the middle of the night. This diurnal variation is not caused by changes in activity level, since it persists during continuous bedrest, and is thus sleep-independent. Indeed sleep deprivation studies indicate the existence of an intrinsic circadian modulation of glucose regulation. Examination of correlations with the variations of the counterregulatory hormones cortisol and GH indicate that the diurnal variation in insulin secretion is inversely related to the cortisol rhythm. Sleep-associated rises in glucose correlated with the amount of concomitant GH secreted. These studies show that glucose regulation is markedly influenced by circadian rhythmicity and sleep and suggest that these effects could be partially mediated by cortisol and GH. Rhythms in insulin secretion also have been found outside the circadian range, demonstrating that human insulin secretion is a complex oscillatory process including rapid pulses recurring every 10–15 minutes superimposed on slower, ultradian oscillations with period in the range of 90–120 minutes. These ultradian oscillations are tightly coupled to glucose.

Alteration of the normal daily profiles of these parameters of glucose tolerance have been identified in normal aging, obesity, and non-insulin- and insulin-dependent diabetes. Observations have indicated that the morning versus evening difference in glucose tolerance observed in normal subjects is abolished in obese subjects and reversed in diabetic subjects who have instead higher glucose responses in the morning than in the evening. Both the rapid and ultradian oscillations of insulin secretion are perturbed in non-insulin-dependent diabetes. While the ultimate cause of these diurnal variations obviously resides in the alteration of wake and sleep states and in the intrinsic effects of circadian rhythmicity, the primary factors responsible for transmitting these modulatory effects of CNS status are still poorly understood.

Asynchronous Circadian Rhythms

Asynchronous circadian rhythms within the vascular system may contribute to underlying cardiovascular disease. Although day–night variations in heart rate and blood pressure have been described since the seventeenth century and are the best-known periodic functions in the cardiovascular system, other parameters have been shown to exhibit circadian variation as well, including stroke volume; cardiac output; blood volume; peripheral resistance; parameters of ECG recordings; plasma concentrations of pressor hormones such as noradrenaline, renin, angiotensin, and aldosterone, atrial natriuretic hormone; plasma cAMP concentrations; blood viscosity; aggregability; and fibrinolytic activity. Although molecular evidence of a vascular clock was not extant, there is circumstantial evidence that it may exist. For example, blood pressure undergoes a marked circadian variability, which is increased in patients with hypertension and coincides with a temporal variability in the incidence of acute vascular events, such as myocardial infarction, sudden cardiac death, and stroke.

In normotensive people as well as in primary hypertension, there is, in general, a nightly drop in blood pressure, whereas in secondary hypertension due to conditions such as renal disease, gestation, and Cushing's disease, the rhythm in blood pressure is in ~70% of cases abolished or even reversed, with highest values at night. This is of particular interest since the loss in nocturnal blood pressure fall correlates with increased end organ damage in cardiac, cerebral, vascular, and renal tissues. Pathophysiological events within the cardiovascular system also display a circadian variability. The onset of nonfatal or fatal myocardial infarction predominates around 6 am–12 noon. A similar circadian time pattern has been shown for sudden cardiac death, stroke, ventricular arrhythmias, and arterial embolism. These early-morning cardiovascular events coincide with the rapid rise in blood pressure, a rapid increase in sympathetic tone and in the concentration of pressor hormones, and the highest values in peripheral resistance. Thus it appears that the early-morning hours are the hours of highest cardiovascular risk.

Using radiotelemetry it is now possible to study cardiovascular parameters in freely moving and undisturbed animals under freerun conditions. In these experiments the rhythms in blood pressure, heart rate, and motor activity were studied in normotensive rats and transgenic hypertensive rats during alternating light–dark cycles (L:D 12:12), during constant darkness (DD) or during constant light (LL). In LD all parameters in both strains exhibited significant and dominant circadian rhythms, which were maintained under freerunning conditions (DD). During LL circadian rhythmicity was almost abolished in all parameters and in both strains, while ultradian components became more dominent. Electrocoagulation of the SCN in normo- and hypertensive abolished circadian rhythms in blood pressure and heart rate. Interestingly, the hypertensive blood pressure values were not reduced by SCN lesion, clearly indicating that the SCN is involved in rhythm generation but not in hypertension. These findings clearly indicate that the rhythms in blood pressure and heart rate must be under the central control of the SCN.

Humoral Signals

Humoral signals can phase-shift and reset peripheral clocks. Molecular clocks similar to those operating in SCN neurons have been uncovered in peripheral tissues and in immortalized rat 1 fibroblast cell lines. In peripheral tissues, such as the liver, kidney, and heart, circadian rhythms in RNA abundance are apparent for each *mPer* gene, although the phase of oscillation is delayed 3–9 hours relative to the oscillation in the SCN. Clock gene oscillations are lost in SCN-lesioned animals, suggesting that the peripheral oscillations may be driven or synchronized by the SCN. It has been suggested that the SCN clock may synchronize peripheral clocks via both neural and hormonal signals. In addition, more recent observations suggest that feeding cycles can entrain peripheral clocks independent of light entrainment.

Observations by our group and others suggest that steroid hormones and vitamins may serve as candidate regulators of peripheral clocks. Circulating concentrations of some steroids undergo circadian variation, and examples of phase shifting of circadian genes in peripheral organs by glucocorticoids has been observed. In addition, we reported a mechanism whereby peripheral circadian oscillators may be regulated by humoral factors. This involves a novel, ligand-dependent interaction between the retinoid receptors, RXRα and RARα, and the bHLH-PAS circadian transcription factors, CLOCK and NPAS2, which directly affects vascular clock function and is of potential importance in both vascular physiology and clinical vascular events.

The mammalian circadian system is organized such that self-sustained oscillators in the SCN entrain peripheral oscillators by releasing a continuous stream of rhythmic signals. The phase and amplitude of peripheral clocks vary between different tissues and organs, in addition to differing from the phase and amplitude in the SCN. The circadian orchestration of relevant targets in these tissues would synchronize the systems-level behavior and physiology in accordance with daily changes in the environment. While in the unicellular algae clock regulation of a large number of genes within the same cell offers an adaptive advantage to the organism, in higher organisms, the clock has evolved to regulate separate sets of genes in different tissue types in a tissue-specific manner. This highlights the importance of the SCN as a master oscillator, which functions to coordinate all peripheral clocks. One function of a peripheral clock might be to regenerate a weak or dampened SCN signal, thus amplifying the oscillation of the signal in that peripheral tissue. A vascular clock may generate an amplified and synchronized vascular rhythm in response to uncoordinated bloodborne signals from central and peripheral clocks. These signals may include hemodynamic stress, the diurnal release of catecholamines and steroids or the periodic bioavailability of signals, such as vitamins, associated with feeding and metabolic cycling.

The existence of multiple oscillators is a common characteristic of all circadian systems described thus far in multicellular organisms. The clock regulates the initiation of locomotor activity in the SCN (and concomitant feeding behaviors elsewhere in the hypothalamus) while it is gearing up for nutrient metabolism in the liver. There is a core set of circadian oscillators in the SCN, liver, and aorta (and presumably all tissue types); some of these are clock components such as *bmal*, *clock*, *npas2*, the *per* genes, and *cry*. Somehow, this core oscillator conscripts the tissue-specific transcriptional machinery to direct circadian oscillation of a subset of those genes—some of these are transcription factors that in turn regulate their targets in a tissue-specific fashion. Peripheral oscillators permit tuning of biological rhythms without difficulty to small, gradual changes in the phase of the input signal. Distorted environmental cues such as transatlantic air travel, the changeover from a day to a night work schedule, or a large hormonal release such as sympathoadrenal activation under conditions of severe stress, may abolish the phase relationships between SCN and peripheral clocks, resulting temporarily in a severely disorganized circadian system.

FURTHER READING

Fisher, DA (1996): Physiological variations in thyroid hormones: Physiological and pathophysiological considerations. *Clin Chem* **42**(1): 135–139.

Forsling, ML (1993): Neurohypophysial hormones and circadian rhythm. *Ann NY Acad Sci* **22**(689): 382–395.

Foulkes, NS, Cermakian, N, Whitmore, D, Sassone-Corsi, P (2000): Rhytmic transcription: The molecular basis of oscillatory melatonin synthesis. *Novertis Found Symp* **227**: 5–18.

Goffin, V, Binart, N, Touraine, P, Kelly, PA (2002): Prolactin: the new biology of an old hormone. *Annu Rev Physiol* **64**: 47–67.

Granner, DK (1996): Pituitary and hypothalamic hormones Murray. In *Harpers Biochemistry*, 24th ed., Granner, DK, Mayes, PA, Rodwell, VW, eds, Appleton & Lange, Stamford, CT.

Kato, Y, Murakami, Y, Sohmiya, M, Nishiki, M (2001): Regulation of human growth hormone secretion and its disorders. *Intern Med* **41**(1): 7–13.

Lemmer, B (1999): Chronopharmacology and its impact on antihypertensive treatment. *Acta Physiol Pharmacol Bulg* **24**(3): 71–80.

McNamara, P, Seo, SP, Rudic, RD, Sehgal, A, Chakravarti, D, FitzGerald, GA (2001): Regulation of CLOCK and MOP4 by nuclear hormone receptors in the vasculature: A humoral mechanism to reset a peripheral clock. *Cell* **105**(7): 877–889.

Mitamura, R, Yano, K, Suzuki, N, Ito, Y, Makita, Y, Okuno, A (1999): Diurnal rhythms of luteinizing hormone, follicle-stimulating hormone, and testosterone secretion before the onset of male puberty. *J Clin Endocrinol Metab* **84**(1): 29–37.

Panda, S, Antoch, MP, Miller, BH, Su, AI, Schook, AB, Straume, M, Schultz, PG, Kay, SA, Takahashi, JS, Hogenesch, JB (2002): Coordinated transcription of key pathways in the mouse by the circadian clock. *Cell* **109**(3): 307–320.

Reppert, SM, Weaver, DR (2002): Coordination of circadian timing in mammals. *Nature* **418**(6901): 935–941.

Sassone-Corsi, P (1995): Transcription facroes responsive to camp. *Annu Rev Cell Dev Biol* **11**: 355–377.

Schaefer, F (2000): Pulsatile parathyroid hormone secretion in health and disease. *Novartis Found Symp* **227**: 225–239.

Van Cauter, E (1990): Diurnal and ultradian rhythms in human endocrine function: A minireview. *Horm Res* **34**: 45–53.

Witte, K, Lemmer, B (1996): Signal transduction in animal models of normotension and hypertension. *Ann NY Acad Sci* **783**: 71–83.

11

HUMAN CIRCADIAN RHYTHMS

Hans P. A. Van Dongen, Gerard A. Kerkhof, and David F. Dinges

While the molecular biology of the circadian clock has been revealed in depth for various animal species, research on humans in this area is still in the early stages. Functional genetic sequences appear to be well preserved across species, suggesting that the molecular biology of human circadian rhythms may be elucidated in the near future. Humans have a full complement of anatomic (e.g., suprachiasmatic nuclei), neuronal (e.g., retinohypothalamic tract), and neuroendocrine (e.g., melatonin) systems involved in circadian rhythm generation and entrainment, strongly suggesting that the circadian biology in humans is similar to that in animals. However, humans express many behaviors for which no clear equivalents are known in the rest of the animal kingdom (e.g., use of clocks and watches; application of artificial light; voluntary night work; self-administration of chronobiotic drugs), which may pose a challenge to generalizing animal phenotypes to humans. Thus, while research on animals continues to provide clues that may apply to humans, there is a need for more research focusing on humans as well. In this chapter, we discuss some successful approaches for laboratory-based research on human circadian rhythms, and the most profound human behavior interfering with this kind of research: voluntary, self-motivated control of sleep and waking activities. We also look at morning-type and evening-type individuals as phenotypes of circadian variability in humans.

■ CIRCADIAN REGULATION AND OVERT CIRCADIAN RHYTHMS IN HUMANS

A Broad Spectrum of Circadian Rhythms in Humans

Human circadian rhythms have been observed in a wide range of variables, including aspects of behavior, physiology,

Molecular Biology of Circadian Rhythms. Edited by Amita Sehgal
ISBN 0-471-41824-2

endocrinology, neurology, and metabolism. Circadian rhythms in humans can be observed in both normal everyday life and highly controlled laboratory experiments; and they are not only present in healthy humans but also implicated in the symptoms of various diseases. Other chapters in this book are filled with examples of circadian rhythms in a variety of animal species—by and large, the same rhythms have been documented for humans as well. Documentation of circadian rhythmicity in humans is often, for practical and ethical reasons, limited to overt circadian rhythms that can be readily observed. The implications of this limitation are discussed below. Table 11.1 gives an overview of the broad spectrum of circadian rhythms documented in humans thus far; this table is not intended to be all-inclusive but serves to get a sense of the pervasiveness of circadian rhythmicity in humans.

The Human Biological Clock Contained in the SCN

It is generally believed that, in healthy humans, at least, the multitude of circadian rhythms is tightly coordinated to create optimal relationships among the various organs and physiological systems and the environment at all times of the day. The master biological clock responsible for this coordination is in the suprachiasmatic nuclei (SCN) of the hypothalamus in the brain. The SCN send out signals to the rest of the brain and to peripheral oscillators and tissues in order to pass on or coordinate the body's "internal" time of day. It is largely unknown what the mechanisms for communication between the SCN and the rest of the brain are; they may involve neuronal firing, neurotransmitter release, and/or humoral cues. Communication with the rest of the body probably involves relay nuclei elsewhere in the brain, but this, too, is mostly undocumented for humans. Nevertheless, because these mechanisms were largely preserved throughout evolution, the information on other organisms given elsewhere in this book is likely to be relevant to humans.

Light: The Dominant Zeitgeber for Humans

There is overwhelming evidence that the most important environmental zeitgeber (i.e., circadian entrainment agent) for humans is the light–dark cycle. Until relatively recently it was believed that only bright light could entrain human circadian rhythms, but it is now known that even ordinary indoor light can serve as a zeitgeber. A phase-response curve and a dose-response curve for humans have been established (see elsewhere in this book for in-depth discussions of response curves). In this age of artificial light, however, ambient light of sufficient strength for entrainment is available all day and beyond the period of natural daylight. The available phase- and dose-response curves were assessed using relatively brief exposures to light, and for this reason cannot be applied to understand entrainment in normal life. Because human civilization tends towards a society active at all 24 hours of the day, animal "wild type" behavior is not entirely representative for human "wild type" behavior, and animal models are not sufficient to fully understand entrainment in humans. Much more work is needed to explain how entrainment by light, and possibly by other zeitgebers, occurs from day to day in humans living under normal (i.e., nonlaboratory) circumstances.

Freerunning Rhythms in Absence of Zeitgebers

Early cave experiments and later isolated laboratory ("bunker") studies have shown that humans, just like other animals, have freerunning rhythms in the absence of zeitgebers. The average period of circadian

TABLE 11.1. Circadian Rhythms Observed in Humans[a]

Category	Example Variables
Cardiovascular system	Heart rate
	Circulating blood volume
	Blood pressure
Respiration	Rate of breathing
	Airway resistance
	Severity of asthma
Endocrine system	Growth hormone
	Cortisol
	Norepinephrine
Gastrointestinal system	Gastric acid secretion
	Hunger
	Excretion
Thermoregulation	Core body temperature
	Rate of sweating
	Vasoconstrictor response
Immune system	Antigen production
	Production of cytokines
	Efficacy of immunosuppressive drugs
Visual perception	Retinal sensitivity
	Pupillary reflex
	Saccadic velocity
Brain function	Cognitive performance
	Melatonin secretion
	Sensitivity to magnetic stimulation
Physical behavior	Sleep–wake cycle
	Athletic performance
	Muscle strength
Disease	Cancer progression
	Occurrence of ischemic stroke
	Efficacy of drug treatment

[a] This table presents an overview of overt circadian rhythms documented for humans, organized by arbitrary categories. Three variables are given as examples for each category. This table is not intended to be complete; rather, it is intended to give a sense of how pervasive circadian rhythms are in humans. All variables in the right-hand column change markedly with a (near-)24-hour periodicity. However, the circumstances under which these periodic changes have been observed (e.g., laboratory- or field-based, with or without sleep deprivation) vary greatly. The implications thereof are discussed in this chapter.

rhythmicity in the bunker experiments was found to be 25 hours. A note of caution is in order for the interpretation of this value, however, as these experiments were not conducted under constant-dark conditions. Thus, feedback on the biological clock from light sources at the disposal of the research subjects probably biased the estimate of the freerunning period in humans. Later laboratory experiments employing a different experimental paradigm, discussed in some detail below, have revealed that the intrinsic period of the human biological clock is much closer to 24 hours than to 25 hours.

Entrainment by Nonphotic Zeitgebers

The cave and bunker experiments done to induce freerunning circadian rhythms raised the question as to what other stimuli, besides light, would be capable of entraining human circadian rhythms. In these studies, steps were taken to remove a wide range of potential zeitgebers, including 24-hour cycles in ambient temperature, noise, electromagnetism, availability of food, and social interactions. There are some indications that meals may serve to entrain the biological clock, and social interactions have long been suspected to serve as zeitgebers as well. Conclusive evidence for the role of social zeitgebers will be difficult to obtain, though, as social stimuli are difficult to standardize for a well-controlled study. Social interaction is also one of the areas where generalization from animal models is difficult, as animal social behavior may differ substantially from human social behavior. In addition, it may well be that the relative role of nonphotic zeitgebers in humans depends on the availability or absence of light for entrainment.

Circadian Control of Sleep–Wake Cycle

Perhaps the most profound circadian rhythm seen in humans—as in other animal species—is the (spontaneous) sleep–wake cycle. As we shall see later, this rhythm is also quite difficult to deal with in humans, because of the unique voluntary control over sleep and waking activities that humans have. Sleep in particular may or may not influence circadian rhythmicity directly, but it clearly limits exposure to other zeitgebers. The role of sleep and waking activity is discussed extensively in this chapter.

■ ENDOGENOUS AND EXOGENOUS DRIVES FOR CIRCADIAN RHYTHMS

Exogenous Drives of Circadian Rhythms

It is of interest to know the origin of the various circadian rhythms demonstrated to exist in humans. It is often implicitly assumed that all overt circadian rhythms originate, through direct or indirect pathways, in the biological clock located in the SCN. Although this assumption is an oversimplification, it has lead to the study of "masking," that is, the effect of exogenous influences that induce, enhance, reduce, or otherwise alter the appearance of circadian rhythmicity independently of the biological clock. A useful working definition of masking in the context of chronobiology is the appearance or modification of a circadian rhythm solely because of the presence of an external influence. A good example is gastrointestinal function. As food is normally taken during the day and not during the night, certain gastrointestinal organs (notably the stomach) are much more active during the day than during the night, creating the appearance of a circadian rhythm. In the absence of food, or if food is distributed evenly across a 24-hour period, however, much of this rhythmicity disappears. Thus, the secretion of gastric acid, for example, is driven primarily by exogenous factors (i.e., food intake) and not so much endogenously.

Light as a Zeitgeber and Masking Factor

Many other circadian rhythms do not depend so heavily on external stimuli, and persist in the absence of these exogenous influences. The circadian rhythm in melatonin secretion by the pineal gland exhibits a regular (near-)24-hour pattern no matter what happens in the environment. Even if the biological clock is not entrained by any zeitgebers, this rhythm persists; thus, the circadian rhythm in melatonin secretion is not a result of masking. It is known, though, that bright-light exposure during the melatonin secretory phase suppresses melatonin secretion; therefore, light can act as a masking factor and alter or even eliminate the melatonin rhythm if exposure is appropriately timed. It should be noted that doing so typically involves keeping a person awake in order to allow for light exposure during the nocturnal period, which is when melatonin is normally secreted. During the day, another masking effect of bright-light exposure can be observed. Bright-light exposure causes an acute, short-lasting enhancement of alertness in humans (and other diurnal species), masking the underlying circadian rhythm of alertness. Thus, light can function as a masking factor as well as a zeitgeber.

Importance of Endogenous and Exogenous Factors in Circadian Function

It has been suggested that distinguishing endogenous from exogenous sources of circadian rhythms is artificial. After all, humans interact with their environment all the time, and the environment is an integral component of human neurophysiology. Furthermore, the effects of masking factors can be of the same order of magnitude as the endogenous circadian rhythmicity. Along the same lines, therefore, it has been suggested that the effects of masking factors are just as important as the effects of zeitgebers for the overall optimization of behavior with respect to the environment. On the other hand, humans have an unparalleled capability to alter their environment (artificial light, temperature control, high-speed transportation to different time zones, etc.), and the various neurophysiological systems cannot always adapt. Even though animal models are excellent tools for understanding endogenous circadian rhythms in humans, it is more difficult to use animal models to study exogenous circadian rhythms and rhythm disruptions that are the result of the interaction with the habitat that humans have created for themselves. Differentiating endogenous from exogenous circadian rhythms helps us understand the human problems of desynchrony encountered during jet lag, shift work, or illness.

Constant Routine and Circadian Rhythms

The gold standard to distinguish endogenous from exogenous drives of circadian rhythmicity is the laboratory experimental protocol called *constant routine*. The objective of this paradigm is to eliminate or keep constant all possible masking factors to expose the underlying endogenous circadian rhythmicity in variables of interest (such as body temperature, heart rate, or cortisol secretion). In a constant-routine paradigm, research subjects are studied in an isolated laboratory for at least 24 hours to allow for the sampling of at least one full circadian cycle. Ambient light and temperature are kept constant, noise is eliminated, and social interactions are minimized. Subjects' posture is held constant, typically near-supine. Food and drink is distributed evenly over the experiment, usually as hourly isocaloric snacks (and not including any caffeine or alcohol). In addition, research subjects in a constant routine are kept awake for at least 24 hours, to eliminate the effects of sleep on the variables measured. Figure 11.1 shows the endogenous circadian rhythms of body temperature, urinary cortisol, and heart rate recorded under constant routine (averages for six individuals). The figure also shows

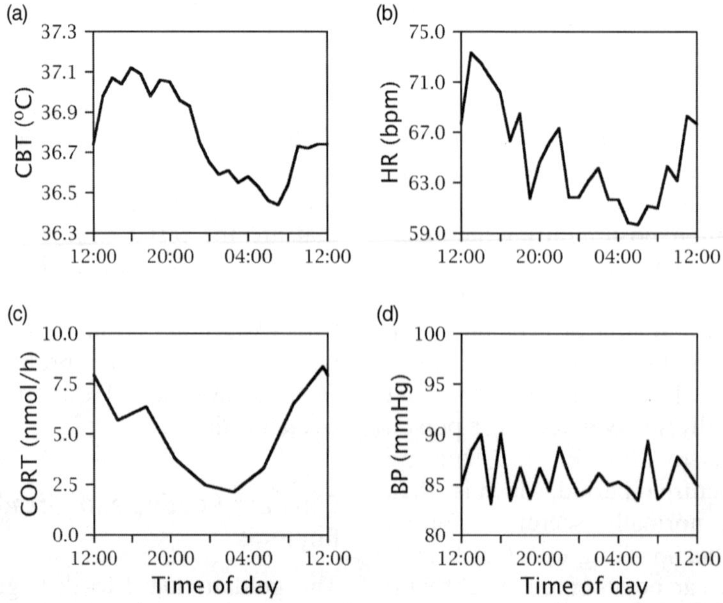

Figure 11.1. Endogenous circadian rhythms. The figure shows endogenous temporal profiles of core body temperature (upper left-hand panel), cortisol excretion (lower left-hand panel), heart rate (upper right-hand panel), and mean arterial blood pressure (lower right-hand panel), recorded during constant routine (averaged over $n = 6$ healthy subjects). Body temperature (CBT) was measured every 2 minutes by means of a rectal probe [hourly averages are plotted in panel (a)]; cortisol excretion (CORT) was derived from urine samples taken every 3 hours; heart rate (HR) and blood pressure (BP) were measured hourly with an oscillometric monitor; and mean arterial pressure (MAP) was calculated as diastolic BP plus one-third of the difference between diastolic and systolic BP. Endogenous circadian rhythmicity is clear in CBT, CORT, and HR. There are fluctuations over time also in BP, which may or may not reflect ultradian rhythms, but there is no endogenous circadian rhythmicity in BP.

the 24-hour profile recorded for blood pressure, which is essentially flat—using the constant routine paradigm, it has been demonstrated that blood pressure does not exhibit any endogenous circadian rhythmicity. It is important to recognize, though, that under ambulatory conditions, blood pressure is not normally found to be constant. There are other mechanisms that affect blood pressure, which would be considered masking from a circadian perspective, but that are no less important for physiological function and health in general. Thus, while circadian rhythms can be studied in isolation to some extent, they must ultimately be considered in the context of the many other aspects of human physiology, endocrinology, neurology, immunology, and behavior.

Voluntary Sleep–Wake Cycle Control

The constant routine involves keeping individuals awake, but the sleep–wake cycle is also under circadian control. Thus, this experimental paradigm exposes endogenous circadian rhythms while eliminating another rhythm (i.e., that of sleep and wakefulness). In animal research, the sleep–wake cycle is often considered an integral part of circadian regulation and is even used as a marker of circadian rhythmicity. In humans, however, this presents a problem due to the unique ability humans have to control their sleep times by self-motivation only. Humans can voluntarily stay awake for long periods of time, during which time they can undertake all kinds of activities, and they can employ numerous environmental (e.g., television) and pharmacological (e.g., coffee) stimuli to help them do so. They can also plan to terminate their sleep at any time by using an alarm or by asking another person to wake them up. The problem lies herein that sleep, or the absence of sleep, affects many of the variables that show circadian rhythmicity. Thus, if sleep is displaced, then the shape of a circadian rhythm can be altered, or masked. This type of masking is not due to exogenous influences per se, and is therefore referred to as *internal masking*. The effect of internal masking is particularly easy to show for body temperature, as illustrated in Figure 11.2. The recognition of internal masking has led to the practice of not considering sleep an intrinsic component of endogenous circadian rhythmicity in humans, even though by itself it is not an exogenous factor. To further understand the special role of sleep, it is necessary to consider the homeostatic mechanism that controls sleep in addition to its circadian rhythm.

THE CIRCADIAN AND HOMEOSTATIC REGULATION OF SLEEP

Homeostatic Drive for Sleep

Sleep is under the simultaneous control of a circadian process (originating in the SCN) and a homeostatic process (the neurophysiological basis of which is as yet to be unraveled). The homeostatic control of sleep serves to balance sleep and wakefulness: Longer periods of wakefulness are followed by longer periods of sleep, and shorter periods of wakefulness require less sleep for "recovery" (it is often implicitly assumed that sleep serves to recover from waking activities, but this is actually an outstanding question). The circadian control of sleep, on the other hand, serves to place wakefulness during the day and sleep during the night. It is generally believed that before the appearance of artificial light, climate control, and 24-hour availability of food, the circadian control of sleep ensured that waking activities and sleep took place at the times optimally suited for survival (i.e., day and night, respectively). Modern technology is often presumed to have eliminated the need for circadian control of sleep, but the evolution of human physiology has not (yet?) been able to catch up with these new developments. Thus, humans desire to stay awake at times when their biological clock tells them they should sleep, and attempt to sleep when the circadian drive for wakefulness is at a maximum. The simultaneous homeostatic control of wakefulness and sleep further complicates this picture; for instance, staying awake at night involves sleep loss, which results in an increased homeostatic drive for sleep during the day at the same time that the circadian drive for wakefulness is high. It is no wonder, then, that many nightshift workers experience problems staying awake on the shift and/or sleeping during the day, and some have serious health problems involving gastrointestinal, cardiovascular, and other pathological conditions.

Exposing Circadian and Homeostatic Regulation

The complex interplay between the homeostatic and circadian regulation of sleep and wakefulness is well understood, and even captured in mathematical models (notably the two-process model of sleep regulation). Two experimental paradigms in particular contributed substantially to our understanding of these processes: sleep deprivation and forced desynchrony. Sleep deprivation experiments typically involve

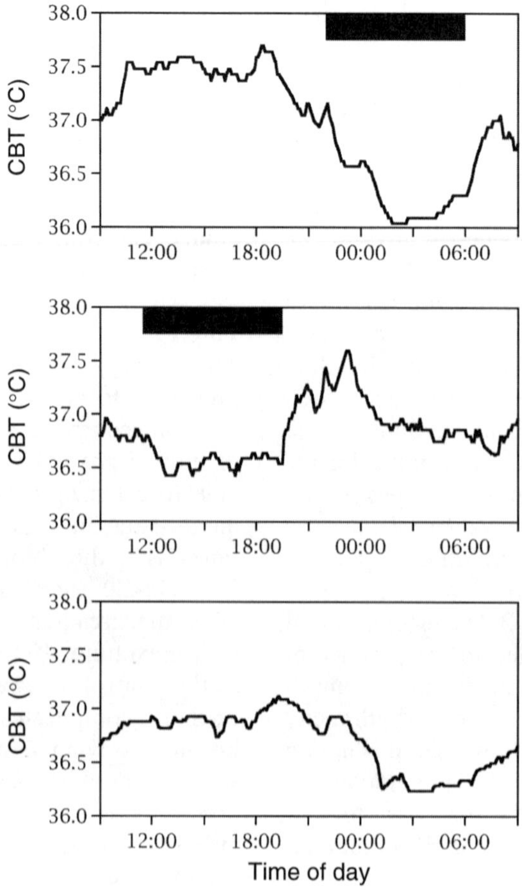

Figure 11.2. Masking of endogenous circadian rhythms by sleep. The figure shows typical circadian profiles of core body temperature (CBT) in humans studied in a laboratory: normal sleep–wake cycle (sleep from 22:00 until 06:00; top panel), displaced sleep–wake cycle (sleep from 11:30 until 19:30; middle panel), and constant routine (no sleep; bottom panel). Sleep periods are indicated with a black bar. The data illustrate that sleep–wake behavior, although under circadian control, can mask endogenous circadian rhythmicity in other variables such as CBT. Therefore, sleep is seldom considered an intrinsic component of endogenous circadian rhythmicity in humans, but is treated as a separate neurophysiological system. The same applies for the expression of (physical) activity during wakefulness. From studies of body temperature profiles obtained under different experimental circumstances, mathematical "purification" procedures have been developed to estimate the endogenous circadian profile of body temperature in the presence of environmental and internal masking factors. Using knowledge of sleep–wake times, activity patterns or light exposure, reasonably accurate estimates of endogenous circadian rhythm parameters can be obtained with such purification methods without the need for laboratory-based constant routine experiments. Note also that even in the constant routine data of individual subjects (bottom panel), recorded under highly controlled laboratory circumstances, there are fluctuations over time that cannot be accounted for by circadian rhythmicity. These fluctuations probably reflect biological or measurement noise, as they are largely averaged out when data for multiple subjects are pooled, leading to a stereotypical sinusoidal profile of CBT circadian rhythmicity.

keeping people awake for extended periods (usually one or two nights and days) and subsequently giving them (unrestricted) recovery sleep. Forced desynchrony involves scheduling wakefulness and sleep (and light and darkness) in cycles deviating considerably from the 24-hour day. The biological clock is not capable of synchronizing to the imposed cycles and resorts to a freerunning state, effectively uncoupling the sleep–wake cycle and the endogenous circadian cycle. Both experimental paradigms expose the interaction of the homeostatic and circadian processes regulating sleep, wakefulness, and other human behaviors.

Preserved Circadian Rhythms during Prolonged Total Sleep Deprivation

The interaction of circadian and homeostatic regulation during sleep deprivation is primarily seen in waking neurobehavioral performance and other variables that track waking behavior (e.g., electroencephalographic spectral power). Figure 11.3 shows psychomotor vigilance performance across 88 hours (3.7 days) of sustained wakefulness. For comparison, the 88-hour profile of melatonin concentration in blood plasma, which reflects circadian rhythmicity but not homeostatic regulation, is shown as well. The same rhythmic pattern seen in melatonin is also observed in psychomotor performance. In addition, however, there is a trend toward poorer performance as sleep loss progresses. The net result is that performance capability is reduced by extended sleep deprivation, but performance during the day is still better than performance during the previous night even after 3 full days of sleep deprivation. Clearly, circadian rhythms are preserved under total sleep deprivation, but the effects of sleep loss also serve to mask the endogenous profile of circadian rhythmicity. On the other hand, waking neurobehavioral performance cannot be measured during sleep, so sleep itself can just as well be a masking factor. Two experimental paradigms have tackled this problem: the forced desynchrony paradigm and the ultradian sleep–wake cycle paradigm.

Forced Desynchrony and Intrinsic Period of Circadian Pacemaker

In the forced desynchrony laboratory paradigm, subjects are exposed to an artificial day–night cycle to which the biological clock is unable to synchronize. Typically a 20-hour or a 28-hour artificial daylength is used. The forced desynchrony paradigm uncouples the circadian and homeostatic regulatory systems because the period of the imposed light–dark cycle is outside the entrainment range of the biological clock. Instead, the biological clock reverts to its intrinsic, freerunning periodicity. The freerunning period of circadian rhythmicity under forced desynchrony is found to be about 24.2 hours on average, which is substantially different from the 25 hours observed in the classical freerun experiments in isolated bunkers (discussed earlier in this chapter). The difference is attributed to the feedback of (self-selected) light exposure on the biological clock in the latter experiments, while this is believed not to contribute overall to the value for the circadian period in forced desynchrony experiments. Thus, the 24.2-hour value of the free-running circadian period found in forced desynchrony studies is thought to be a better reflection of the intrinsic period of the biological clock. However, in his early studies of circadian rhythms, Aschoff already noted that any estimate of circadian period reflects, at least partially, the experimental or environmental circumstances. Thus, it may not be possible to give a universal estimate of the intrinsic period of human circadian rhythmicity without the connotation of the experimental conditions yielding the estimate.

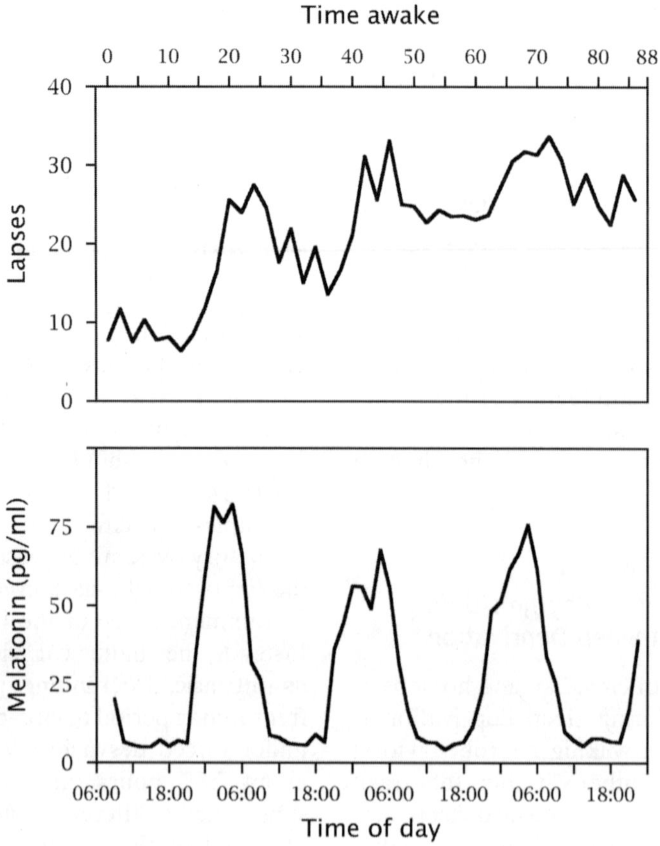

Figure 11.3. Circadian rhythms during total sleep deprivation. The figure shows the temporal profiles of psychomotor vigilance performance lapses (top panel) and plasma melatonin concentration (bottom panel) across 88 hours of total sleep deprivation (for $n = 10$ subjects). Melatonin concentrations are under circadian control only, while performance capability shows the interaction of circadian rhythmicity and the buildup of homeostatic pressure for sleep. Thus, sleep deprivation experiments serve to demonstrate the interaction of circadian and homeostatic regulation for variables affected by both regulatory systems. However, disentangling their respective contributions to the overall temporal profile presents a mathematical problem.

Interaction of Circadian and Homeostatic Processes

Besides its utility for estimating the intrinsic circadian period, the forced desynchrony paradigm has another important use for the study of circadian rhythms—it makes it possible to average out the "masking" influences of sleep and wakefulness. With each circadian cycle, the phase of the freerunning circadian rhythm shifts with respect to the imposed sleep–wake (i.e., light–dark) cycle. By collapsing the data of interest over the free-running circadian period, therefore, the cyclicity of sleep and wakefulness is averaged out, resulting in an average profile with no net effects of sleep and/or wakefulness. Similarly, by folding the data over the imposed light–dark cycle, circadian changes can be averaged out. It turns out that the same variables that display an interaction of circadian and homeostatic influences under total sleep deprivation also show this interaction under forced desynchrony. However, by folding the data obtained under forced desynchrony, the circadian and homeostatic processes can be separated easily and without the mathematical problems encountered for total sleep deprivation. Interestingly, this led to the discovery that the interaction between the circadian and homeostatic mechanisms is nonlinear. This means that the contribution of the homeostatic process depends to some extent on the state of the circadian process, and vice versa. (It should be noted that the consequence of this discovery is that accurately separating the circadian and homeostatic systems is not so simple after all in the forced desynchrony context, as data folding and averaging yields only a first-order approximation of the relative contributions of the two processes.) This underlines the inherent interrelationship between circadian rhythmicity and homeostatic regulation—they influence each other and manipulating one process will have consequences for the other.

Ultradian Day and Sleep Propensity

Another human experimental paradigm that recognizes the importance of sleep–wake regulation for the study of circadian rhythms exists, namely, the "ultradian day." This refers to a research paradigm using artificial sleep–wake cycles that are extremely short in duration—typically 20–90 minutes—with about one-third of each cycle reserved for sleep. Such schedules permit the study of sleep and wakefulness at many different circadian phases while, at least conceptually, leaving the overall 24-hour homeostatic balance between sleep and wakefulness intact. This is another paradigm that permits the measurement of endogenous circadian rhythms in variables that are affected by sleep and sleep loss and that cannot, therefore, be measured under constant routine. The interaction between circadian and homeostatic regulation profoundly affects the ultradian day paradigm, however. One-third of each sleep–wake cycle is reserved for sleep, but depending on the time of day, not all this time is actually spent asleep. Of course, the circadian drive for wakefulness is the primary factor responsible for this variation. To what extent this phenomenon reduces the usefulness of the ultradian day paradigm for the study of circadian rhythms in variables also affected by sleep has not been determined.

Jet Lag

The knowledge gained in sleep deprivation, forced desynchrony, and ultradian day studies about the combined circadian and homeostatic influences on sleep, wakefulness, and neurobehavioral performance helps explain the phenomenon of jet lag. This is the malaise typically experienced after transmeridian travel (i.e., rapid travel to a different time zone) and caused by circadian misalignment and desynchrony. It takes some time (up to a week, depending on the direction of travel) for the biological clock to synchronize with the light–dark cycle in the new time zone, which causes the transient circadian misalignment. Circadian rhythms that are (partially) driven by exogenous factors (in the new time zone) appear to adjust more quickly, resulting in transient internal desynchrony. In addition, a shift of the sleep–wake cycle is needed for realignment to the light–dark cycle in the new time zone, leading to a transient disturbance of the homeostatic regulation of sleep and wakefulness. Jet lag is often thought of as a purely circadian phenomenon, but clearly the homeostatic process is involved as well, and contributes to the symptoms of jet lag.

Shiftwork

During shiftwork, the same mechanisms are at work to cause problems focusing on the job, sleeping at unconventional times, and a number of other work- and health-related problems. The involvement of homeostatic regulation is more readily apparent in this case, as shiftwork, and nightwork in particular, often involves a period of acute sleep loss. Furthermore, the circadian misalignment may not be as transient as for jet lag, for often there are no zeitgebers to help align the circadian rhythm with the work–rest schedule. On the other hand, it is worth noting that problems of circadian misalignment can be partially overcome, even if no realignment takes place. The temporary alerting effect of bright light, for instance, may be used to compensate for a lack of circadian drive for wakefulness during the night. Various pharmacological countermeasures (e.g., caffeine) are available for the same purpose. From a chronobiological research point of view, these constitute masking factors. But in real life, these "masking" factors are an intrinsic part of how humans deal with, and control, the cycle of day and night (see Fig. 11.4).

MORNINGNESS/EVENINGNESS AND CIRCADIAN PHENOTYPES

Differences in Circadian Phase Position of Biological Clock

One noteworthy illustration of human circadian rhythmicity is the existence of morning-type and evening-type individuals ("larks" and "owls"). Morning types prefer to be awake and active early during the day, at which time they experience peak alertness, and they retire early in the evening.

Figure 11.4. Humans have voluntary, self-motivated control over sleep and wakefulness. The development of intelligence and the use of tools has allowed the human species to step beyond the influence of daily light–dark and other environmental controls of behavior. Paradoxically, the rising of societies operating 24 hours a day, 7 days a week (i.e., 24/7) has led to a lifestyle that is cognitively impairing, because circadian and homeostatic processes regulating sleep also appear to be critical for waking cognitive ability. The response to this dilemma has been the invention and social acceptance of countermeasures serving to sustain wakefulness, alertness, and cognitive capability. As a result, human "wild type" behavior nowadays tends to ignore the endogenous regulatory neurobiology for sleep and waking activities. (Artwork by Erik Timmers.)

Evening types, on the other hand, favor the afternoon and evening for activities, as they are more alert then, and they prefer to sleep in the early morning. For a long time, it was believed that morningness reflected an aspect of personality, similar to introversion and neuroticism. However, it was since demonstrated that morning and evening types differ in the phase of their circadian rhythmicity. In other words, the biological clock of evening types is delayed with respect to morning types, as illustrated in Figure 11.5. Moreover, this difference is at least partially endogenous in nature. The endogenous circadian rhythmicity of (extreme) evening types, measured with core body temperature under constant routine, runs behind by approximately 2 hours when compared to (extreme) morning types. Oscillator theory would predict that morning types, therefore, would also have a somewhat shorter circadian period than would evening-types under freerunning conditions, and some evidence to this extent has been documented in the literature. It may also be that morning and evening types differ in the entrainment properties of the biological clock, such that they synchronize differentially to the natural light–dark cycle regardless of the period of their free-running circadian rhythms.

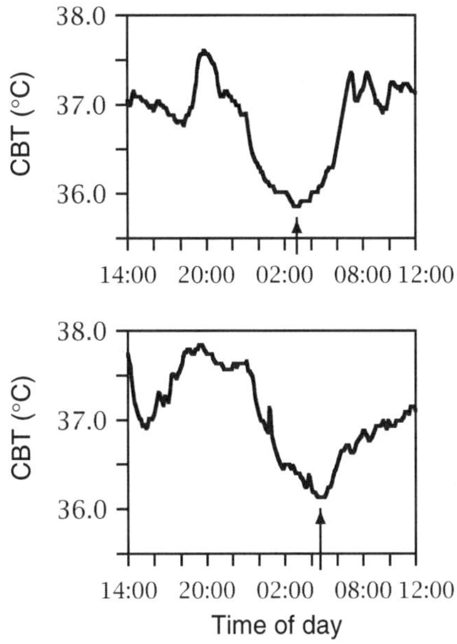

Figure 11.5. Core temperature rhythms in a morning-type individual and an evening-type individual. The figure shows the difference between a randomly selected morning-type individual (top panel) and a randomly selected evening-type individual (bottom panel), in the circadian rhythm of core body temperature (CBT) recorded under field conditions. The arrows indicate the minimum of CBT, the traditional phase marker for CBT data, which is reached 1.8 hours earlier (i.e., relatively phase-advanced) in the morning-type than in the evening-type individual.

Trait Variability in Circadian Rhythms

Although morningness/eveningness is partly affected by developmental changes (puberty in humans is accompanied by a tendency toward eveningness, and advanced aging is accompanied by a tendency toward morningness), overall it appears to be an enduring trait. For this reason, it has been suggested that morningness/eveningness may be genetically determined, and that morning types and evening types may be considered phenotypes of human circadian variability. In support of this idea comes the finding that monozygotic twins have more similar scores on a morningness/eveningness questionnaire than do dizygotic twins. In one twin study, the total heritability for morningness/eveningness was found to be approximately 45%. Further supporting a genetic basis for morningness/eveningness is some preliminary evidence that a polymorphism located in the 3′ flanking region of the human *Clock* gene corresponds to a tendency toward eveningness. This discovery led to the development of a *Clock* mutant mouse model of morningness/eveningness. Another polymorphism identified in the genomic machinery underlying human circadian rhythms, the A2634G substitution in the *Timeless* gene, appears to have no influence on morningness/eveningness.

Advanced and Delayed Sleep Phase Syndromes

The tendency toward morningness or eveningness can be so extreme in an individual that the misalignment with the rest of society is disabling, resulting in phenotypes referred to as advanced sleep phase syndrome (ASPS) or delayed sleep phase syndrome (DSPS), respectively. Autosomal semidominant mutations in rodents with short or long circadian periods have been shown to be associated with similarly advanced or delayed sleep–wake rhythms. These animal models predicted the existence of familial circadian rhythm disorders in humans, a variation of which has been documented: familial advanced sleep phase syndrome. In three families, ASPS segregated as an autosomal dominant trait, and the disorder appeared to be caused by a missense mutation in the *hPer2* gene, which shortens the circadian period. Indeed, a shorter circadian period would be expected to result in an earlier phase position under entrained circumstances, resulting in the extreme morning-type behavior of familial ASPS patients. The Mendelian inheritance pattern of familial ASPS makes it unlikely, though, that the exact same genes are involved in this syndrome as in the morningness/eveningness trait. DSPS has also been associated with a genetic abnormality. In this case, a structural polymorphism in the *hPer3* gene has been implicated in the pathogenesis of the disorder.

■ CONCLUSION

The documentation of human familial advanced sleep phase syndrome—in the light of earlier, comparable findings for mutations in rodents—strengthens the notion that animal models can inform the investigation of circadian rhythms in humans. Given the strict practical and ethical limitations of research involving human beings, it is clear that animal models will continue to play an important role in the investigation of the molecular biology of human circadian rhythmicity. When interpreting animal data in the context of humans (or vice versa), however, care must be taken also to consider the differences that exist between humans and other animal species. From a circadian perspective, the most important difference may well be the human's unique, voluntary control over sleep and waking activities. Circadian rhythms and sleep–wake patterns are strongly related in humans as they are in animals—but human behavior clearly shows that they can be easily dissociated. Thus, sleep–wake patterns should not be indiscriminately used as a surrogate measure of circadian rhythms in humans.

FURTHER READING

Carskadon, MA, Dement, WC (1977): Sleepiness and sleep state on a 90-min schedule. *Psychophysiology* **14**: 127–133.

Czeisler, CA, Khalsa, SBS (2000): The human circadian timing system and sleep–wake regulation. In *Principles and Practice of Sleep Medicine*, 3rd ed, Kryger, MH, Roth, T, Dement, WC, eds, Saunders, Philadelphia, pp. 353–375.

Jones, CR, Campbell, SS, Zone, SE, Cooper, F, DeSano, A, Murphy, PJ, Jones, B, Czajkowski, L, Ptácek, LJ (1999): Familial advanced sleep-phase syndrome: A short-period circadian rhythm variant in humans. *Nature Medicine* **5**: 1062–1065.

Kerkhof, GA, Van Dongen, HPA (1996): Morning-type and evening-type individuals differ in the phase position of their endogenous circadian oscillator. *Neuroscience Lett.* **218**: 153–156.

Mrosovsky, N (1999): Masking: History, definitions, and measurement. *Chronobiol. Int.* **16**: 415–429.

Van Dongen, HPA, Dinges, DF (2000): Circadian rhythms in fatigue, alertness, and performance. In *Principles and Practice of Sleep Medicine*, 3rd ed, Kryger, MH, Roth, T, Dement, WC, eds, Saunders, Philadelphia, pp. 391–399.

Vink, JM, Groot, AS, Kerkhof, GA, Boomsma, DI (2001): Genetic analysis of morningness and eveningness. *Chronobiol. Int.* **18**: 809–822.

Wever, RA (1979): *The Circadian System of Man*, Springer-Verlag, New York.

INDEX